Borderlands in World History, 1700

Borderlands in World History, 1700–1914

Edited by

Paul Readman
Senior Lecturer in Modern British History, King's College London

Cynthia Radding
Distinguished Professor, The University of North Carolina at Chapel Hill

and

Chad Bryant
Associate Professor, The University of North Carolina at Chapel Hill

First published 2014 by
PALGRAVE MACMILLAN

Palgrave Macmillan in the UK is an imprint of Macmillan Publishers Limited,
registered in England, company number 785998, of Houndmills, Basingstoke,
Hampshire RG21 6XS.

Palgrave Macmillan in the US is a division of St Martin's Press LLC,
175 Fifth Avenue, New York, NY 10010.

Palgrave Macmillan is the global academic imprint of the above companies
and has companies and representatives throughout the world.

Palgrave® and Macmillan® are registered trademarks in the United States,
the United Kingdom, Europe and other countries.

ISBN 978–1–137–32057–5   hardback
ISBN 978–1–137–32056–8   paperback

This book is printed on paper suitable for recycling and made from fully
managed and sustained forest sources. Logging, pulping and manufacturing
processes are expected to conform to the environmental regulations of the
country of origin.

A catalogue record for this book is available from the British Library.

A catalog record for this book is available from the Library of Congress.

Typeset by MPS Limited, Chennai, India.

Contents

List of Maps and Figures

Maps

Figures

Acknowledgements

This book represents just one of many efforts underway between our two institutions, King's College London and the University of North Carolina at Chapel Hill, which in 2005 formally launched a strategic partnership that has grown to include student exchanges, joint PhD committees, student-led workshops, collaborative research, conferences, and other combined efforts among members of our respective intellectual communities. Thus, the editors would like to begin by thanking the members of the University of North Carolina–King's College London Collaboration Committee, the University of North Carolina–King's Strategic Alliance Fund, and the King's College Department of History for generously supporting the 'Borderlands as Physical Reality: Producing Place in the Eighteenth and Nineteenth Centuries' conference. This event was held at King's in October 2011, and provided the basis for the book. Our efforts owe much to the encouragement and advice of David Ricks, Keith Hoggart, Jan Palmowski, Jim Bjork, Bob Miles, Jonathan Hartlyn, Lloyd Kramer, James LeLoudis, John McGowan, and the staff at the University of North Carolina's Winston House in London, all of whom have been key contributors to our universities' ambitious partnership as well. The Center for European Studies, the Department of History, the Institute for the Arts and Humanities, and the University Research Council at the University of North Carolina Chapel Hill generously contributed towards meeting the costs of this project, as did the Department of History and School of Arts & Humanities at King's.

Many more individuals helped to bring this book to its successful completion. Arthur Burns, Bernard L. Hermann, Jan Rüger, and the numerous other participants of the conference collectively refined and developed many of the themes and questions that informed this book. Jenny McCall, Clare Mence, and Holly Tyler at Palgrave offered enthusiastic support for our project, and the two anonymous readers provided thoughtful and constructive comments that improved the volume as a whole. Philip Schwartzberg at Meridian Mapping in Minneapolis worked creatively with the editors' numerous requests and questions while composing the book's maps. Audra Yoder took time out from other commitments to compile the index. Finally, we thank all the contributors for engaging with our editorial interventions and meeting the publishing deadlines. We truly enjoyed reading these innovative and original works of research and trust that you will as well.

Paul Readman, Cynthia Radding, and Chad Bryant
Chapel Hill and London
July 2013

Notes on Contributors

Timothy P. Barnard is an associate professor in the Department of History at the National University of Singapore. His research focuses on the cultural and environmental history of island Southeast Asia, with a particular focus on the Straits of Melaka during the early modern era. His publications include *Multiple Centres of Authority: Environment and Society in Siak and Eastern Sumatra, 1684–1827* (2003) and the edited volume *Contesting Malayness: Malay Identity across Boundaries* (2004). He is currently working on a social history of the Komodo dragon.

Jim Bjork is a senior lecturer in the Department of History at King's College London. His book, *Neither German nor Pole: Catholicism and National Indifference in a Central European Borderland* (2008), examines the role of the Roman Catholic Church in blurring national identifications and facilitating national ambiguity in Upper Silesia between the late nineteenth century and the frontier plebiscite of 1921. He remains interested in the tensions between religion and nationality and is working on an article engaging recent debates about the concept of 'national indifference.' His current research is focused on the local, national, and transnational dimensions of the reconstruction of the Roman Catholic Church in Poland after the Second World War.

Frank Bongiorno is Associate Professor of History at the Australian National University in Canberra. He has previously held academic appointments at Griffith University in Brisbane, the University of New England in Armidale, New South Wales, and King's College London. In 1997–98 he was Smuts Visiting Fellow in Commonwealth Studies at the University of Cambridge. He is the author or co-author of two books on Australian labor politics and his most recent book is *The Sex Lives of Australians: A History* (2012).

Chad Bryant is an associate professor at the University of North Carolina at Chapel Hill who studies the social and cultural history of Central and Eastern Europe from the eighteenth century to the present. His research has focused on the Bohemian Lands, most of which now constitute the Czech Republic. His book, *Prague in Black: Nazi Rule and Czech Nationalism* (2007), examined how Nazi rule radically transformed nationality politics and national identities in the Protectorate of Bohemia and Moravia. His most recent publications include 'Into an Uncertain Future: Railroads and Vormärz Liberalism in Brno, Vienna, and Prague,' *Austrian History Yearbook* (2009); 'After Nationalism? Urban History and Eastern European History,' *East European Politics and Societies* (2011); and 'Zap's Prague: The City, the Nation, and Czech Elites before 1848,' *Urban History* (2013). He is currently writing a history of modern Prague and its inhabitants.

Benjamin H. Johnson is Associate Professor of History at the University of Wisconsin, Milwaukee. His research interests include environmental history, North American borders, Texas history, and western history. He is author of *Revolution in Texas: How a Forgotten Rebellion and Its Bloody Suppression Turned Mexicans into Americans* (2003) and the prize-winning *Bordertown: The Odyssey of an American Place* (2008). His edited volumes include *Steal This University: The Labor Movement and the Corporatization of Higher Education* (2003), *The Making of the American West* (2007), *Bridging National Borders in North America* (2010), and *Major Problems in North American Borderlands History* (2011). He is currently researching a book on American environmentalism in the early twentieth century.

Lloyd Kramer is Professor of History at the University of North Carolina, Chapel Hill. His research and teaching focus on modern European intellectual history, the history of cross-cultural exchanges in the Atlantic World, and the history of modern France. He is the author of several books, including *Threshold of a New World: Intellectuals and the Exile Experience in Paris, 1830–1848* (1988), *Lafayette in Two Worlds: Public Cultures and Personal Identities in an Age of Revolutions* (1996), and *Nationalism in Europe and America: Politics, Cultures, and Identities since 1775* (2011).

Lisa A. Lindsay is an associate professor in the History Department at the University of North Carolina, Chapel Hill, where she teaches African history and the history of the Atlantic slave trade. She is the author of *Working with Gender: Wage Labor and Social Change in Southwestern Nigeria* (2003) and *Captives as Commodities: The Trans-Atlantic Slave Trade* (2008), and is co-editor, with Stephan F. Miescher, of *Men and Masculinities in Modern Africa* (2003) and, with John Wood Sweet, of *Biography and the Black Atlantic* (2013). She is currently at work on the contextualized biography of a South Carolina freedman who in the 1850s migrated to modern-day Nigeria, making trans-Atlantic connections that his descendants and their American relatives maintain to this day.

Oksana Mykhed is a doctoral candidate in history at Harvard University. Her research interests include early modern and modern East European and Russian history. In her current work, she explores the growth of empires and the formation of imperial boundaries, and focuses on the history of migration control, public health, and commerce in the Russian Empire and Poland-Lithuania in the seventeenth and eighteenth centuries.

Roland Quinault was educated at Oxford, where he was a Junior Research Fellow at Merton College. He has been Honorary Secretary of the Royal Historical Society, a Reader in History at London Metropolitan University, and the Fulbright–Robertson

Visiting Professor in British History at Westminster College, Missouri. He is currently a Senior Research Fellow at the Institute of Historical Research, University of London. His main field of research is British political and social history from the Victorian period to the later twentieth century. He is the author of *British Prime Ministers and Democracy from Disraeli to Blair* (2011) and co-editor of *William Gladstone: New Studies and Perspectives* (2012).

Cynthia Radding is the Gussenhoven Distinguished Professor of Latin American Studies and Professor of History at the University of North Carolina, Chapel Hill. Her research focuses on imperial borderlands in both North and South America, with methodological emphases on ethnohistory and environmental history. She is the author of four books, including the prize-winning *Wandering Peoples: Colonialism, Ethnic Spaces and Ecological Frontiers in Northwestern Mexico, 1700–1850* (1997) and *Landscapes of Power and Identity: Comparative Histories in the Sonoran Desert and the Forests of Amazonia from Colony to Republic* (2005). Her more recent publications include a prize-winning article, 'The Children of Mayahuel: Agaves, Human Cultures, and Desert Landscapes in Northern Mexico,' *Environmental History* (2012). Her current book project is *Bountiful Deserts and Imperial Shadows: Seeds of Knowledge and Corridors of Migration in Northern New Spain*. Radding is past president of the American Historical Association Conference on Latin American History (2011–13).

Daren Ray is a Visiting Assistant Professor at the American University in Cairo. His research draws on historical linguistics, ethnography, and colonial studies to examine the historical development of East African communities from early times to the present. He has received funding for his research as a Fulbright–Hays Fellow (2009) and as a Dissertation Proposal Development Fellow with the Social Science Research Council (2008). He received his PhD in African history from the University of Virginia in 2014.

Paul Readman is Senior Lecturer in Modern British History at King's College London. He is the author of *Land and Nation in England: Patriotism, National Identity and the Politics of Land 1880–1914* (2008). His other publications include (as co-editor with M. Cragoe) *The Land Question in Britain, 1750–1950* (2010) and (as co-editor with T.G. Otte) *By-elections in British Politics, 1832–1914* (2013), as well as articles in journals such as *Past & Present, English Historical Review, History, Historical Journal,* and *Journal of British Studies*. He is Principal Investigator on a major Arts and Humanities Research Council project, 'The Redress of the Past: Historical Pageants in Britain, 1905–2016.' His present research focuses on historical pageants and the place of the past in modern Britain, and on meanings of landscape in England between c. 1750 and c. 1950.

Michael Rowe is Senior Lecturer in Modern European History at King's College London. Prior to moving to King's in 2004, he held a Research Fellowship in

Nuffield College, Oxford and taught in Queen's University Belfast. His major publications include the prize-winning monograph *From Reich to State*, published by Cambridge University Press, and the edited volume *Collaboration and Resistance in Napoleonic Europe*, published by Palgrave. He is currently working on a comparative history of the Napoleonic Empire.

Matthew Salafia earned his PhD in American history from the University of Notre Dame and currently teaches at North Dakota State University. His research interests are early American history, western borderlands, American slavery and the American Revolution. His book, *Slavery's Borderland: Freedom and Bondage along the Ohio River* (2013), is a study of the Ohio River as a fluid boundary between slavery and freedom in antebellum America.

Nina Vollenbröker is a teaching fellow at the Bartlett School of Architecture, University College London. She holds a Diploma in architecture and a Masters in architectural history and theory. Her PhD research is funded by the Arts and Humanities Research Council and focuses on spatializations of home and rootedness in people who do not have a long-term attachment to a single geographical location.

Jason M. Yaremko is an associate professor with the Department of History at the University of Winnipeg, and History Program Coordinator with the Faculty of Education's Bachelor of Arts/Bachelor of Education Access Program. His current research examines indigenous diaspora, Aboriginal transculturation, identity and cultural persistence, and Aboriginal and non-Aboriginal representations of Aboriginal identity. His most recent publications include '"Obvious Indian"—Missionaries, Anthropologists and the "Wild Indians" of Cuba: Representations of the Amerindian Presence in Cuba,' *Ethnohistory* (2009), and 'Colonial Wars and Indigenous Geopolitics: Aboriginal Agency, the Cuba–Florida–Mexico Nexus, and the Other Diaspora,' *Canadian Journal of Latin American and Caribbean Studies* (2011).

Introduction: Borderlands in a Global Perspective

Paul Readman, Cynthia Radding, and Chad Bryant

In the winter of 1716–17 Lady Mary Montague crossed the border dividing the Habsburg Empire and the Ottoman Empire. As she noted in a letter to Alexander Pope, dated February 12, 1717, decades of conflict between the two empires had caused the belligerents to police the border with particular care. The Habsburg governor and the Ottoman Bassa negotiated, via courier, a place along the border frontier where Montague and her husband, Richard Wortley, could cross. A convoy of Habsburg soldiers then escorted the couple and their entourage to a small village on the border, where they were met by Ottoman Janissaries and regular soldiers. From there the couple traveled to Belgrade, then heavily fortified and filled with the tension of war. The previous year Prince Eugene of Savoy had defeated the Grand Vizier Damat Ali Pasha's two hundred thousand-strong army near the spot where Montague had crossed into the Ottoman Empire. Eugene of Savoy had now set his sights on Belgrade, which he successfully captured a year after Montague's journey from England to Constantinople had ended.

The Habsburg–Ottoman conflict had devastated the countryside on both sides of the border. Passing over the fields near Sremski Karlovci,[1] Montague described for Pope a land still 'strewed with the skulls and carcasses of unburied men, horses, and camels.' 'I could not look without horror,' she continued,

> on such numbers of mangled human bodies, nor without reflecting on the injustice of war, that makes murder not only necessary but meritorious. Nothing seems to be plainer proof of the irrationality of mankind ... than the rage with which they contest for a spot of ground, when such vast parts of fruitful earth lie quite uninhabited.[2]

And yet, Montague wrote, this was not a simply a place where empires battled to draw and redraw borders as part of a maddeningly irrational international

1

contest of power and prestige. Habsburg soldiers, without pay and forced to arm themselves, had turned to plunder. 'They rather look like vagabond gypsies, or stout beggards, than regular troops,' she observed.[3] On the other side of the border, Janissaries had recently murdered their Bassa for attempting to restrict their plundering. Peasants on both sides of the border lacked for food and clothing, yet they managed to eke out a simple life amidst the destruction.

Perhaps most surprisingly, rather than dividing populations the Habsburg–Ottoman border formed an axis around which various languages, cultural practices, and religious traditions turned. The priests on the Habsburg side of the border combined elements of Catholic and Greek Orthodox religions while 'letting their hair and beard grow inviolate, [and thus] make exactly the figure of Indian Bramins.'[4] As she traveled south Montague noted further cases of this religious interblending in a letter of April 1, 1717 to Abbé Conti, a favorite in George I's court. She took special note of the Arnouts, later called Albanians, who,

> living between Christians and Mahometans, and not being skilled in controversy, declare that they are utterly unable to judge which religion is best; but to be certain of not entirely rejecting the truth, they prudently follow both. They go to the mosques on Fridays, and to church on Sunday, saying for their excuse, that at the day of judgment that they are sure of protection from the true prophet; but which that is, they are not able to determine in this world.[5]

Later into the nineteenth century one local described for Edith Durham a similar practice and the 'light way religion hangs on the Albanian.'[6]

Montague, like most travel writers, was keen to note the different and exotic for her readers. (She spent considerable time describing the Austrian court's strange custom of employing personal dwarves, for example.[7]) Yet her observations, for the most part sober and penetrating, neatly point to a number of ways that historians have come to understand borderlands in the eighteenth and nineteenth centuries. Borderlands are places where states, empires, and other sources of governing institutional authority demarcate, expand, and protect territories under their control. As such, they are historically constructed spatial entities incorporating political borders or boundaries of some kind. They have often been fluid or dynamic (as in the case of the borderlands of the nineteenth-century western United States, conceptualized by the federal government of the time as an 'open frontier'). They have formed the basis of cultural interactions, exchanges, and admixtures. They have been claimed, defined, and contested by social, ethnic, and national groups, as well as by institutions, the presence of the latter or their agents—in some form—being essential for any given tract of territory to be meaningfully described, experienced, and understood as a borderland.

In an influential article, Jeremy Adelman and Stephen Aron have suggested that this institutional presence has been normatively imperial in form: borderlands, in their account, being 'contested boundaries between colonial domains,' sites 'of intense imperial rivalry and of particularly fluid relations between indigenous peoples and imperial interlopers.'[8] Yet whatever the heuristic merit of such a definition for the understanding of colonial North America, it is of limited utility for other historical contexts. Indeed, it is unduly restrictive as a definition, as borderlands could and did exist in non-imperial/colonial contexts. The institutions of empire and colonialism provided just one means by which institutional presences acted in the creation of borderlands. Such presences could be those associated with nation-states as well as empires; they could be those of supra-national entities (such as federations), of local governing institutions, of small social organizations (such as tribal or kinship groups). But in all cases, these presences necessarily wielded significant political authority over the groups with which they were associated, and also acknowledged the existence of borders (even in cases where they challenged their legitimacy).[9] These institutional presences might not in all contexts have intruded very obviously into everyday life in borderlands, but they were there, for without them, no borderland could exist.

Borders and authoritative institutional presences are thus necessary conditions for the existence of borderlands. Yet they are not sufficient on their own. Various institutions—empires, nation-states, and their agents—have divided the continent of Antarctica into separate territories, demarcated by boundary lines. But the borders of Antarctic territories have never been associated with borderlands, since institutional presence and the drawing (and contesting) of boundaries have not brought borderlands into existence.[10] This is because borderlands cannot exist in the absence of significant human exchange and interaction: as Antarctica is very largely uninhabited and has no indigenous population, its borders do not define borderlands. In this sense, following the lead of anthropologists, borderlands can be understood as ecumenes—that is to say, areas or 'regions of persistent cultural interaction and exchange.'[11] Borderlands are thus associated with boundaries recognized and contested by ordinary people on the ground. They are places defined by cultural admixture and transnationalism. They are places where—the best efforts of states aside—ideas, goods, and people move among various contact zones. Without such cross-cultural interaction, borderlands cannot be said to exist.

Borderlands, then, were worldwide phenomena during the modern era in which various authoritative institutional presences—many of them new to world history—attempted to establish borders, thus forming the basis for a myriad of reactions, counter-reactions, and interactions. Yet, despite the significance of borderlands to human experience across the globe, their study has largely remained confined within the circles of various regional specializations.

Over the course of a long career, beginning in the 1920s, Owen Lattimore produced a series of enormously erudite studies of the Inner Asian 'frontiers' of China—work which emphasized the importance of the borderland experiences and interrelationships of settled and nomadic peoples to the history of that country over the *longue durée*, and which continues to influence present-day scholarship.[12] Work by A.I. Asiwaju and others in the 1970s and 1980s stimulated research into cross-border social, cultural, and economic interactions in Africa in the context of nineteenth- and twentieth-century colonization and decolonization.[13] In more recent years, a good deal of attention has been paid to the Caucasus and the 'steppe frontier' by historians of Russia, while historians of South Asia, notably Willem Van Schendel, have begun to examine the volatile and contested borderlands of India, Bangladesh, and Burma.[14] The richest literature, however, concerns the study of the Americas, whose history cannot adequately be understood without extensive reference to the warfare and cultural exchange among settlers, states, or imperial powers and native peoples which took place in its shifting borderlands. Indeed, such is the centrality of borderlands to North American history that scholarly writing on the subject has now assumed diluvial proportions, constituting an area of specialism in itself. More recently historians of Europe, often embarking on their studies with distinctly different research questions and agendas, have made important contributions to the field of borderland studies. These two regional areas of specialization, the Americas and Europe, point the way to exciting new approaches to the study of borderlands, yet scholars in each of these fields only rarely share their ideas and research. This volume draws inspiration from each of these fields to present a global approach to the study of borderlands.

In adopting this integrative agenda, the book—as its title suggests—seeks to make a contribution to world history, a field of scholarly inquiry which in recent years has developed rapidly in the context of late twentieth- and early twenty-first-century globalization.[15] The essays that follow reflect the challenge posed by world history to the Eurocentricity and adherence to national historiographical paradigms which still characterize much historical writing in other fields.[16] They also chime with the emphasis placed by many current practitioners of world history on cross-cultural interaction and exchange, as well as on transnational history and entangled histories/*histoire croisées*.[17] (This emphasis is now well established in the scholarship, being evident in major textbooks aimed at undergraduate students.[18]) As Jerry Bentley has put it, 'particularly since the 1980s, the new world history has focused attention on comparisons, connections, networks, and systems rather than the experiences of individual communities or discrete societies.'[19] Indeed, though the study of borderlands has yet to receive much sustained attention from scholars of world history, doing so offers an excellent means of exploring these interrelations. This is because borderlands—as ecumenes—are key sites of intercultural contact, conflict, exchange, and identity formation. As such, they offer a way of giving locational specificity to the doing of world history, of rooting it in particular places rather than seeking, somehow, to encompass the whole globe—an approach that has usually led to diffuse, patchy, and surface-level treatment.[20] The approach taken here, by contrast, seeks to contribute to the understanding of the history of the world by studying it thematically, by examining a particular subject that speaks to wider questions.[21] While the range of topics for such a project is inexhaustible, borderlands as key features of the modern world offer notably good potential for inquiry in this vein.

Borderlands in the Americas

Borderlands as a concept in the history of the Americas developed at the interstices of the principal European imperial spheres in greater North America: the Spanish, arising from the Caribbean, Central America, and Mexico; the French, in both Louisiana and Canada; and the British, extending along the Atlantic seaboard and spreading westward to the Appalachian Mountains. Its imperial context flowed from the distinct but entwined themes of frontiers, understood as the limits of European settlement and influence, and of wilderness, a somewhat mythical space dominated by the forces of nature and inhabited by unconquered indigenous peoples. The conventional understanding of *frontiers* as the movable and contested limits of European expansion framed the initial meaning of *borderlands* conveyed by nineteenth- and early twentieth-century historians Frederick Jackson Turner, Francis Parkman, and Herbert Eugene Bolton.[22] Scholars working in this tradition tended to envision European expansion as an inevitable process and portrayed borderlands not so much as wilderness, but as contested territories on the edges of effective governance by any one imperial power.

Borderlands histories of this genre produced epic narratives of European exploration, conquest, and settlement. At the same time they established strong archival bases for the analysis of the imperial economies and institutions which, in their view, shaped frontier regions. Thus, this foundational phase of borderlands historiography focused on religious missions, mining centers, and military presidios as the poles of Spanish presence on the northern fringes of the viceroyalty of New Spain. Bolton's legacy, in particular, weighed heavily on Spanish-American historical traditions through his own prodigious scholarship and his students who continued this line of research. Subsequent generations of historians trained in both Latin America and the US West took the Boltonian and Turnerian foundations as a point of departure to advance borderlands scholarship in new directions, beginning in the last third of the twentieth century. Eschewing to a degree the Eurocentric and imperialist frameworks that had defined the early borderlands narratives, and informed by the work of cultural geographers, archaeologists, and anthropologists, historians expanded the range of historical actors to foreground the indigenous peoples and local, ethnically mixed populations. This new phase of borderlands scholarship continued the archivally based research of its founding generation, focusing on some of the same institutions, but used them as windows through which to imagine the communities that lived within them and to pose questions about changing ethnic identities over time.[23]

Borderlands developed as a recognized field of history largely in North America—in the colonial territories that would become the United States and Canada, as is profiled in the chapter by Benjamin H. Johnson in this volume. Borderlands scholarship has developed parallel themes for South America in the many contested boundaries across Spanish and Portuguese dominions and in the vast expanses of grasslands, wetlands, mountain steppes, and rain forests where indigenous bands and tribal federations held sway. The majestic Andean cordilleras define a vertical backbone to the South American continent, and their eastern piedmont and tropical lowlands constitute ecological, ethnographic, and political borderlands. In the same vein, the greater Amazonian and Paraguayan river basins trace geographical borderlands that have marked the historical processes of their peoples over centuries. Imperial maps of South America emphasized the official, but shifting, boundary between Portuguese America (Brazil) and Spanish America; yet, broad territories within the subcontinent remained under the control of equestrian and riverine indigenous peoples, such as the multiple bands of Gê-speakers of the interior of Brazil, and the Guaraní, Guaycurú, Charrúa, and Mapuche of Paraguay, Argentina, Uruguay, and Chile. As in North America, mission histories have guided the historiography of the South American borderlands in important ways, but the Spanish system of presidios did not develop in the same way as in the northern Mexican borderlands, and the impressive profiles of nomadic and semi-nomadic peoples as powerful traders and livestock breeders in the pampas have, to a certain extent, dominated the field.[24]

Current modes of inquiry among borderlands historians have expanded the range of borderlands themes, framing their work in the light of anthropological and environmental points of view. In North America Richard White's now famous formulation of the 'middle ground' to characterize the *pays d'en haut* surrounding the Great Lakes in the wetlands and grasslands of both Canada and the US emphasized the processes of transculturation that were set in motion by repeated imperial encounters.[25] For South America, anthropologist Thierry Saignes shaped scholarly views on the complex ethnic mosaics of indigenous borderlands and 'forgotten peoples' in the *selva* east of the Andes.[26] In both subcontinents the pervasive history of enslavement of both indigenous and African captives has heightened the sense of tension and violence in the concept of borderlands and in the narratives of specific regions. In the historiography now spanning nearly a century, the meaning of borderlands has shifted from the notion of Europeans 'taming' a wild and dangerous frontier to a zone of cultural commingling, and—more recently—to contested spaces marked by violent encounters among multiple European, Afro-descendant, and indigenous groups.[27]

In the context of the Americas, borderlands histories may be characterized as a counterpoint that recognizes both imperial structures of power and the durability of indigenous territories. We find a creative tension within this counterpoint between the dual emphases on the agency and resilience of different peoples living in the borderlands and histories of their displacement and loss. Recent comparative histories and collective volumes highlight this tension at the same time that they excavate a historical archaeology of layered territorial spheres of power.[28]

New directions in borderlands studies in the Americas incorporate spatial theories of history and employ interdisciplinary methods of research and analysis. Thematic emphases in the scholarship point to the construction of networks of trade and of protocols for negotiating and sharing power among historical actors albeit across gendered and racialized demarcations that signal shifting degrees of inequality.[29] Yet in order to advance the conceptual frameworks for this field even further, we must recognize the need for histories of borderlands that expand the conversation beyond the Americas to other geographical spheres.

Historians and Borderlands beyond the Americas

Compared to historians of the Americas, scholars of European history came later to the study of borderlands. Such tardiness was in part a function of the hold of the Turnerian 'frontier' paradigm. While this interpretive framework seemed applicable to the North American context and also, perhaps, to British and Russian imperial expansion,[30] it did not appear relevant to the explication of developments elsewhere. Lattimore was an early and effective critic of its applicability to the Chinese case.[31] It never really seemed suited to mainland Europe, either. One key reason for this was the long-assumed naturalness of European borders (and thus borderlands), which were seen almost as ontological givens, derived from the physical geography of the continent itself. Enlightenment rationalism had defined rivers, mountain ranges, and coastlines as natural delimiters of sovereignty, the corollary of this being that political borders should coincide with them where possible. This way of thinking helped justify state expansion, as in the French case, as well as resistance to changes in the geopolitical status quo; it also helped to establish a more general conception of borders—and by extension borderlands—as natural entities whose existence was to be taken as read, part of the normative order of things.[32] This was a fallacy, of course, all borderlands being human constructions and as such subjective, negotiated, and contested; but its strength in public discourse across the modern period was unquestionable, and persists today as a still-commonplace unspoken assumption. Indeed its power and persistence may well form part of the explanation for why the study of European borderlands was for many years neglected by historians, at least by comparison with those of North America. In contrast to North American borderlands, most European borderlands seemed to tell no especially dramatic or significant stories, at best appearing as the backdrop or context for larger narratives, as places where wars were fought and on which nationalists focused their irredentist grievances. At least in west-central Europe, there was no equivalent to Frederick Jackson Turner's western frontier.[33]

Despite the academic discrediting of the idea of 'natural' borders,[34] many European borderlands remain relatively understudied: notwithstanding path-breaking work by Peter Sahlins and others,[35] the history of borderlands in Europe can still be described as a developing field. The continued efflorescence of work on borderlands elsewhere in the world, North America in particular, has done something to remedy this, but has not supplied anything like a comprehensive corrective. Indeed, in one sense this scholarship presents problems for the historical study of Europe's borderlands. While extremely sophisticated and productive of many illuminating findings, the methodologies developed by scholars of North American borderlands—White's 'middle ground' is a notable example[36]—are typically of limited utility when applied to the European context. This is because on the European continent the historical experience of the eighteenth and nineteenth centuries did not generally feature dynamically changing borderlands (or frontiers), bound up with imperial encounters with indigenous peoples and large-scale settlement projects, as was the case in North America. Russia has provided a partial exception to this rule (the process of colonization having long been regarded as central to any understanding of that country's history),[37] even if the older scholarship tended towards the Turnerian 'march of civilization' thesis.[38] Influenced by the work done by 'new western historians' such as White, nuanced understandings of Russian expansion have now been developed, though the focus has typically been on Siberia, the Caucasus, and other Asian rather than European borderlands of Russia—as it is for these places that the approaches first developed for the North American context have most explanatory power.[39]

As the Russian case illustrates, historical approaches to the study of borderlands have been powerfully conditioned by the North American example. While this has certainly enriched the scholarship, it is potentially problematic as most European borderlands were quite unlike those found on the opposite side of the Atlantic. This is not to say other European borderlands have attracted no historical attention, but coverage has been patchy until comparatively recently, and significant gaps remain. The borderlands of France have probably received the most sustained examination, largely on account of the important work of Sahlins, whose study of the Cerdanya valley in the Pyrenees has been influential (though arguably less with historians than with anthropologists).[40] In addition, recent years have seen significant work on Eastern Europe, with Omer Bartov and Eric D. Weitz's *Shatterzone of Empire* being one notable landmark.[41] Yet, it remains fair to say that whole geographical regions have been largely ignored in a still-developing historiography of European borderlands. For example, despite the advance of a 'four nations' paradigm in the writing of modern British history, the borderlands of the United Kingdom have received little attention, especially for the modern period.[42] Moreover, notwithstanding the work of Jim Bjork on Upper Silesia and Caitlin Murdock on Saxony/Bohemia,[43] research on borderlands has been less extensive than might be expected even in the literature on modern Germany, despite that country's eighteenth- and nineteenth-century transformation from a constellation of states to an expansive empire.

While explicable, this incomplete coverage of eighteenth- and nineteenth-century European borderlands is surprising in the light of the prominence of nationalism and national identities in recent historical scholarship. Indeed, it can be described as something of an anomaly (notwithstanding the fact that the development of national borders was not necessarily connected to the rise of the modern-day nation-state). As anthropologists and political geographers have been at pains to emphasize, borderlands can be and have been important in the construction of national cultures and identities, functioning as what Hastings Donnan and Thomas M. Wilson have called 'symbolic territories of state image and control.'[44] International borderlands are often the first and last areas of a state seen by travelers, and as such are places over which government authorities are particularly keen to establish markers of differentiation. Not the least reason for this is because in many cases other markers of differentiation are lacking: landscape features, culture, patterns of settlement and social life, and economic organization are often strikingly similar on both sides of any given borderline.[45] Indeed, historians have argued that for many European states in the eighteenth and nineteenth centuries, the apparently uncertain loyalties and identities of inhabitants of borderlands made them problematic places, potentially subversive of national cohesion and security. According to this logic it followed that their liminality had to be managed, even suppressed, by central state action, a process perhaps best described in Eugene Weber's classic study, *Peasants into Frenchmen*,[46] but one that was also powerfully evident in the modern state's monopolization of legitimate means of movement by way of passport and other documentary controls. Typically imposed with most force and visibility at borders, and in borderlands, this imposition of identity controls was not only a process through which states asserted the right to regulate the movement of individuals, it was also (through the assertion of this right) 'a central feature of their development *as* states,' as John Torpey has persuasively argued.[47]

Now, while it is true that increasingly assertive state authorities certainly sought to tame and control borderlands in this way, a top-down heuristic is of limited utility. Indeed, the drawing or reassertion of borders often brought monarchs, municipalities, nobles, church leaders, and nationalists, among others, into conflict during the modern period.[48] As Anthony D. Smith and others have suggested, national identities were not merely imposed from above on a passive populace; their ideologies were consonant—necessarily consonant—with myths and traditions cherished by the common people of localities and regions, and moreover were often shaped by input from below.[49] This could be the case in borderlands as elsewhere. As Frederick Barth demonstrated long ago, national and ethnic identities often depend upon persistent contact, interaction, and rivalry among competing groups of people.[50] In his study of the Pyrenees, Sahlins rejected the idea that borderlands and borderlanders were places and peoples to whom states did things, instead arguing for their significance as independent agents of change. Specifically, Sahlins suggested that the process of nationalization could proceed from the periphery rather than just from the center, the people of the Cerdanya mobilizing the French and Spanish states to intervene in local disputes. In this way, he argued, 'The Cerdans came to identify themselves as French or Spanish, localizing a national difference and nationalizing local ones, long before such differences were imposed from above.'[51]

Yet while Sahlins's approach suggests one way of answering Donnan and Wilson's complaint that there has been 'a relative dearth of historical studies of borders, border peoples and borderlands as motive forces in the development of ... nations and states,'[52] it too is problematic. Sahlins was right to insist on the importance of research that gives an active role to borderlands and their populations in eighteenth- and nineteenth-century European projects of nation-state-building. Our understanding of these projects (and the construction of national identities generally) remains incomplete, one-sided, without this perspective, and this alone is sufficient justification for the more extensive study of European borderlands. Yet Sahlins's approach—or too uncritical an adoption of it—risks understating the extent to which borderlands could be places of autochthonously generated peaceable transnational interactions (as opposed to germinators of conflictive national differences), and could remain so even in the context of the increasingly insistent nationalizing imperatives of European modernity. As William Douglass has shown in his work on the Basque country, Pyrenean communities on either side of the Franco–Spanish borderline were less inclined to define themselves in confrontational opposition to their 'French' or 'Spanish' counterparts than a reading of Sahlins might suggest. These Basque communities, Douglass notes, made transnational agreements with each other concerning such matters as the access of livestock to scarce pasture land, and continued to operate these agreements even in times of war—when they functioned as de facto local peace treaties.[53] And more recent scholarship has shown that similar arrangements existed in other borderland regions too. The fishing communities on the French and English sides of the Channel, for example, also entered into truces that cut across presumed national allegiances, Anglo–French divisions not being as sharply defined in this maritime borderland as the influential work of Linda Colley might suggest.[54] As late as the end of the nineteenth century, it would seem that many of the ethnic Poles and Germans who lived in Upper Silesia displayed a persistent ambivalence about exclusionary nationalistic categories of identity; in Jim Bjork's formulation, they were 'neither German nor Pole.'[55] And as Pieter Judson shows in his study of the borderlands of Imperial Austria, nationalists first created the idea of 'language frontiers' dividing, for example, Czechs and Germans in Bohemia. Yet, despite the nationalists' hard work of building schools, constructing tourist sites, and demonizing the opposite nationality, most locals remained frustratingly indifferent to nationalism.[56]

These recent critiques of what Bjork has helpfully described as 'the muscular teleology powering the modernist theory of nationalization'[57] suggest a vital lesson that can be drawn from the study of borderlands, and one especially germane to Europe, as it is in their explaining of the European experience that historians have applied this teleology with most force. Against that teleology the history of borderlands demonstrates the limitations, unevenness and variegated character of the nationalizing process identified as so central to any understanding of European modernity. The study of national identities, of course, is still all the rage with historians of modern Europe, but many analyses are predicated on unhelpfully binary assumptions about the character of these identities, the accent often being on the development of mutual antagonisms, divisions, and 'Othering.'[58] The social experience of borderlands suggests that the reality was more complex, that national allegiances could be ambivalent, weak, trumped by other loyalties, and reflexively adjustable depending on circumstances. This was a function of the fact that borderlands were as much zones of interaction as they were of demarcation and division. Indeed, they are best understood, as Van Schendel has argued in his work on Bengal, not as regions within nation-states, but as transnational spaces: proper study of any given borderland means studying both sides of the borderline, or at any rate acknowledging that no borderland is delimited by a national boundary.[59] As Caitlin Murdock writes,

> Thus modern borderlands—the territorial and cultural zones adjacent to these lines of division—are defined not by barriers by but movement. It is the mobility of populations, political and cultural ideas and material goods that creates lived frontier zones in places otherwise distinguished only by a few territorial markers.[60]

For historians of modern Europe, this means accepting that while borderlands can tell us a great deal about state governance strategies and ideologies, any true appreciation of their significance can only be garnered by moving beyond what in another context Neil Brenner has called 'state-centrism.'[61] Research that does this, recognizing that borderlands are transnational zones, helps undermine 'lazy assumptions that state and society, state and nation, or state and governance are synonymous or territorially coterminous'– assumptions that continue to be implicit in much writing on modern European history. In this way the study of borderlands offers a means of writing history free from what Van Schendel has labelled 'the iron grip of the nation-state.'[62] The potential of such an approach has been demonstrated in recent scholarship that uses geography rather than states or nation-states as framing devices for its analyses. Charles King's masterfully wide-ranging study of the Black Sea—by any definition a region of liminality, cross-cultural interconnection, and conflict—presents one particularly good example.[63]

Of course, being open to going beyond state-centric interpretations in this way does not mean consigning the state, as a paradigmatic reference point, to a dustbin marked error: states were crucial sources of institutional presence in many borderland contexts. However, it does involve adopting a transnational approach that does not begin with the assumption that, when it comes to writing the history of a borderland, the state (or states) should necessarily be assigned interpretive priority. After all, in some borderlands, the state receded far into the background—a point even Van Schendel seems reluctant to concede, defining borderlands as 'areas that are bisected by a state border' (a definition that itself seems predicated on a form of state-centrism).[64] Borderlands could be *intra*state as well as *inter*state. Some time ago, John Cole and Eric Wolf's pioneering anthropological study of two villages in the Italian province of Tyrol showed that despite great ecological similarities between the two places, social and ideological differences abounded to the extent that, in the authors' terminology, a 'hidden frontier' could be said to have existed.[65] And there are more obvious examples. One that is especially telling is provided by the United Kingdom after 1707, whose internal borderlands stand as a particularly good indictment of state-centric approaches. Wales and Scotland might have been parts of the same polity as England, that is, the United Kingdom, yet they also shared borderlands with England. These borderlands were independent of state boundaries, being contained within in a union of multiple national identities.[66] Indeed, in Britain an ideology of Unionism predicated on a shared acceptance of the legitimacy of the UK state proved able—at least before the First World War—to coexist in a mutually supportive or at least accommodative relationship with most forms of Welsh and Scottish national identity, even nationalism. This occurred to the extent that 'Unionist nationalism,' a term first coined by Graeme Morton to describe the mid-nineteenth-century Scottish case,[67] seems more generally applicable, with the Welsh and Scots borderlands with England being important sites for its negotiation and expression across the period as a whole. In the rest of Europe, too, national identities and nationalism found powerful expression within multinational states. The Habsburg Monarchy presents the classic example, where late nineteenth- and early twentieth-century nationalist movements 'from below' forced central state authorities to recognize the existence of separate nations, and also to define their geographical extent.[68] The contestation and establishment of internal national boundaries within these states, a process most palpably visible in borderland areas, serves as a further illustration of the limitations of state-centric approaches: for all that they were shaped by institutional presences of various kinds, very many borderlands, national and non-national alike, were not bisected by state borders.[69]

Towards a Global Perspective

Broadly speaking, then, recent years have seen the study of European and American borderlands draw away from state-centered approaches, instead giving more emphasis to the resilience of indigenous peoples and/or the transculturation that borderlands engender. This volume takes as its starting point the assumption that borderlands are not just constructions but places where states, individuals, and various groups interact within the contexts created in part by institutionally defined borders. In doing so, it aims to provide a worldwide perspective, offering carefully crafted primary source-based studies of a wide variety of borderlands across five continents. The intention is that the rich texture of borderlands and the social experience of the inhabitants of those lands, as well as the liminality of transnational and trans-imperial spaces, will emerge through the essays that follow. These essays are rooted in particular regions, but collectively seek to transcend the boundaries of their scholarly traditions. They also seek to push the study of borderlands in new directions. For example, a common theme running through many of them is an emphasis on materiality and the ways in which this materiality made possible—or hindered—the making and unmaking of borderlands as experienced by inhabitants of borderland environments. This emphasis on materiality and lived experience does not preclude discussion of borderlands' various and contested meanings, but—in contrast to much recent scholarship—the accent is as much on the 'how' and the diachronic as the 'what' and the synchronic. In other words, the essays do not only discuss what borderlands were, or were imagined to be, but how they were imagined, and how they came into being as places that were lived in, encountered, negotiated, blurred, and erased.

To bring out this and other common themes and issues central to the study of borderlands in a global perspective, the essays have been grouped into six conceptually based parts. Part I offers critical reflections on the ways in which scholars in two different continents have approached borderland histories.

In Chapter 1, Benjamin H. Johnson presents a critique of the influential interpretation, derived from Adelman and Aron's seminal *American Historical Review* article of 1999,[70] that after the middle of the nineteenth century North American national borderlands can be described, very largely, as settled and restrictive zones. On the contrary, Johnson argues, the new boundaries between the United States and Mexico and Canada remained significantly porous, open and contested places for many years after their creation, with inhabitants' relationship with them—and acceptance (or not) of them—often being markedly at variance with the assumptions of state administrations.

Johnson, of course, is intervening in an extraordinarily dense and rich scholarly literature, one that—in part thanks to his own contributions[71]—disdains the previously widespread concept of the 'frontier' as unhelpfully one-sided, predicated on an expansionist perspective. Such revisionism has not yet taken place in the writing of Australian history, where a nationalist 'frontier'-based historiography still exerts considerable influence despite the recent and considerable impact of transnational history. In fact, as Frank Bongiorno notes in Chapter 2, 'borderland' is a term that seldom appears in Australian historical literature. Yet, Bongiorno argues, a borderland heuristic offers considerable potential for understanding the dynamics of settler–Aborigine relations, and more generally reconceptualizing Australian history as the history of a continent of multiple nations and sovereignties.

Part II of the book is concerned with territoriality and landscape. The interrelationship between borderlands and physical landscape has occupied the attention of geographers for some time,[72] though historians have been relatively slow to explore it in detail. Offering three case studies from very different parts of the world, the chapters in Part II emphasize human interaction with borderland landscapes, and—echoing the preoccupations of much recent environmental history—the impact of these landscapes on social experience. In her chapter, Cynthia Radding explores the relationship between the cultivated agrarian landscapes distinctive to northern Mexico and the economic and political transformations experienced by the people of that borderland in the nineteenth century. As Radding shows, borderlands in this region were ecological and cultural transitional zones that intersected with the economic and political spheres of influence that had developed in the colonial frontiers of New Spain and the emerging republic of Mexico.

Ecological–cultural connections are also discussed in Timothy P. Barnard's essay on the Siak polity of eighteenth-century eastern Sumatra. As Barnard demonstrates, the region's natural environment—its status as a borderland of eco-niches and various forms of localized rule between the ocean and the mountains—influenced the development of the remarkably flexible yet powerful Siak state. Here, as in the case of many of the other places featured in essays in this volume, the physical character of the borderland landscape did much to shape human history. But territoriality could be very consciously politically constructed too, of course, as Daren Ray shows in his analysis of the colonial-era Mombasa region. The British reconfiguration of administrative boundaries in and around Mombasa was influenced by the perceptions of local residents, but at the same time overlaid earlier, more complexly variegated territorial demarcations distinguishing one community from another—and in particular, as Ray writes, 'assumed a degree of cultural homogeneity that local communities had never previously articulated.' In doing so, this process helped forge a new, cruder set of identities based on a binary distinction between the 'Islamic' coast and the 'native' interior.

The role of authoritative agency is explored further in the third group of essays in Part III, which focus on state action in particular. The intention here is not to reassert any totalizing state-centric paradigm but rather to demonstrate the importance—at least in some contexts—of state policy in the shaping of borderland identities and social experience. In Chapter 6, Oksana Mykhed describes the techniques employed by Russian state authorities to control, manage, and discipline their eighteenth-century borderland with Poland. Here, the extension of state control after c. 1770 into a disease-blighted and lawless region seems to have won over local inhabitants, reconciling them to imperial rule and so abetting the more complete integration of the Polish Palatinate into the Russian Empire. A similar desire to control potentially troublous borderlands and their populations was evinced by the state apparatus of Revolutionary and Napoleonic France, Michael Rowe tells us in his essay. Under the impress of war with Britain, French national borders hardened, this hardening effected and reflected by the imposition of state controls on, around, and pertaining to the boundary line itself: customs posts, territorial markers, barriers, tariff controls, and so on. In this way the nation-building and national security concerns of the late eighteenth- and early nineteenth-century French state provided the prototype for more modern forms of international border arrangements.

Yet not all borderlands were international in this sense. Or, put another way, not all were bisected by boundaries of nation-states. This can be seen through examination of other European borderland zones, which forms the focus of Part IV of the book. Borderlands can be *intra-* as well as *inter*state. One telling example of this is provided by the Anglo-Scottish case, the subject of Paul Readman's contribution. In this essay, Readman highlights the complex and multiple nature of British national identities. As Colley aptly remarked, 'identities are not like hats. Human beings can and do put on several at a time.'[73] And the English–Scottish borderland functioned as an important context for the articulation of these plural identities—identities which were not incompatible with that of a wider British Unionism, based on an acceptance of the constitutional status quo. Indeed, the plurality of and interplay between national and other identities—a point recently insisted on by Peter Mandler[74]—is well demonstrated through the study of borderlands, which were after all important zones of cultural interaction and intermingling throughout the modern period. Jim Bjork's chapter illustrates this, further problematizing the relationship between borderlands and national identities. Using a micro-historical methodology, Bjork shows that communitarian disputes in the Roman Catholic parish of Siemianowitz on the Germany–Russia border were far less about national divisions than might be expected. Participants in this church fight (over how the parish would divide into two) defined themselves and acted politically in ways that were often in complex tension with pre-sumed ethnic and national allegiances.

Part V considers the theme of labor relations in borderlands, with a view to contributing to the understanding of the variety of social experience in these contexts. A good sense of the richness of this variety is given by Jason M. Yaremko's study of the lived experience of Amerindian peoples as indentured, enslaved, and free-wage workers in Cuba. Often overlooked in Cuban histori-ography, with its stress on the process of racial integration, the experience of these migrants was multileveled and fluid, their story of adaptation and per-sistence serving as an excellent illustration of the heterogeneity of borderland societies. Yaremko's chapter elaborates on the theme of spatial dislocations in the historical processes of crossing borders, as his subjects migrated either freely or under coercion from mainland Florida and, later, Mexico to the island of Cuba. In addition, his chapter exemplifies newer interpretations of border-lands in the historiography on the Americas, linking continental territories and peoples to the maritime borderlands of the Caribbean islands with their long, dramatic history of forced labor.

Such ambiguity, fluidity, and uncertainty were all common to life in border-lands generally, a fact that runs counter to the assumption that borders neatly divide one group, with one set of experiences, from another, with quite different experiences. Borderlines have, for example, often seemed to demar-cate free from unfree (or less free) systems of work and labor. The Ohio River border, the subject of Matthew Salafia's essay, is one such case in point. Yet as Salafia shows, paralleling some of the arguments of classical and medieval historians,[75] the distinction between a slave and a free man or woman was a slippery one. The Ohio River supposedly divided the one from the other, but the labor mobility the river facilitated served to blur the distinction between waged employment and chattel slavery for African Americans, in whose eyes the difference between the two was in practice less one of kind than one of degree. Similar ambiguities were at play in borderlands in other parts of the world, such as those of Africa. In the third chapter of Part V, Lisa A. Lindsay shows how the borders of Liberia did not strictly delineate a space of freedom from surrounding spaces of slavery, and this despite the fact that the colony was founded as a place of liberty (hence its name) for African American settlers. Ironically, in acting to defend their own freedom and attack the slave trade, mid-nineteenth-century Americo-Liberians inflicted conditions akin to slavery on indigenous Africans inside the colony.

Part VI shifts the focus to individual experience. Roland Quinault discusses three prominent Victorians' observations on the Welsh–English borderlands of the nineteenth century. As with the Anglo-Scottish example, this was an *intra*- rather than *inter*state boundary zone, but one that was nonetheless experienced as a lived reality as well as a historical memory. Yet, borderlands could of course be deeply meaningful to the lives and worldviews of non-elite as well as elite individuals. More specifically, they could offer rootedness and a sense of home even in very fluid and transient settings. In her essay, Nina Vollenbröker reveals how Overland travelers in the nineteenth-century American West could, through their interaction with an ever-changing and unfamiliar physical environment, still maintain a stable sense of rootedness and personal identity. The borderland condition, Vollenbröker reminds us, need not unsettle or discomfort; indeed it may be that just such a way of being is immanent, to a greater or lesser extent, in all human experience.

Borderlands, then, complicate the histories of empires, peoples, and nation-states evermore so as they are interpreted as transnational spaces and contested grounds. Borderlands may be viewed from a continental lens, as in the broad sweep of North and South America or the dramatic regional diversity of Africa, or through the changing cartographies of Europe, as evidenced in the fraught boundary between the Habsburg and Ottoman empires described by Lady Mary Montague in the early eighteenth century. But it bears repeating that borderlands are not imposed from above, as imperial or national borderlines; for all that authoritative institutional presences are necessary for their existence, borderlands are vitally shaped at the level of local communities through short- and long-term migrations, trading networks, and various assertions of group identities. Most poignantly, individuals have given meaning to borderlands and borderland experiences. The efforts of these numberless and now often nameless persons form the basis of any scholarly understanding of human experience in these unique landscapes and waterways.

Notes

1. Carlowitz in Lady Montague's account.
2. *Letters of the Right Honourable Lady Mary Montague* (3 vols, London, 1763), I, p. 141.
3. *Letters*, I, p. 140.
4. *Letters*, I, p. 140.
5. *Letters*, II, p. 11.
6. M. Mazower, *The Balkans* (London, 2000), p. 73.
7. *Letters*, I, pp. 114–15.

8. J. Adelman and S. Aron, 'From Borderlands to Borders: Empires, Nation-States, and the Peoples in between in North American History,' *American Historical Review*, 104 (1999), 814–41, at 816–17.

9. By 'political authority' here, we mean 'the authority attached with the function of governing'; see C.W. Cassinelli, 'Political Authority: Its Exercise and Possession,' *The Western Political Quarterly*, 14 (1961), 635–46.

10. For the history and political geography of Antarctica's borders, see P. Beck, *The International Politics of Antarctica* (London, 1986).

11. This is the definition offered by Igor Kopytoff in his work on Africa: see his 'The Internal African Frontier: The Making of African Political Culture,' in Kopytoff (ed.), *The African Frontier* (Bloomington, 1987), p. 10. For other ecumenical anthropological approaches, see in particular the work of Ulf Hannerz, to which his *Transnational Connections: Culture, People, Places* (London and New York, 1996) is a good introduction (see esp. pp. 6–7, 172 n. 4). For some commentary on the utility of ecumene as a heuristic device for scholars of world history, see A. Dirlik, 'Performing the World: Reality and Representation in the Making of World Histor(ies)', *Journal of World History*, 16 (2005), 407–10. Also useful on this theme is D.J. Weber and J.M. Rausch, 'Introduction,' in Weber and Rausch (eds), *Where Cultures Meet: Frontiers in Latin American History* (Wilmington, 1994), pp. xii–xli.

12. See esp. O. Lattimore, *Inner Asian Frontiers of China* (Oxford, 1988 [1940]); Lattimore, *Pivot of Asia: Sinkiang and the Inner Asian Frontiers of China and Russia* (Boston, 1950); and Lattimore, *Studies in Frontier History: Collected Papers 1928–1958* (London, 1962). For Lattimore's continuing relevance to more recent scholarship, see J. Cotton, *Asian Frontier Nationalism: Owen Lattimore and the American Policy Debate* (Manchester, 1989), and W.T. Rowe, 'Owen Lattimore, Asia, and Comparative History,' *Journal of Asian Studies*, 66 (2007), 759–86.

13. A.I. Asiwaju, *Western Yorubaland under European Rule, 1889–1945* (London, 1976); A.I. Asiwaju (ed.), *Partitioned Africans: Ethnic Relations across Africa's International Boundaries, 1884–1984* (London, 1985).

14. See M. Khodarkovsky, *Russia's Steppe Frontier: The Making of a Colonial Empire, 1500–1800* (Bloomington, 2002); W. Sunderland, *Taming the Wild Field: Colonization and Empire on the Russian Steppe* (Ithaca, NY, 2004); T.M. Barrett, 'Lines of Uncertainty: The Frontiers of the North Caucasus,' *Slavic Review*, 114 (1995), 578–601; Barrett, *On the Edge of Empire: The Terek Cossacks and the North Caucasus Frontier, 1700–1860* (Boulder, 1999); N.B. Breyfogle, *Heretics and Colonizers: Forging Russia's Empire in the South Caucasus* (Ithaca, NY, 2005); W. Van Schendel, *The Bengal Borderland: Beyond State and Nation in South Asia* (London, 2005); S. Misra, *Becoming a Borderland: The Politics of Space and Identity in Colonial Northeastern India* (New Delhi, 2011).

15. For a useful recent survey of the field, see J. Osterhammel, 'World History,' in A. Schneider and D. Woolf (eds), *The Oxford History of Historical Writing, vol. 5: Historical Writing since 1945* (Oxford, 2011), pp. 93–112. See also J.H. Bentley (ed.), *The Oxford Handbook of World History* (Oxford, 2011).

16. See D. Sachsenmaier, 'World History as Ecumenical History?,' *Journal of World History*, 18 (2007), 465–89.

17. See, for example, 'AHR Conversation: On Transnational History,' *American Historical Review*, 111 (2006), 1440–64; M. Werner and B. Zimmermann, 'Beyond Comparison: Histoire Croisée and the Challenge of Reflexivity,' *History and Theory*, 45 (2006), 30–50; E.H. Gould, 'Entangled Histories, Entangled Worlds: The English-Speaking Atlantic as a Spanish Periphery,' *American Historical Review*, 112 (2007), 764–86; M. Seigel, 'World History's Narrative Problem,' *Hispanic American Historical Review*, 84 (2004), 431–46.

18. For the pre-eminent example of a textbook of this kind, see R. Tignor et al., *Worlds Together: Worlds Apart: A History of the World from the Beginnings of Humankind to the Present* (3rd ed., New York, 2011), the 'primary organizing framework' of which, according to its authors, 'is the theme of interconnection and divergence' (p. xxxiv).

19. J.H. Bentley, 'The Task of World History,' in Bentley (ed.), *Oxford Handbook of World History*, p. 2.

20. As in the especially notable case of the UNESCO-commissioned *History of Mankind* (6 vols, London, 1963–69).

21. For suggestive comments along these lines, see Seigel, 'World History's Narrative Problem.'

22. F.J. Turner, *The Frontier in American History* (Huntington, 1976 [1920]); F. Parkman, *The Battle for North America*, ed. J. Tebbel (London, 2001 [1889]); H.E. Bolton, *Spanish Exploration in the Southwest, 1542–1706* (New York, 1963 [1908]).

23. The University of Arizona Office of Ethnohistorical Research, formerly the Documentary Relations of the Southwest, is one of the most influential academic institutions that has consolidated multiple archival holdings through microfilm and published annotated texts in both the original languages (mainly Spanish) and English translations. The following bibliography is but a representative sample of the expansive borderlands historiography produced from the latter third of the twentieth century onward: A.T. Bushnell, *Situado and Sabana: Spain's Support System for the Presidio and Mission Provinces of Florida* (Athens, GA, 1994); E.J. Burrus and F. Zubillaga (eds), *El noroeste de México: Documentos sobre las misiones jesuíticas, 1600–1769* (México, 1986); J.F. Brooks, *Captives and Cousins: Slavery, Kinship, and Community in the Southwest Borderlands* (Chapel Hill, 2002); S.M. Deeds, *Defiance and Deference in Mexico's Colonial North: Indians under Spanish Rule in Nueva Vizcaya* (Austin, 2003); B. DeLay, *War of a Thousand Deserts: Indian Raids and the US–Mexican War* (New Haven, 2008); W.B. Griffen, *The Apaches at War and Peace: The Janos Presidios, 1750–1858* (Albuquerque, 1988); S. Ortelli, *Trama de una guerra conveniente: Nueva Vizcaya y la sombra de los apaches (1748–1790)* (México, 2007); C. Radding, *Landscapes of Power and Identity: Comparative Histories in the Sonoran Desert and the Forests of Amazonia from Colony to Republic* (Durham, NC, 2005); M. Rodríguez, *Historias de resistencia y exterminio. Los indios de Coahuila durante el siglo XIX* (México, 1995); D.J. Weber, *The Spanish Frontier in North America* (New Haven, 1992); D.J. Weber, *Bárbaros: Spaniards and Their Savages in the Age of Enlightenment* (New Haven, 2005).

24. The network of Jesuit missions for both the Spanish *selva* running north–south on the eastern margins of the Andes and in Brazil has a voluminous history; two foundational works are D. Alden, *The Making of an Enterprise: The Society of Jesus in Portugal, Its Empire, and Beyond, 1540–1750* (Stanford, 1996); S. Leite, *História da Companhia de Jesus no Brasil* (Lisboa, 1938–50). Representative newer works include D. Block, *Mission Culture on the Upper Amazon: Native Tradition, Jesuit Enterprise, and Secular Policy in Moxos, 1660–1880* (Lincoln, NE, 1994); J.S. Saeger, *Chaco Mission Frontier: The Guaycuruan Experience* (Tucson, 2000); B. Ganson, *The Guaraní under Spanish Rule in the Río de la Plata* (Stanford, 2003); G. Boccara, *Los Vencedores. Historia del pueblo mapuche en la época colonial* (San Pedro de Atacama, 2007); H. Langfur, *The Forbidden Lands: Colonial Identity, Frontier Violence, and the Persistence of Brazil's Eastern Indians, 1750–1830* (Stanford, 2006); R.J. Manrini (ed.), *Vivir entre dos mundos: Las fronteras del sur de la Argentina. Siglos XVIII y XIX* (Buenos Aires, 2006); M. de Fátima Gomes Costa, *História de um país inexistente: O Pantanal entre los séculos XVI e XVIII* (São Paulo, 1999).

25. R. White, *The Middle Ground: Indians, Empires, and Republics in the Great Lakes Region, 1650–1815* (Cambridge, 1991).
26. T. Saignes, *Los Andes orientales: Historia de un olvido* (Lima, 1985); L. Gutiérrez Brockington, *Blacks, Indians and Spaniards in the Eastern Andes: Reclaiming the Forgotten in the Colonial Mizque, 1550–1782* (Lincoln, NE, 2006).
27. Langfur, *Forbidden Lands*; Brooks, *Captives and Cousins*; P. Hämäläinen, *The Comanche Empire* (New Haven, 2008); N. Blackhawk, *Violence Over the Land: Indians and Empires in the Early American West* (Cambridge, MA, 2006).
28. C. Daniels and M.V. Kennedy (eds), *Negotiated Empires: Centers and Peripheries in the Americas, 1500–1820* (London, 2002); D. Guy and T.E. Sheridan (eds), *Contested Ground: Comparative Frontiers on the Northern and Southern Edges of the Spanish Empire* (Tucson, 1998); J.F. de la Teja and R. Frank (eds), *Choice, Persuasion and Coercion: Social Control on Spain´s North American Frontiers* (Albuquerque, 2005); E. Langer and R.H. Jackson (eds), *The New Latin American Misión History* (Lincoln, NE, 1995); R.J. Mandrini and C.D. Paz (eds), *Las fronteras hispanocriollas del mundo indígena latinoamericano enlos siglos XVIII–XIX: Un estudio comparativo* (Tandil, 2003).
29. P.N. Limerick, *The Legacy of Conquest: The Unbroken Past of the American West* (New York, 1987); J. Barr, *Peace Came in the Form of a Woman: Indians and Spaniards in the Texas Borderlands* (Chapel Hill, 2007); DeLay, *War of a Thousand Deserts*; I. Combès (ed.), *Definiciones étnicas, organización social y estrategias políticas en el Chaco y la Chiquitanía* (La Paz, 2006); Combès, *Etno-Historias del ISOSO. Chané y chiriguanos en el Chaco boliviano (siglos XVI a XX)* (La Paz, 2005).
30. For a classic British imperialist statement of 'frontier' ideology, see Lord Curzon of Kedleson, *Frontiers: The Romanes Lecture 1907* (Oxford, 1907).
31. Lattimore, 'The Frontier in History,' in *Studies in Frontier History*, pp. 469–91, esp. 489–91; Rowe, 'Owen Lattimore,' 771–2, 780–2.
32. P. Sahlins, 'Natural Frontiers Revisited: France's Boundaries since the Seventeenth Century,' *American Historical Review*, 95 (1990), 1423–51; A.C. Diener and J. Hagen (eds), *Borderlines and Borderlands* (Lanham, 2010), p. 7.
33. For the significance of Turner in American history, see J.M. Faragher, *Rereading Frederick Jackson Turner* (New Haven, 1998).
34. Geographers have been particularly insistent that all borders are human constructions rather than natural givens. For a recent statement of this view, see A.C. Diener and J. Hagen, 'Borders, Identity, and Geopolitics,' in Diener and Hagen (eds), *Borderlines and Borderlands*, p. 3.
35. The seminal work is P. Sahlins, *Boundaries: The Making of France and Spain in the Pyrenees* (Berkeley, 1989).
36. White, *Middle Ground*.
37. For a useful overview, see D. Moon, 'Peasant Migration and the Settlement of Russia's Frontiers, 1550–1897,' *Historical Journal*, 40 (1997), 859–93.
38. See, e.g., D.W. Treadgold, 'Russian Expansion in the Light of Turner's Study of the American Frontier,' *Agricultural History*, 26 (1952), 147–52; G.V. Lantzef and R.A. Pierce, *Eastward to Empire: Exploration and Conquest on the Russian Open Frontier to 1750* (Montreal, 1973); J.L. Wieczynski, *The Russian Frontier: The Impact of Borderlands upon the Course of Early Russian History* (Charlottesville, 1976). For a more recent example of the use of a Turnerian frontier in Russian history, see J. Pallot and D.J.P. Shaw, *Landscape and Settlement in Romanov Russia, 1613–1917* (Oxford, 1990), pp. 13–32.
39. See, e.g., Khodarkovsky, *Russia's Steppe Frontier*; Sunderland, *Taming the Wild Field*; Barrett, 'Lines of Uncertainty'; Barrett, *On the Edge of Empire*; Breyfogle, *Heretics and Colonizers*; N.B. Breyfogle, A. Schrader, and W. Sunderland (eds), *Peopling the Russian*

Periphery: Borderland Colonization in Eurasian History (London, 2007); B.J. Boeck, *Imperial Boundaries: Cossack Communities and Empire-Building in the Age of Peter the Great* (Cambridge, 2009). The major studies, in English, of the European borderlands of Russia remain E.C. Thaden, *Russia's Western Borderlands, 1710–1870* (Princeton, 1984), and T.R. Weeks, *Nation and State in Late Imperial Russia: Nationalism and Russification on the Western Frontier, 1863–1914* (DeKalb, 1996).

40. Sahlins, *Boundaries*. Some indication of Sahlins's influence on anthropologists can be inferred from his inclusion as a contributor in Thomas Wilson and Hastings Donnan's influential collection of 'ten anthropological case studies': T.M. Wilson and H. Donnan (eds), *Border Identities: Nation and State at International Frontiers* (Cambridge, 1998), at p. [i]; and see p. 5 for explicit acknowledgment of Sahlins's influence on anthropologists. See also H. Donnan and T.M. Wilson, *Borders: Frontiers of Identity, Nation and State* (Oxford, 1999), esp. pp. 50–3, and H. Donnan and T.M. Wilson, 'An Anthropology of Frontiers,' in Donnan and Wilson (eds), *Border Approaches: Anthropological Perspectives on Frontiers* (Lanham, 1994), esp. p. 5. For a more recent book-length study of a French borderland, see T. Baycroft, *Culture, Identity and Nationalism: French Flanders in the Nineteenth and Twentieth Centuries* (Woodbridge, 2004).

41. Omer Bartov and Eric D. Weitz (eds), *Shatterzone of Empire: Coexistence and Violence in the German, Habsburg, Russian, and Ottoman Borderlands* (Bloomington and Indianapolis, 2013).

42. The only work of any real significance for the eighteenth and nineteenth centuries being R. Morieux, *Une mer pour deux royaumes: La Manche, frontière franco-anglaise XVIIe-XVIIIe siècles* (Rennes, 2008), and his 'Diplomacy from Below and Belonging: Fishermen and Cross-Channel Relations in the Eighteenth Century,' *Past and Present*, 202 (2009), 83–125. For the 'four nations' approach see H. Kearney, *The British Isles: A History of Four Nations* (Cambridge, 1995 [1989]; J.G.A. Pocock, 'British History: A Plea for a New Subject,' *Journal of Modern History*, 47 (1975), 601–28; J.G.A. Pocock, 'The Limits and Divisions of British History: In Search of an Unknown Subject,' *American Historical Review*, 137 (1982), 311–36. Premodern British borderlands have been better served in the scholarship. For some recent examples, see A. Groundwater, *The Scottish Middle March, 1573–1625* (Woodbridge, 2010), and M. Lieberman, *The Medieval March of Wales: The Creation and Perception of a Frontier 1066–1283* (Cambridge, 2010).

43. J.E. Bjork, *Neither German nor Pole: Catholicism and National Indifference in a Central European Borderland* (Ann Arbor, 2008); C. Murdock, *Changing Places: Society, Culture, and Territory in the Saxon–Bohemian Borderlands, 1870–1946* (Ann Arbor, 2011).

44. Donnan and Wilson, *Borders*, p. 13.

45. Having been commissioned by the Swedish government to travel to North America on a botanical fact-finding mission, the eighteenth-century Linnaean naturalist Pehr Kalm traveled through the Pas de Calais in 1748. In high seas, his ship was buffeted back and forth between the English and French coasts: 'the land on both sides has the same *facies* and appearance,' he remarked, 'so that if one who had seen the coast of England should get to see the coast of France here, and did not know it was such, he would certainly believe that it was the English coast … and English hills': *Kalm's Account of his Visit to England on His Way to America in 1748* (London, 1892), p. 455.

46. E. Weber, *Peasants into Frenchmen: The Modernization of Rural France, 1870–1914* (London, 1977).

47. J. Torpey, *The Invention of the Passport: Surveillance, Citizenship and the State* (Cambridge, 2000), esp. ch. 2–4, at p. 3. For an interpretation emphasizing the persistence of identity

controls in Europe across the whole of the nineteenth century, even after the supposed relaxation of passport requirements in most European states in the 1860s, see L. Lucassen, 'A Many-Headed Monster: The Evolution of the Passport System in the Netherlands and Germany in the Long Nineteenth Century,' in J. Caplan and J. Torpey (eds), *Documenting Individual Identity: The Development of State Practices in the Modern World* (Princeton, 2001), pp. 235–55.

48. R.J.W. Evans, 'Frontiers and National Identities in Central-European History,' in R.J.W. Evans (ed.), *Austria, Hungary, and the Habsburgs: Essays on Central Europe, c. 1683–1867* (Oxford, 2006), pp. 114–33.

49. Smith's 'ethno-symbolic' interpretation of national identity seeks 'to go beyond the "top-down" approaches of modernism in order to bring the popular, emotional, and moral dimensions of national identity back into focus.' See A.D. Smith, *The Nation in History: Historiographical Debates about Ethnicity and Nationalism* (Cambridge, 2000), at p. 77; A.D. Smith, *The Antiquity of Nations* (Cambridge, 2004), esp. pp. 17–23; A.D. Smith, *Myths and Memories of the Nation* (Oxford, 1999). See also J. Hutchinson, *The Dynamics of Cultural Nationalism: The Gaelic Revival and the Creation of the Irish Nation State* (London, 1987). For interpretations stressing the symbiosis between local, regional, and national identities, see C. Applegate, *A Nation of Provincials: The German Idea of Heimat* (Berkeley, 1990); A. Confino, *The Nation as Local Metaphor: Württemburg, Imperial Germany, and National Memory, 1871–1918* (Chapel Hill, 1997); P. Readman, 'The Place of the Past in English Culture, c.1890–1914,' *Past and Present*, 186 (2005), 147–200.

50. F. Barth, *Ethnic Groups and Boundaries: The Social Organization of Culture Difference* (Oslo, 1969).

51. Sahlins, *Boundaries*, p. 286.

52. Donnan and Wilson, *Borders*, p. 49.

53. W.A. Douglass, 'A Western Perspective on an Eastern Interpretation of Where North Meets South: Pyrenean Borderland Cultures,' in Wilson and Donnan (eds), *Border Identities*, pp. 62–95, esp. pp. 70–1.

54. Morieux, 'Diplomacy from Below and Belonging'; cf. L. Colley, *Britons: Forging the Nation 1707–1837* (2nd ed., New Haven, 2009 [1992]).

55. Bjork, *Neither German nor Pole*.

56. P. Judson, *Guardians of the Nation: Activists on the Language Frontiers of Imperial Austria* (Cambridge, MA, 2006). See also Mazower, *Balkans*.

57. Bjork, *Neither German nor Pole*, p. 5.

58. For a classic and seminal account in this vein, see L. Colley, 'Britishness and Otherness: An Argument,' *Journal of British Studies*, 31 (1992), 309–29, and also her *Britons*.

59. Van Schendel, *Bengal Borderland*, and see also M. Baud and W. Van Schendel, 'Toward a Comparative History of Borderlands,' *Journal of World History*, 8 (1997), 211–42.

60. Murdock, *Changing Places*, p. 10.

61. N. Brenner, 'Beyond State-centrism? Space, Territoriality, and Geographical Scale in Globalization Studies,' *Theory and Society*, 28 (1999), 39–78.

62. Van Schendel, *Bengal Borderland*, at p. 8.

63. C. King, *The Black Sea: A History* (Oxford, 2004).

64. Baud and Van Schendel, 'Toward a Comparative History of Borderlands,' 226.

65. J.W. Cole and E.R. Wolf, *The Hidden Frontier: Ecology and Ethnicity in an Alpine Valley* (2nd ed., Berkeley, 1999 [1974]).

66. L.W.B. Brockliss and D. Eastwood (eds), *A Union of Multiple Identities: The British Isles c.1750–c.1850* (Manchester, 1997).

67. G. Morton, *Unionist-Nationalism: Governing Urban Scotland, 1830–1860* (East Linton, 1999).
68. J. King, *Budweisers into Czechs and Germans: A Local History of Bohemian Politics, 1848–1948* (Princeton, 2002), esp. pp. 114–52.
69. For recent research taking issue with the approach of historians who have 'tended to elide the work of *nation* building with the particular requirements of modern *state* building,' see P.M. Judson and M.L. Roszenblit (eds), *Constructing Nationalities in East Central Europe* (Oxford, 2005), at p. 2.
70. Adelman and Aron, 'From Borderlands to Borders.'
71. See, most recently, B.H. Johnson and A.R. Graybill (eds), *Bridging National Borders in North America: Transnational and Comparative Histories* (Durham, NC, 2010), and P. Hämäläinen and B.H. Johnson (eds), *Major Problems in the History of North American Borderlands* (Boston, 2012).
72. See, e.g., D. Rumley and J.V. Minghi (eds), *The Geography of Border Landscapes* (London, 1991).
73. Colley, *Britons*, p. 6.
74. P. Mandler, 'What Is "National Identity"? Definitions and Applications in Modern British Historiography,' *Modern Intellectual History*, 3 (2006), 271–97.
75. See, e.g., A. Rio, 'Self-Sale and Voluntary Entry into Unfreedom, 300–1100,' *Journal of Social History*, 45 (2012), 1–25, and 'Freedom and Unfreedom in Early Medieval Francia: The Evidence of the Legal Formulae,' *Past and Present*, 193 (2006), 7–40.

Part I
Writing Borderlands

1

Negotiating North America's New National Borders

Benjamin H. Johnson

The introductory essay of this volume refers to the 'diluvial proportions' of the literature on borders and borderlands in North American history, a description that might with equal justice be applied to the entire historiography of the United States and the imperial currents from which it emerged in the late eighteenth century. Like US historiography as a whole, the body of work on North American borderlands simultaneously benefits and suffers from its richness and size: its students have the privilege of joining a scintillating and vibrant conversation, but at the same time its volume can all too easily crowd out the discussions emanating from other rooms in the mansion of history. Borderlands, as the introduction also notes, 'were worldwide phenomena during the modern era,' yet North Americanists have rooted their accounts in the distinctive regional, colonial, and national histories most clearly shaping their subjects.[1] We have thereby left largely unexplored the question of whether or not there are important enough dynamics of borders and borderlands *in general* to warrant the kind of broadly comparative approach undertaken in this book, and locked ourselves in too close a conversation to learn from those examining similar developments elsewhere.

To historians interested in borders and borderlands, the replacement of empires by nation-states might appear as the key turning point in the history of North America, for nations seem invested in their boundaries in ways that empires are not. Jeremy Adelman and Stephen Aron emphasize just this viewpoint. They argue that in places where European empires competed for control—the Great Lakes, the Mississippi Valley, and Spanish Texas and New Mexico—native peoples often maintained broad autonomy and power. But in the nineteenth century, when European powers gave way to the nation-states of Canada, the United States, and Mexico, natives ended up as conquered peoples forced to live in the context of national, not borderland, societies. 'Hereafter,' they write, 'the states of North America enjoyed unrivaled authority to confer or deny rights to peoples within their borders.' These new borders

28

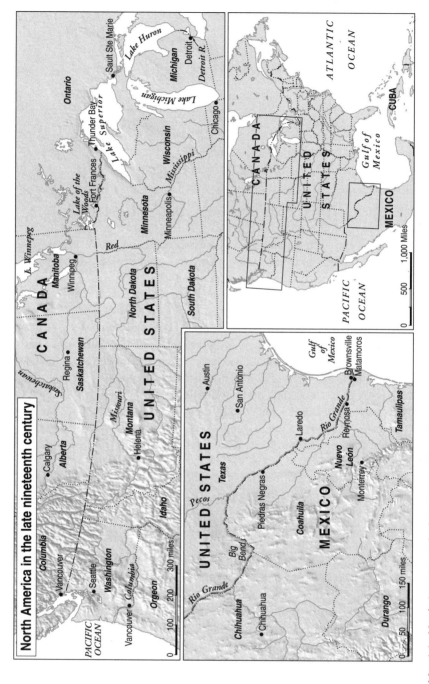

Map 1.1   North America in the late nineteenth century

divided North American peoples in new ways, but also had ramifications 'for internal membership in the political communities of North America.' 'The rights of citizens—never apportioned equally—were now allocated by the force of law monopolized by ever more consolidated and centralized public authority.' 'With the consolidation of the state form of political communities,' they conclude, 'borderland peoples began the long political sojourn of survival within unrivaled polities.'[2]

Adelman and Aron were criticized, sometimes harshly, for downplaying the power of native peoples, for minimizing the ways in which nation-states were continuations of colonial processes rather than breaks from them, and for assuming that nations were more powerful than they were in actuality.[3] Yet whatever the inadequacies of their depiction of different North American regions, the focus on the territorial ambitions of nation-states raises productive questions for scholars examining North America and suggests some ways to relate North American history to global history. Indeed, scholars of empire more generally point to the importance of similar shifts as those noted by Adelman and Aron. 'Imperial and national frontiers,' notes Charles Maier,

> usually enclose different processes of governance and institutional structuration within their respective territories. The nation-state will strive for a homogenous territory. It imposes taxation, not equally on all classes, but more equally than an empire on all districts ... eventually it strives for internal improvements and developments. Because of their size, and their assumption of power over old states and communities, empires possess a far less administratively uniform territory. They accommodate enclaves with local liberties and charters.[4]

Other scholarship points to a similarly more exclusive concept of territory, but links it less to the shift from empire to nation than to the rise of agrarian and industrial capitalism, which thrived in bounded (if linked) national economies where states guaranteed property rights and created accessible national markets. The most dynamic polities of the nineteenth century envisioned territory 'not just as an acquisition or as a security buffer but as a decisive means of power and rule.'[5]

In the case of liberal Republics such as the United States, Mexico, and ultimately Canada, territory was particularly tightly bound up with the question of sovereignty and national identity. As President Abraham Lincoln told his Congress in 1862,

> A nation may be said to consist of its territory, its people, and its laws. The territory is the only part which is of certain durability. 'One generation passeth away, and another generation cometh, but the earth abideth forever.'

That portion of the earth's surface which is owned and inhabited by the people of the United States is well adapted to be the home of one national family, and it is not well adapted for two, or more. Its vast extent, and its variety of climate and productions, are of advantage ... for one people, whatever they might have been in former ages.[6]

This chapter uses the early history of the international borders that divided and linked Mexico, the United States, and Canada to argue that focusing on state power *and its limits* in this period would let North Americanists write accounts of these places that simultaneously reflect their historical specificity and are more open to comparison with similarities elsewhere. Specifically, it examines efforts to map the newly drawn borders and to police them against other claims to sovereignty mounted by native groups and border leaders. The unevenness and paradoxes of the state-building and territorial consolidation that redrew the continent's map in the mid-nineteenth century not only make it helpful for historians to compare the two international borders to one another (rarely enough done), but also to make their studies part of a worldwide conversation about borderlands and borders.[7]

Modern maps tell us that borders are international spaces—zones made by the encounter of empires and nation-states. And yet even with the most powerful nation-states, they remained in critical ways contested, open, and permeable, to the frustration of metropolitan dreams of discrete sovereign spaces. In North America as elsewhere, non-national geographies, economies, and identities persisted into the twentieth century, and in some ways up to our own time. This tension is perhaps the great central theme of modern borderlands history.

On the Ground

The modern map of North America took shape with the 1848 Treaty of Guadalupe Hidalgo between Mexico and the United States and the extension of the US–Canada border from the Rocky Mountains to the Pacific along the 49th parallel after American saber-rattling. But maps are abstractions, and the first efforts to accurately survey the borders suggested just how weak was the grasp of central governments.

The joint US–Mexico survey provided for in the Treaty of Guadalupe Hidalgo was to begin near the Spanish Mission at San Diego, California. But because the land route across the continent was so grueling and dangerous, the US party instead traveled by boat to Panama, crossed the narrow land bridge there, and sailed to California. The journey to Central America was uneventful, but once the US commissioners arrived there in March 1849, the flood of traffic prompted by the California gold rush delayed them for two full months, exhausting much of their funding in the process. Quarreling over finances was joined by a deeper split within the US side of the commission, with the northerners suspicious that southerners were intent on finding a southwestern route for a transcontinental railroad, and thus ensuring the spread of slavery westward. When the principal commissioners left the surveyors near the Gila River in Arizona for what was supposed to be a short resupply trip to Sonora, they became lost, spent several weeks in the desert, and became gravely ill. Pedro García Conde, the Mexican commissioner, died, while his US counterpart spent several months recuperating.[8]

It was not until four years later, in the summer of 1853, that mapping of the Rio Grande section of the border even began. Even this was hard to pull off: Yellow Fever killed the party's doctor—and nearly one of the US head commissioners—while they awaited transport in Florida. Hurricanes turned the journey across the Gulf of Mexico, which usually took five days, into an eighteen-day ordeal. Once on land, the party depended on the protection of the US Army from the Apache, Comanche, and other Indian peoples who seemed not to know or care that their lands were now split between the United States and Mexico. Some of the river proved simply impossible to survey. The rugged terrain of the Big Bend country struck William Emory with its desolate beauty. 'No description,' he wrote, 'can give an idea of the grandeur of the scenery through these mountains. There is no verdure to soften the bare and rugged view; no overhanging trees or green bushes to vary the scene from one of perfect desolation.' The qualities that made the area visually striking precluded its adequate mapping, however. 'Rocks are here piled one above another,' continued Emory,

> over which it was with the greatest labor that we could work our way. The long detours necessarily made to gain but a short distance for the pack-train on the river were rapidly exhausting the strength of the animals, and the spirit of the whole party began to flag. The loss of the boats, with provisions and clothing, had reduced the men to the shortest rations, and their scanty wardrobes scarcely afforded enough covering for decency. The sharp rocks of the mountains had cut the shoes from their feet, and blood, in many instances, marked their progress through the day's work.

In the face of such hardship, the Commission headed south deep into Mexico, leaving this stretch of the border unsurveyed.[9]

Similar challenges confronted the United States, Canada, and the United Kingdom as they parsed out the northern border. The 1846 treaty between Britain and the United States ended conflict—and perhaps avoided a war—by setting the 49th parallel as the land border between the Pacific and the Rocky Mountains. (In 1818 they had agreed on the 49th parallel as the border from Lake of the Woods to the Rockies.[10]) The 1856 Fraser River gold rush prompted colonial authorities to propose a joint survey of the still-unmarked line separating Washington

Territory and British Columbia. Like their counterparts to the south, the difficulties of a land crossing required the surveying parties to travel across Panama and head up the Pacific coast by ship, a journey of some three months.[11] The survey of the border between the ocean and the Rockies was completed in 1861. Again conditions on the ground hampered the work and suggested the limits of state power. The governor of British Columbia repeatedly requisitioned the British surveying party to keep the civil peace in his raucous gold-rush territory.[12] In the absence of any real infrastructure, the region's topography posed considerable challenges. The intention was to map the line and to leave occasional markers, particularly where rivers and other notable natural landmarks intersected the parallel. This could require risky and dramatic efforts, as in one instance recalled by the British surveyor Charles Wilson:

> We have made a bridge over the Chilukweyuk river here, which caused some trouble, we managed it however by felling two trees each about 150 ft long on opposite sides of the river so that their ends rested on a small island in the middle of the river & over this we can now pack mules. Two of our men nearly lost their lives in the operation, having fallen into the rapids from which they were drawn out almost miraculously ...[13]

The mosquito—an animal with its own territoriality—became a major player in this survey. At first diarists simply noted their abundance and aggressiveness with some sense of amazement, but things soon became more ominous. 'Found poor Buckner in bed,' noted Wilson in July of 1859, 'his mind is rather irritable & the mosquitoes have worked him so that he has scratched into a vein in his neck which bled a good deal & we had to keep bathing it during the day.' The next day, 'Buckner was very weak from loss of blood & could hardly stand.' The British party suspended work for a day. Five days later, the axemen hired to blaze a clearing along the border walked off the job in disgust at the mosquitoes. Wilson began to suffer considerably: 'My hands, during the last few days, have been so swollen & stiff that I could hardly bend my joints & have had to wrap them in wet towels to be ready for the next day's work.' The expedition's pack animals joined him in misery: 'two of [our] mules have been blinded & 6 of our horses were so reduced that we had to turn them out on the prairie & let them take their chance of living. I never saw anything like the state of their skins, one mass of sores.' By the end of the month work was suspended and the party split between Vancouver Island, already surveyed, and higher portions of the route where the bugs were not unbearable.[14]

The American Civil War derailed any intention to complete the survey from the Rockies to Lake of the Woods. This effort did not begin until 1872, more than fifty years after Britain and the United States agreed upon that portion of the border. The more easily traversed topography of the Plains and the presence of railroad lines in Minnesota and Dakota Territory made this survey more easily conducted than its western counterpart. But here too it was difficult for governments simply to transport dozens of men and their surveying equipment the length of the line. Much of the Dawson Route, the name for the 480-mile 'road' from Winnipeg to the Lake of the Woods, was barely a trace across the land. Travelers heading west on the only overland route in Canada connecting the Great Lakes and Manitoba were put on wagons that traversed corduroy roads that linked 'a succession of more than a dozen lakes and rivers.' As a recent account relates,

> Baggage would be thrown from the wagons and jammed onto a small, usually overloaded and dangerously under-maintained steamboat or tug. At the far end of the lake, baggage and passengers would again take to wagons for another bone-rattling ride along a collapsing corduroy road, followed by another lake crossing, another wagon ride and so on for up to two weeks.

The stretch of the road from Winnipeg to the Lake of the Woods, where the survey was to begin, horrified Samuel D. Anderson, head of the Canadian commission, as it ran 'through a most desolate swamp, a bottomless bog over which it is dangerous to walk as a man sinks into it to his waist. The road was supported on the branches and stems of trees and on each side there was the brown marshy water with grass and reeds growing in it.'[15]

The exhausting labor of surveying these borders resulted in minimal demarcations and control. For decades after the borders had been surveyed, they remained largely unmarked; the physical reality of the borders did not reflect the theoretical claims to territorial sovereignty asserted by Ottawa, Washington DC, and Mexico City. The work of the surveys themselves was problematic and led to several recalculations, moving the line hundreds of yards, with consequent problems to title. One such resurvey in 1870 seemed to place a Hudson Bay Company post, built shortly north of what the Company initially thought the border, hundreds of yards within US soil. A zealous American customs inspector promptly seized the post and inventoried its contents to assess the customs duties on them.[16]

States did go to considerable efforts to mark their territory. But these markings were dwarfed by the landscape, and often collapsed back into it. In the 1870s, after the incident with the trading post, Canadian and American authorities remarked their shared border between Lake of the Woods and the Continental Divide, leaving 135 iron pillars, 129 stone cairns, 113 earth mounds, three stone and earth mounds, and eight timber poles along a stretch of nearly one thousand miles.[17] The situation on the southern boundary was similar. An American resurvey of the border with Mexico in the 1890s found that many of the stone markers used two generations before had been moved or had simply disappeared.[18] Remarking the land border aided in the enforcement of customs duties—at the time, still the primary source of state revenue.[19] The first fences demarcating North American borders, to my knowledge, appeared in 1909 to stop livestock movements across the California–Baja California line.[20] It was in the same decade that the US–Canada border was resurveyed and enough markers erected to make the longest gap between them 1.5 miles rather than 25.[21]

The difficulties in conducting these surveys and their limitations did not mean that they did not matter or ultimately happen. But if the modern state seeks to control territory and hold it against other claimants, it must first map it and create some kind of infrastructure for its forces to travel across it. This was no easy task, and one barely completed by the dawn of the twentieth century.

Native Sovereignties

The challenges of mapping the physical terrain of the borderlands became particularly manifest when nation-states sought to extinguish the military power of independent Indian peoples, perhaps the most obvious threat to their claims to exclusive territorial sovereignty. This is a familiar if morose story: the process was complete by the 1880s, ushering in a period rightly considered to be the nadir of native history in all three nations. In the north, Plains peoples were defeated by the combination of the collapse of the buffalo herds, extended US military campaigns, and Canadian withholding of rations they obligated themselves to provide by treaty. In the south, Apaches proved to be the last holdouts, delaying the settlement and economic development of a vast area straddling Sonora and Arizona, and ultimately only defeated by a cooperative effort between Mexican and American forces.

The brutality and long-term consequences of these conquests can obscure some of what they reveal about national borders in this period. One easily forgotten aspect is the extraordinary difficulty that states had in waging these campaigns, given the enormous advantage in population, economic resources, and political centralization that they enjoyed. Here geography and climate worked to the advantage of natives. Mexican leaders had long been aware of the military might of Comanches, Apaches, and others, and in the Guadalupe Hidalgo negotiations cleverly insisted that US authorities commit to suppressing 'incursions into the territory of Mexico.' Only three years after the signing of the treaty, the US government abandoned this effort and sought release from this treaty provision. As Secretary of State Daniel Webster wrote,

> The hostile acts of the Indians whose homes are in the territory ceded to the United States by the Treaty of Guadalupe Hidalgo, have not been confined

to Mexican citizens only, but have probably been as frequent, as destructive and as barbarous on citizens of the United States, especially of North Western Texas, New Mexico, and California ... It is obvious that along a frontier of such an extent, most of it a rugged wilderness, without roads of any kind and impassable, not only by wheeled vehicles but perhaps even by horses, no means which could have been employed since the Treaty of Guadalupe Hidalgo went into operation, would have sufficed to prevent incursions of United States Indians into Mexican territory. The subsistence, forage, and ammunition of the troops must necessarily have been conveyed from one or the other extremity of the line of boundary, and without roads, this would have been impracticable. It is also notorious that that part of the boundary which extends from the Rio Grande to the Gila [roughly modern-day El Paso to central Arizona], and which is not a natural line, such as those rivers afford, has not yet even been marked. This would in any event have rendered it uncertain where a road for the conveyance of our military stores ought to have been constructed or where our troops should have been posted.[22]

The climate and weather challenges of the northern plains posed similar challenges for American authorities. As with the boundary surveys, military commanders confronted the difficulties of moving larger numbers of men, supplies, and equipment across a landscape with little to no modern transportation infrastructure. One enterprising commander charged with pursuing Sioux forces in the early 1870s found repeated frustrations with river crossings, an obstacle that he overcame by improvising pontoon bridges out of empty whiskey kegs roped to upside-down wagon beds.[23] Here the frigid winters were the analogue to the desert heat and aridity of the southern borderlands. Ultimately it was winter campaigns that led to the military defeat of the Plains Indian peoples, but winter campaigning could be fatal to the American military even in the absence of actual fighting. 'The ground was so hard that driving a tent pin, which had to be iron, was almost impossible, and the removal of it was so difficult that we had often to tie our tent ropes to trees or bushes,' recalled a cavalry captain of moving across the landscape in the winter of 1875. 'All food was frozen solid and had to be thawed before cooking, bits had to be warmed before being placed in the mouths of the horses, and any teamster who touched a trace chain or iron part with bare hands would quickly drop it or be blistered.' The cold soon became enough of an enemy:

Our trail was lost, or obliterated by the snow; our eyes were absolutely sightless from the constant pelting of the frozen particles. And thus we struggled on. A clump of trees or a hill for shelter from the killing and life-sapping wind would have indeed been a sweet haven. With frozen hands and faces, men becoming weaker and weaker, many bleeding from the nose and the ears, the weakest lying down and refusing to move, a precursor of death; with them the painful stinging bite of the frost had been succeeded by the more solid freezing, which drives the blood rapidly to the center and produces that warm, delightful, dreamy sensation, the forerunner of danger and death. They had to be threatened and strapped to their saddles, for if they were left behind, death would follow ... Ours now was a struggle for life, to halt was to freeze to death, to advance our only hope ...[24]

The second easily overlooked point is that, in the arena of Indian fighting, the three nation-states of North America came to operate as allies and collaborators as much as antagonists. This was perhaps epitomized in the 1880s campaigns against the Apache, in which nearly a quarter of the American standing army was deployed. US General George Crook determined that the only way to defeat Geronimo and his followers was to pursue them deep into Mexico. In 1882, the United States and Mexico signed an agreement providing 'that the regular federal troops of the two republics may reciprocally cross the boundary line of the two countries when they are in close pursuit of a band of savage Indians.'[25]

To the north, the dynamic was more complicated. Canada did not grant the US army similar latitude in its campaigns against the northern Plains people after their shocking defeat of the 7th cavalry in 1876, and the military might of the United Kingdom kept American commanders from simply crossing the border with impunity.[26] Nor did it wage its own all-out warfare against natives. Perhaps this relative freedom accounts for the 1880 report of a Canadian officer that Indians 'call the boundary the "Medicine Line," because no matter what they have done upon one side they feel perfectly secure after having arrived upon the other.'[27] On the other hand, Canadian authorities were not above the use of hunger or withholding rations to bend native peoples to their will. Their vision of economic development and settlement required a similar pacification of Indians and the conversion of their territory into private property.[28]

Much of the literature on borders, especially that of a cultural orientation, and especially that of the US–Mexico border, stresses the idea of difference and conflict. And this is true even when the ultimate point is about the cultural hybridity of borderlands, which in these accounts emerges from the meeting of two supposedly very different cultures and peoples.[29] The theme of difference and conflict is particularly strong in the case of US–Mexico borderlands, given the two nations' conflictual histories. But what we see from the vantage of Native American history is the deep similarity between Canada, the United States, and Mexico, as settler-states and liberal republics insistent on extinguishing native sovereignties (and at many points, native identities themselves). International borders may be the products and instigators of

international conflict, but they are also the reflections of international cooperation to extinguish other sovereignties.

Border Rebellions

Indians were not the only groups whose assertions of sovereignty and people-
hood conflicted with the central states' assertions of territorially exclusive
sovereignties. Indeed, nineteenth-century North America witnessed numerous political projects whose survival would have resulted in a very different
map than the one that we have today. To name a few: William Augustus
Bowles' Creek state of Muskogee, the Lone Star Republic of Texas, the Mormon
Kingdom of Deseret, New Mexico's 1837 Chimayó Rebellion, the Canadian
revolts of 1837 and 1838, California's Bear Flag Republic, Antonio Canales's
Republic of the Rio Grande, the Yucatán peninsula's independence movement,
the Fenian 'Dominion of the Brotherhood North of the St. Lawrence,' and of
course, the largest of them all, the Confederate States of America. If today we
easily assume that the three nation-states were destined to control all of the
continent along more or less the lines that they did, then this incomplete list
suggests that people in the nineteenth century assumed otherwise.

The careers of two mid-century rebels at either border, Juan Cortina and
Louis Riel, capture some of the contingency of continental nation-building.
In 1859, Cortina, the scion of a prominent ranching family near the mouth of
the Rio Grande, led a paramilitary force that captured Brownsville, Texas, and
did battle with the Texas Rangers and both US and Mexican troops for much of
the next two years in a territory about half the size of the state of Connecticut.
Cortina and his followers feared for the safety, property, and autonomy of
Hispanics under Anglo-American rule.[30] Similarly, in 1869, Métis (mixed
European and native peoples) residents along the Red River of the North under
the leadership of Louis Riel prevented the Canadian Confederation's appointed
governor from entering their territory and created their own provisional government.[31] For a time, this resistance resulted in the protection of a bilingual
society and the preservation of Métis landholding, but many Métis moved
further west, where they again joined a rebellion lead by Riel in 1885 after his
return from the United States, for essentially the same reasons.[32]

Both Cortina and Riel acted in the name of borderlands peoples with abundant reason to fear their incorporation into the United States and Canada.
These peoples had their own homelands, which in both cases straddled the
new international borders. Cortina's can be seen as the old Mexican northeast,
whose settlement by Hispanic frontiersmen predated the border by a century
and whose trade and cattle-ranching economy continued after the Treaty
of Guadalupe Hidalgo. Cross-river commerce in livestock, saddles, blankets,
silver, and manufactured goods grew, often to the irritation of officials on

both sides, from Texas independence in 1836 through the political tumult and sporadic warfare of the Lone Star Republic, whose 1845 annexation to the United States prompted the US–Mexico war (1846–48). Borderlands merchants could put a break on racial animosity; as Corpus Christi's founder, Henry Kinney, observed, 'When Mr. Mexican came, I treated him with a great deal of politeness, particularly if he had me in his power; when Mr. American came, I did the same with him; and when Mr. Indian came, I was also very frequently disposed to make a compromise with him.'[33] And this was a political community, too: the northeast was a bastion of liberalism within Mexican politics, and after the US military forced him into Mexico, Cortina remained a key player, running guns to Union partisans in Texas and fighting as part of a politically and militarily coherent northeastern bloc against a French invasion and conservative forces in Mexico later that decade. Cortina's political and military movements in the Mexican northeast suggest that Hispanic south Texas in this period continued to be part of a regional economic and political geography that transcended the new border. Moreover, it could mobilize itself militarily, as much of it would under an 1891 revolt in which a cross-border coalition led by Catarino Garza challenged the rule of the Porfirio Díaz regime, only to be defeated by the Mexican and US armies, Texas rangers, and local militias.[34]

Similarly, in these decades, the Métis led a cross-border existence. Their community on the Plains was a creation of these decades rather than a continuation of an earlier society, as in the Mexican northeast. Starting in the 1830s, Métis living in the upper Red River Valley began trading with native peoples to their west. In the 1860s and 1870s, their trips had become frequent enough to leave distinctive trails west from the Red River, later to be used by British and American surveyors and the North-West Mounted Police.[35] In these decades, Métis founded numerous communities deep in the northern plains, on either side of the 49th parallel. Here their economy rested on buffalo hunting and trading liquor and weapons with the Sioux, Assiniboine, Gros Ventres and other Indian peoples. Much of this trade was illegal in both nations, conducted in violation of general tariffs and specific provisions limiting trade with natives. As was the case 'along the Texas–Mexico border,' notes historian Michel Hogue, 'contraband helped stitch borderland populations together across the international boundary.'[36] The Métis attempted to use this border to their advantage. When American officials seized trade goods and arrested Métis traders near Frenchman's Creek in 1874, for example, Métis complained to Canadian authorities that they had been north of the border and thus that the American actions were illegal and a violation of Canadian sovereignty. Canadian authorities ultimately accepted that the raid had taken place in US territory, but the incident 'exposed the limits of official knowledge about the Plains borderlands.'[37]

The rebellions led by Juan Cortina and Louis Riel shared important characteristics. Both invoked liberal principles in defense of their actions. In Cortina's words, 'To defend ourselves, and making use of the sacred right of self-preservation, we have assembled in a popular meeting with a view of discussing a means by which to put an end to our misfortunes.'[38] In Riel's words from the first revolt, 'a people, when it has no government, is free to adopt one form of government in preference to another, to give or refuse allegiance to that which is proposed.'[39] Militarily, both took advantage of their followers' intimate knowledge of the landscape to fight much larger and better-armed professional armies. Modern technologies marshaled by the states overwhelmed this advantage, as when Canada used its infant transcontinental railroad to mobilize forces against the second revolt.

Neither Cortina nor Riel absolutely rejected the nation against which he contended: Cortina expressed his admiration for the US constitution, pointed out that he and his followers had not renounced their citizenship rights, and condemned US authorities for allowing Mexican military authorities to cross to the north. The Métis rebels declared themselves willing to negotiate with the young Canadian government. This point is particularly important. It was not simply that the international borders of North America had yet to be solidified, but that other borders had not yet been drawn either: the borders between white and Mexican, between white and Indigenous, between Anglophone and Francophone Canada, and the general questions of who would count as citizens of the young nations, and on what terms.[40] The United States and Mexico ripped themselves apart over precisely these questions in the 1860s (and Mexico did so again in the 1910s), and the Quebec question endures into the twenty-first century. So it is no wonder that actors in the nineteenth century thought that so much was up for grabs. Just what the change from imperial borderlands to national borderlines meant was not predetermined.

Crossing to Freedom

Most of the groups discussed in this essay had reason to believe that the international borders had taken on heightened importance by the turn of the century. It is easy to overstate this transition (in Adelman and Aron's schema, 'from borderlands to bordered lands'). Borders were lightly patrolled by the measure of the later twentieth century, and contraband goods and people routinely crossed both. (Contemporary American political discourse includes calls for reestablishing 'control' of borders against sanctioned goods and people, but this is yearning for a past that never was.) But there were changes nonetheless: the borders were finally well mapped and somewhat monitored by customs and immigration agents of the three nation-states. Indian peoples on the northern plains had become largely dependent on rations provided by the Canadian and American governments, and were heavily monitored and their mobility restricted on reservations on both sides of the line.[41] After the failure of the second Northwest Uprising and execution of Louis Riel in 1885, the Métis and other Plains indigenous people ceased armed resistance to their incorporation into Canada. In the same decade in Mexico, Juan Cortina languished in federal prison in Mexico City, one of many victims of the Porfirio Díaz regime's subordination of formerly powerful northern leaders. Only one major uprising aimed at redrawing the border—the 1915–16 Plan de San Diego revolt in South Texas—would occur.

Ironically, however, the consolidation of borders into meaningful demarcations of the territory of different sovereign states created opportunities for uses of these borders that governments had never anticipated. Even those most disadvantaged by border creation—native peoples—could use them for their own ends. American and Canadian authorities insisted that Indians were political subordinates and properly 'belonged' to one nation or another—that they were 'American' or 'Canadian.' At the same time, they had great difficulty in identifying native bands, distinguishing them from one another, and tracking their whereabouts consistently. Officers and Indian agents were consequently left dependent on the assurances of other Indians, who found that they could reward their indigenous friends and punish their enemies. In the late 1870s, for example, Assiniboine and Gros Ventre residing on a Montana reservation insisted that some Métis residing with them—but not all—be included in their tribal ranks. The American army pursued those for whom they did not vouch.[42]

Unfree laborers across North America found much greater freedom from the establishment of meaningful international borders. The different status of chattel slavery in the nineteenth century—outlawed in the British Empire in 1834, increasingly restricted in the American North, increasingly entrenched in the American South, and outlawed if still practiced in some regions of Mexico by 1829—meant that the simple crossing of the border could dramatically change one's legal status and social opportunities. It is well known that there was extensive slave flight from the United States to Canada, where former bondsmen and bondswomen could be free from the long arm of fugitive slave hunters. As William Wells Brown wrote in 1847, 'I would dream at night that I was in Canada, a freeman, and on waking in the morning, weep to find myself so sadly mistaken.'[43] Contemporaries estimated that from 15,000 to 75,000 fugitive slaves found sanctuary in Canada West (later Ontario), with modern scholarship encountering considerable difficulty in arriving at a more precise measure.[44]

Less well known is the story of slave flight from Upper Canada (present-day Ontario) to Michigan Territory. When the Jay Treaty of 1796 turned over control of Detroit to the United States, the hundreds of slaves within striking distance of the easily crossed Detroit River could find freedom simply by crossing the river, where US law banned the importation of slaves. (Their American counterparts could do the same, because Upper Canada was governed by a similar proscription.) By 1807, their flight put at risk the entire institution of slavery on the Canadian side of the river. Enslaved 'people in the Detroit borderland,' concludes Gregory Wigmore, 'acquired their freedom long before those held elsewhere in Upper Canada and Michigan—jurisdictions where slavery persisted until the mid-1830s.'[45]

Unfree laborers in the Rio Grande/Río Bravo borderlands created similar opportunities for themselves. Mexico abolished slavery in 1829, within a decade of gaining its independence, while slaveholders still dominated the highest offices in the United States. If most Americans thought of Mexico as a backwards nation hampered by primitive Catholicism and burdened by the weight of its past, slaves had reason to see it in a different light: they would belong to nobody but themselves if they could get to the other side of the Rio Grande. 'In Mexico you could be free,' remembered one former slave decades later. 'We would hear about [those who fled] and how they were going to be Mexicans. They brought up their children to speak only Mexican.' This was a much more arduous journey than crossing the Detroit river. Most of Texas's slaves lived hundreds of miles from the border, in the humid cotton-growing regions of east and central Texas. The open plains, numerous rivers, and chaparral thickets that lay between plantation districts and the border both hampered slave flight and ensured that they were generally safe from slave-catchers once in Mexico. This promise made slave flight much more common in Texas than in the South as a whole. Texas slave masters understood this threat; a group complained that 'something must be done for the protection of slave property in this state. Negroes are running off daily. Let the frontiers of slavery begin to recede and when or where the wave of recession may be arrested God only knows.' Escaped slaves found refuge in Mexico, prompting occasional slave-raiding expeditions by Texas authorities, or on the Texas side of the borderlands, where some settled down and married into local families.[46]

Fleeing north held a similar promise for some Mexicans. At the same time as slaves fled south from Texas, servants exploited or held in debt peonage in the Mexican Northeast found refuge by crossing the border into the United States. This exodus continued for several decades; a Mexican government commission conservatively estimated that more than 5000 fled Nuevo León and Coahuila for Texas in the first 15 years of the border's existence (1848–63). Much as fugitives undermined slavery in Texas, the peasants of Mexico's northeast edge forced hacienda owners to abandon debt peonage and offer better wages and working conditions. 'Nobody changes nationalities to assume a worse condition,' wrote a Mexican senator of these migrants, 'and it is very dangerous to see just beyond the arbitrary line prosperity and wealth, and on this side destitution and poverty.'[47] Mexican commissioners studying the issue were less charitable, complaining of the 'loss of labor to places where the population is sparse' and condemning the fugitive workers in terms that Texas slave masters used for their fled bondsmen: 'for the most part criminals, for they always steal before fleeing or have already been prosecuted for other crimes … These criminals and others of another class, especially the cattle thieves who have managed to escape, all reside in Texas.'[48]

Conclusion

The frustration of Mexican elites and American slave-masters at the ability of their laborers to take advantage of young national borders suggests the complexity of the emergence of 'bordered lands' in North America and beyond. By the twentieth century, over much of the globe consolidated states had surveyed and demarcated their borders, extinguished the military power of native peoples and other rival sovereignties, and regulated the crossing of goods and people across the lines. These processes happened differently in different places. The actions of settler-states such as Canada and the United States looked to native peoples like continuations of the colonial processes set in motion by European expansion. In places like Asia that faced intensified European pressures, polities such as Siam and Meiji Japan embraced a similar conflation of territory and sovereignty in order to resist further European expansion.[49]

Yet nowhere did states achieve all of their ambitions. The searing deserts, swarming mosquitos, rugged mountains, and frigid plains that posed such daunting challenges to the surveyors and militaries of North America had their counterparts in the mountains of upland southeast Asia, the islands and channels of Melaka Straights, the swamps and forests of Right Bank Ukraine.[50] The people of these and countless other places cooperated with, ignored, defied, and sometimes took advantage of borders. They ran from states, made war on them, and carried goods and people across the borderlines when there was a profit in doing so. And they moved themselves when a change of sovereignty was advantageous. National borders remained contested and 'fuzzy' in ways that the term 'hardening' obscures: states had trouble 'seeing' their borders and border peoples, even as subaltern populations made unexpected uses of hardened borders to serve their own ends. The physical and social reality of borderlands ensured a degree of openness, one that lasts to today and characterizes other places beyond North America.

Notes

1. S. Truett and E. Young (eds), *Continental Crossroads: Remapping US–Mexico Borderlands History* (Durham, NC, 2004).
2. J. Adelman and S. Aron, 'From Borderlands to Borders: Empires, Nation-States, and the Peoples in between in North American History,' *American Historical Review*, 104 (1999), 840.

3. E. Haefeli, 'A Note on the Use of North American Borderlands,' *American Historical Review*, 104 (1999), 1222–6; J. Wunder and P. Hämäläinen, 'Of Lethal Places and Lethal Essays,' *American Historical Review*, 104 (1999), 1229–35.

4. C. Maier, *Among Empires: American Ascendancy and Its Predecessors* (Cambridge, MA, 2006), p. 102.

5. C. Maier, 'Consigning the Twentieth Century to History: Alternative Narratives for the Modern Era,' *American Historical Review*, 105 (2000), 818. C. Hill, *National History and the World of Nations: Capital, State, and the Rhetoric of History in Japan, France, and the United States* (Durham, NC, 2009); C. Bright and M. Geyer, 'Where in the World Is America? The History of the United States in the Global Age,' in T. Bender (ed.), *Rethinking American History in a Global Age* (Berkeley, 2002), pp. 63–99.

6. A. Lincoln, 'Annual Message to Congress,' 1 Dec. 1862, in *Lincoln: Speeches and Writings, 1859–1865* (New York, 1989), p. 403.

7. B.H. Johnson and A. Graybill (eds), *Bridging National Borders in North America: Transnational and Comparative Histories* (Durham, NC, 2010).

8. R. St John, *Line in the Sand: A History of the Western US–Mexico Border* (Princeton, 2011), pp. 12–13.

9. W.H. Emory, *Report on the United States and Mexican Boundary Survey* (Washington DC, 1857), pp. 14–15, 23, 58, 84–5. More generally, see P. Rebert, *La Gran Línea: Mapping the United States–Mexico Boundary, 1849–1857* (Austin, 2001).

10. G.F.G. Stanley, *Mapping the Frontier: Charles Wilson's Diary of the Survey of the 49th Parallel, 1858–1862, While Secretary of the British Boundary Commission* (Toronto, 1970), p. 6.

11. Stanley, *Mapping the Frontier*, p. 23.

12. See, for instance, Stanley, *Mapping the Frontier*, pp. 27, 32.

13. Stanley, *Mapping the Frontier*, 68.

14. Stanley, *Mapping the Frontier*, 60–3.

15. T. Rees, *Arc of the Medicine Line: Mapping the World's Longest Undefended Border across the Western Plains* (Lincoln, NE, 2007), pp. 44–5.

16. Rees, *Arc of the Medicine Line*, pp. 35–6.

17. Rees, *Arc of the Medicine Line*, pp. 343–4.

18. T. Torrans, *Forging the Tortilla Curtain: Cultural Drift and Change along the United States–Mexico Border from the Spanish Era to the Present* (Ft. Worth, 2000), p. 33.

19. G.M. Colín, 'The Political Economy of Mexican Protectionism, 1868–1911,' PhD dissertation, Harvard University, 2002; P.W. Magness, 'From Tariffs to the Income Tax: Trade Protection and Revenue in the United States Tax System,' PhD dissertation, George Mason University, 2009.

20. St. John, *Line in the Sand*, pp. 103, 203.

21. Rees, *Arc of the Medicine Line*, p. 357.

22. Daniel Webster to Robert P. Letcher, 19 Aug., 1851, in W.R. Manning (ed.), *Diplomatic Correspondence of the United States: Inter-American Affairs, 1831–1860, vol. IX, Mexico, 1848 (mid-year)–1860*, pp. 89–92 (Washington DC, 1937). The United States secured its release from Article 11 in 1853, as part of the Gadsden Purchase, which in 1854 pushed a disputed portion of the boundary line south and delivered much of present-day southern New Mexico and Arizona to the US. See B. DeLay, *War of a Thousand Deserts: Indian Raids and the US–Mexican War* (New Haven, 2008), p. 303.

23. S.J. Barrows, 'Crossing the Big Muddy,' *Army and Navy Journal*, 10 (1873), 328–9, reprinted in P. Cozzens (ed.), *Eyewitnesses to the Indian Wars, 1865–1890 vol. IV: The Long War for the Northern Plains* (Mechanicsburg, 2004), pp. 150–3.

24. G.V. Henry, 'A Winter March to the Black Hills,' *Harper's Weekly*, 39 (27 July 1895), reprinted in Cozzens, *Eyewitnesses to the Indian Wars, vol. IV*, pp. 183–7.

25. 'Mexico: Reciprocal Right to Pursue Savage Indians across the Boundary Line,' in P. Cozzens (ed.), *Eyewitnesses to the Indian Wars, 1865–1890 vol. I: The Struggle for Apacheria* (Mechanicsburg, 2001), pp. 343–5.
26. D.G. McCrady, *Living with Strangers: The Nineteenth-Century Sioux and the Canadian–American Borderlands* (Toronto, 2010 [2006]), p. 97.
27. Quoted in B. LaDow, *The Medicine Line: Life and Death on a North American Borderland* (New York, 2001), p. 41.
28. LaDow, *Medicine Line*, pp. 59, 21; McCrady, *Living with Strangers*, p. 112. For a more general treatment of the parallels and differences in the US and Canadian incorporation of the Great Plains, see A.J. Graybill, *Policing the Great Plains: Rangers, Mounties, and the North American Frontier, 1875–1910* (Lincoln, NE, 2007).
29. The classic and widely influential statement of this idea in a North American context is G. Anzaldúa, *Borderlands: The New Mestiza, La Frontera* (San Francisco, 1987).
30. J. Thompson, *Cortina: Defending the Mexican Name in Texas* (College Station, 2007).
31. For a comparison of Riel and Cortina, see Graybill, *Policing the Great Plains*, pp. 71–90.
32. T. Flanagan, *Louis 'David' Riel: Prophet of the New World* (Toronto, 1978); J.H. Kinsey, *Strange Empire: A Narrative of the Northwest* (Toronto, 1952).
33. Quoted in M.Á. González-Quiroga, 'Conflict and Cooperation in the Making of the Texas–Mexico Border Society, 1840–1880,' in Johnson and Graybill (eds), *Bridging National Borders*, p. 44.
34. E. Young, *Catarino Garza's Revolution on the Texas–Mexico Border* (Durham, NC, 2004).
35. McCrady, *Living with Strangers*, p. 3.
36. M. Hogue, 'Between Race and Nation: The Creation of a Métis Borderland on the Northern Plains,' in Johnson and Graybill (eds), *Bridging National Borders*, pp. 60, 63, 67. On smuggling more generally, see P. Andreas, *Smuggler Nation: How Illicit Trade Made America* (Oxford, 2014).
37. Hogue, 'Between Race and Nation,' pp. 69–70.
38. US Congress, House, *Difficulties on the Southwestern Frontier*, 36th Congress, 1st Session, 1860 H. Exec. Doc. 52, p. 70.
39. 'Declaration of the People of Rupert's Land and the Northwest,' in *History of the North-West*, vol. II (Toronto, 1894–95), p. 416.
40. For works that contrast a later socially bordered world with a previously more open one, see E.R. Schlereth, 'Privileges of Locomotion: Expatriation and the Politics of Southwestern Border Crossing before Manifest Destiny,' *Journal of American History*, forthcoming; D. Montejano, *Anglos and Mexicans in the Making of Texas, 1836–1986* (Austin, 1987); K. Benton-Cohen, *Borderline Americans: Racial Division and Labor War in the Arizona Borderlands* (Cambridge, MA, 2009); A. Graybill, *The Red and the White: A Family Saga of the American West* (New York, 2013).
41. McCrady, *Living with Strangers*, p. 3.
42. Hogue, 'Between Race and Nation,' pp. 74–5.
43. W.W. Brown, *Narrative of William W. Brown, a Fugitive Slave, Written by Himself* (Boston, 1847), p. 50.
44. R.W. Winks, *The Blacks in Canada: A History* (Montreal, 1997 [1971]); M. Wayne, 'The Black Population of Canada West on the Eve of the American Civil War: A Reassessment Based on the Manuscript Census of 1861,' *Histoire Sociale/Social History*, 56 (1995), 465–85.
45. G. Wigmore, 'Before the Railroad: From Slavery to Freedom in the Canadian–American Borderland,' *Journal of American History*, 98 (2011), 450–1, 453.
46. G. Horne, *Black and Brown: African Americans and the Mexican Revolution* (New York, 2005), pp. 14, 17; W. Carrigan, *The Making of a Lynching Culture: Violence and*

Vigilantism in Central Texas, 1836–1916 (Urbana, 2004); S. Kelley, '"Mexico in His Head": Slavery and the Texas–Mexico Border, 1810–1860,' *Journal of Social History*, 37 (2004), 709–23; B.H. Johnson, *Revolution in Texas* (New Haven, 2003), p. 16. For contemporary estimates of general slave flight, see J.H. Franklin and L. Schweninger, *Runaway Slaves: Rebels on the Plantation* (New York, 1999), p. 279.

47. J. Mora-Torres, *The Making of the Mexican Border: The State, Capitalism, and Society in Nuevo León, 1848–1910* (Austin, 2001), p. 28.

48. *Reports of the Committee of Investigation Sent in 1873 by the Mexican Government to the Frontier of Texas* (New York, 1875), p. 403.

49. T. Winiachakul, *Siam Mapped: A History of the Geo-Body of a Nation* (Honolulu, 1994); Hill, *National History and the World of Nations*.

50. J.C. Scott, *The Art of Not Being Governed: An Anarchist History of Upland Southeast Asia* (New Haven, 2010); E. Tagliacozzo, *Secret Trades, Porous Borders: Smuggling and States along a Southeast Asian Frontier, 1865–1915* (New Haven, 2005); K. Brown, *A Biography of No Place: From Ethnic Borderland to Soviet Heartland* (Cambridge, MA, 2004).

2

'The Men Who Made Australia Federated Long Ago': Australian Frontiers and Borderlands

Frank Bongiorno

Alfred Deakin, a leading federalist and later Australian prime minister three times, ended an 1898 speech delivered in the Victorian town of Bendigo advocating federation of the Australian colonies, by quoting a local poet, William Gay:

> Her seamless garment, at great Mammon's nod,
>   With hands unfilial we have basely rent,
> With petty variance our souls are spent,
> And ancient kinship under foot is trod:
> O let us rise, united penitent,
> And be one people,—mighty, serving God![1]

This chapter will suggest that not only did the garment have seams, but that these mattered enormously to the people affected by their presence. On the eve of federation, which occurred on January 1, 1901, Australia comprised six self-governing colonies, five of them on the mainland, the sixth the island of Tasmania. What is now the Northern Territory was then administered by South Australia. The British founded Australia as a penal colony in January 1788, initially establishing a settlement at Port Jackson, now Sydney. Another settlement soon followed on Norfolk Island, about 1000 miles east of Sydney in the Pacific Ocean. In the early years of the nineteenth century, penal establishments would also be initiated at present-day Hobart in Tasmania, and Newcastle, in the Hunter Valley north of Sydney. In the mid-1820s, a convict settlement was established at Moreton Bay, the site of what is now Brisbane, and at the end of the decade a private venture initiated a free settlement on the Swan River, an area now occupied by Perth, Western Australia. The mid-1830s saw further territorial expansion; a private company commenced the colony of South Australia, while ambitious settlers from Van Diemen's Land (Tasmania) moved across Bass Strait with their sheep in order to establish properties in

Map 2.1 Australia, 1900

Port Phillip, thereby founding the city of Melbourne and the future colony of Victoria. By this time, the rapid expansion of the fine wool industry had resulted in the occupation of large tracts of territory, in many cases by settlers known colloquially as squatters since they occupied land unlawfully. Convict labor underpinned the development of the wool industry. The effect of this expansion on traditional Aboriginal society was devastating, with many dying of disease or through settler violence.

The middle decades of the century saw the winding down and eventual abolition of convict transportation to Eastern Australia (although it would be introduced in Western Australia between 1850 and 1868), and the outbreak of a gold rush in the early 1850s that greatly stimulated both immigration and the desire for self-government and democracy. By 1859 the original colony of New South Wales had shrunk to the mere proportions of a large European nation, as new territories were carved out of it and in each case (except Western Australia, which was not self-governing until 1890) given the right to rule itself.

While most settlers felt a wider loyalty to the empire and understood themselves as a British people, each colony had a sense of itself as a 'nation.' In the nineteenth century, according to Alan Atkinson, Australia was a 'multiplicity of nations,' and 'nationhood ... a variously defined, mutually overlapping, frequently evanescent sense of community and place.'[2] In the 1890s the federal movement found it necessary to ensure that, in the making of a new set of national political institutions, the political and cultural integrity of the individual colonies would not be destroyed. Recent scholarship has underlined the nationalist fervor of the federalists, some of whom nurtured an image of an island continent made by God for nationhood.[3] There was also a clear sense that certain fields of activity, such as external affairs, tariffs, defense and immigration, were best handled by a new and wider political entity whose jurisdiction would stretch across the continent. All the same, the federalists were sufficiently pragmatic, and also sufficiently attached to their own respective antipodean homelands, to ensure that the sovereignty of the colonies, which became states under the federal constitution of 1901, would still be respected under the new arrangements.

In a recently published collection of academic essays with the title *Frontier Skirmishes*, Australia was described as 'a nation whose borders are coeval with its gigantic coastline, thereby naturalizing them to the point of invisibility. Borders are something that Australians, if they are not immigrants, have never had to live with.'[4] There is an overwhelming historical amnesia contained in this sweeping claim. Australian historians, following Russel Ward's influential *The Australian Legend* (1958), have mainly ignored internal borders and instead turned to the concept of the frontier—understood in often vague terms as a place of remote rural settlement by Europeans—in order to chart the course of their national history and, even more ambitiously at times, to explain the shape of Australian national identity itself.[5] Indeed, the versatility and durability of the frontier concept in Australian historiography is suggested in a recent study by Penelope Edmonds, who reminds us that colonial cities, too, were 'frontiers,' not just 'the bush, backwoods, or borderlands' to which earlier historians gave the bulk of their attention.[6]

Yet notwithstanding Edmonds's passing reference, Australian historians have made insubstantial use of 'borderlands' terminology in their work. Their neglect may well be indebted to a longer tradition of continental national-ist contempt for the supposed provincialism of allegiance to a mere 'colony' or 'state.' Certainly, some of the most famous nationalist writers of the late nineteenth century reinforced a sense of the divisions between the colonies being artificial, created by a petty and squabbling humanity. The Henry Lawson poem from which the title of this chapter has been borrowed was com-posed in 1901, on the occasion of the visit of the Duke and Duchess of York to open the first federal parliament in May:

> There'll be royal times in Sydney for the Cuff and Collar Push,
>> There'll be lots of dreary drivel and clap-trap
> From the men who own Australia, but who never knew the Bush,
>> And who could not point their runs out on the map.
> O the daily Press will grovel as it never did before,
>> There'll be many flags of welcome in the air,
> And the Civil Service poet, he shall write odes by the score—
>> But the men who made the land will not be there.

For Lawson, the real Australia had already been made by the bushmen of the interior: they had 'federated long ago.'[7]

Lawson's short story 'Hungerford' (1893) is even more evocative of a conti-nental nationalist contempt for internal borders. Hungerford is a small town on the border of New South Wales (NSW) and Queensland. The borderland is presented in this brief sketch not as a place of social interaction or 'fluid' identities, but a microcosm of the hopelessness of the outback, the boredom relieved only by drunken brawling outside one of the pubs on the Queensland side. The irrationality of the boundary itself is evoked at the outset, where the narrator reports that an intercolonial rabbit-proof fence runs straight through the town—but 'with rabbits on both sides of it.' Law and order also prove problematic: 'The police cannot do anything if there's a row going on across the street in NSW, except to send to Brisbane and have an extradition warrant applied for.' The narrator asks a man minding some goats and sheep which of the two colonies he preferred. After scratching his head for a while, and

with the bored air of a man who has gone through the same performance too often before, he stepped deliberately up to the fence and spat over it into New South Wales. After which he got leisurely through and spat back on Queensland.

'That's what *I* think of the blanky colonies! he said ...

And if I was at the Victorian and South Australian border I'd do the same thing.'

He let that soak into our minds, and added: 'And the same with West Australia—and—and Tasmania.' Then he went away.

The last would have been a long spit—and he forgot Maoriland.[8]

Australian historians have been implicated in this continental version of nationalism, and, rather like the authors and poets of the federation era such as Lawson, they have sometimes understood their efforts as part of a nation-building enterprise.[9] It is nonetheless possible that a historiographical stress on borders and borderlands might be better able than the nearly ubiquitous 'frontier' to capture a sense of Australia as a region comprising nations, borders, and sovereignties subject to ongoing redefinition and renegotiation through the colonizing process. Some of these formations were the result of settler occupation; others were unknown or unrecognized by most Europeans but meaningful to Aboriginal people—as they sometimes remain today.

For reasons to be explained later in this essay, the settler communities in borderlands strongly favored the federal union achieved on 1 January 1901, after which the distinctiveness of borderland experience was less marked. Following federation, these areas largely conformed to what Michiel Baud and Willem van Schendel have called declining borderlands, areas that lose their political significance and, in a peaceful process, gradually wither away.[10] Australia's settler borderlands, in fact, have never quite withered away, but the establishment of a customs union in the early twentieth century as a part of the federal settlement, and the central government's rapid assumption of immigration control, ensured that borderlands were of declining distinctiveness in political and economic terms. All the same, in the 1940s and again in the 1970s, some borderlands—and notably the Albury-Wodonga area on the New South Wales–Victoria border—were seen by centralist Labor governments in Canberra as ripe for government-sponsored regional development. Borderland areas such as the Riverina and New England areas of (respectively) southern and northern New South Wales generated movements for the creation of new states until the 1960s, although in each case it was distance from Sydney rather than proximity to a border that mattered most.

Indigenous Borderlands

What might an Australian history of borderlands look like? Australia's settlers have spent much of the last couple of centuries creating boundaries and borders but these overlaid and ignored existing indigenous understandings of country. From at least the early nineteenth century, the British settlers of the Australian colonies showed an interest in delineating the boundaries separating Aboriginal nations, although maps were a rarity before the late nineteenth century.[11] When such maps have been drawn—even in quite recent times—they have been controversial. In particular, there has been much confusion between different types of group in Aboriginal society, with the result that some maps have unconsciously and confusingly incorporated diverse categories. Similar problems have occurred in the mapping of tribal territories among nomadic or semi-nomadic peoples in North and South America. In the Australian case, the conflation of land-usage with land-holding is one prominent example. This kind of confusion might be likened to a British map that contained, say, boundaries delineating counties, House of Commons constituencies and local government areas, without distinguishing between them.[12]

Maps of Aboriginal nations, with their alluringly tidy borders, fail to capture the complexity of either classical Aboriginal society or the colonizing process. Indeed, we might see the creation of such documents as an effort to produce a colonial order with the kinds of clearly defined borders and territories with which Europeans were increasingly comfortable as the nineteenth century unfolded. But as Elspeth Young has argued, whereas the 'significance and strength of Aboriginal territoriality and its expression in cultural and economic terms is undeniable … the exact delineation of that territoriality through publicly displayed maps is highly problematic.' The Aboriginal relationship to the land, she suggests, is expressed not through clearly demarcated territories and borders 'but rather through responsibility for sections of Dreaming tracks representing the travels of ancestral beings.' Some of the more important points at which these tracks cross—such as the location of waterholes—are shared between different groups.[13]

Jurists would eventually draw on the legal fiction of *terra nullius* (land belonging to no one) in order to explain how an Aboriginal ownership and sovereignty that had endured over tens of thousands of years was extinguished in Australia by the arrival of the British in 1788. Yet some recent scholarship on the emergence of European ideas and practices of sovereignty in Australia indicates how this recognition of the contingency of the borderland survived European arrival. Lisa Ford has argued that in the earliest years of colonial New South Wales, the dominant approach to sovereignty was a legal pluralism. British settlement clung to a coastline here and there, and officials assumed Aboriginal sovereignty had survived 1788. Indigenous people were not subjected to the common law when they committed 'crimes' because they were not seen to be subjects. From the 1820s, however, there was a drive towards a more modern form of territorial sovereignty that would soon come to treat Aborigines as subjects, if inferior ones, instead of people of another nation with its own laws.[14] In practice, even in the absence of formal recognition by the colonizing power, Aboriginal people would continue to exercise effective sovereign rule over much of the land mass of Australia for decades afterwards. For instance, warfare between Aboriginal groups, without interference by white authorities, continued well into the twentieth century—as did massacres by Europeans.[15]

By the early twentieth century, the colonizers had come to exercise a despotic authority over many Aboriginal people through a system of missions and reserves. Virtually every aspect of the lives of Aboriginal people who came under the protection acts, including their movement around the country, was now subjected to government control. Nonetheless, divided authority between the states sometimes allowed Aborigines living near the state borders to slip across them in the hope of escape from a dire situation, or of better treatment from another government.[16] The issue of Aboriginal people crossing from one jurisdiction to another 'to avoid disciplinary action ... or what they might consider painful medical treatment' was considered sufficiently important to attract the attention of the first national native administrators' conference in 1937. Some, but not all, state and territory laws allowed Aboriginals to be returned across a border to their homeland. The Victorian representative reported that during his state's centenary celebrations (1934), a troupe of Aboriginal minstrels had visited Melbourne, resulting 'in a drunken melee in front of one of the city's leading hotels ... Of course, if a native visitor is a decent fellow, we do not want to send him back.' By the 1930s, to seek to exercise a similar control in the ordinary course of events over the movement across borders of non-indigenous people would have been unthinkable (as well as unconstitutional).[17]

Settler Borderlands

The internal borders of Australia have been fairly stable since 1863. By that time, South Australia (1836), Victoria (1851), Queensland (1859), and the Northern Territory (1863) had been carved out of New South Wales. Western Australia's eastern boundary had been set earlier, in 1829, and Queensland's western border was extended to incorporate the gulf country in 1862. In the following decade, Queensland's northern maritime border would be extended twice, in 1872 and 1879, to cover the Torres Strait islands. The latter annexation, made at the behest of the Colonial Office in London, would have dramatic consequences for the land law of the entire nation 120 years later, providing the occasion for recognition of native title.[18]

It is in the Torres Strait, and elsewhere along the northern coastline, that one perhaps finds a 'borderland' closest in character to the conventional understanding of that term. Even before the arrival of European settlers, these were zones of contact between Aborigines and peoples living in the lands to the north. The Aboriginal people were not totally isolated from the rest of the world. Northern Australia might be seen as part of a regional community stretching northward into what is now the Indonesian archipelago and Malaysia.[19] Long before (and after) European settlement, Macassan trepang fisherman made regular voyages to northern Australia, engaging in a complex set of economic, cultural, and sexual interactions with local indigenous people. The Aboriginal people they encountered accommodated these sojourners into their lifestyles and traditions.[20] The arrival of the dingo (native dog) from Asia perhaps four thousand years ago also attests to continuing contact between Australia's indigenous people and the outside world. Aboriginal people in northern Queensland, moreover, appear to have had regular contact with the inhabitants of what is now New Guinea, yet they did not emulate their agricultural practices.

The area immediately to Australia's north also became a site of imperial rivalry. In the early 1880s, a French nobleman and conman, Marquis de Rays, attempted to found 'New France' by settling French, Italian, and German colonists in unclaimed territory extending from eastern New Guinea to the Solomon Islands. Around 570 people joined the four expeditions. Many died of malaria and the survivors had to be resettled elsewhere, the Italians in northern New South Wales. The only truly pleasing aspect of this tragic tale is that King Charles of New France, who had the sense never to visit his dominion, went to prison.[21] Whitehall also repudiated the Queensland government's attempted annexation of eastern New Guinea in 1883; the territory was carved up between Britain and Germany the following year. In 1906 British New Guinea came under Australian administration and, on the seizure of the territory by an Australian expeditionary force in September 1914, so did German New Guinea. After Versailles, the former German colony became a Mandate administered by Australia under the League of Nations. So, as a result of Australia's acquisition of a colonial domain to its north, it gained a land border with the Dutch East Indies, of which West New Guinea was a part. And in the 1960s and early 1970s—much to its initial alarm—Australia shared a land border with Indonesia.

Australia's internal boundaries have not gone uncontested.[22] There have been occasional disputes between governments over territory, such as in the isolated wheat-growing area around Serviceton on the border of Victoria and South Australia. The Privy Council in London eventually resolved the dispute in Victoria's favor but as recently as the 1970s I recall my father, a native of a nearby town on the Victorian side of the border, still referring to Serviceton as 'The Disputed Territory.'[23] Victoria also made occasional, and unsuccessful, claims to territory between the Murray and Murrumbidgee rivers. There have been demands for new colonies and states stretching back to the 1850s, with the most vigorous movements in northern Queensland, and northern and southern New South Wales. The Australian constitution contains provision for new states and as late as 1967, the New South Wales state government, in fulfillment of an election promise, agreed to put to a popular vote the creation of a state of New England based in the north.[24] In 1933 Western Australia voted to secede from the Commonwealth, but an incoming state Labor government ignored this expression of the popular will and the imperial government similarly refused to act. The mining boom of the 1960s and 1970s once again provoked talk of Western Australian secessionism.

In the nineteenth century, the borderlands of the Australian colonies were places where the movement of goods and people was restricted. With self-government in the 1850s, they gained the power to regulate their own trade and immigration. This authority was used to levy duties on imported goods and restrict the entry of people considered undesirable. These were exercises in the expression of sovereignty as well as 'practical' measures to raise revenue and preserve the 'British' character of their society. While considerable historiographical attention has been paid to the history of Australian coastal border control,[25] there has been little recognition that the colonies themselves developed a regime for the restriction of intercolonial movement of goods and people. The men who made Australia might have federated long ago, but someone forgot to tell their governments.

Goods

Tariff duties were an important source of revenue for colonial governments. Victoria and South Australia went further and used tariffs to develop local industry. The colonies imposed such duties not only on the world outside Australia but on each other. Tariffs applied to a vast array of goods, including grain and livestock. The flashpoint for conflict over this issue was the Murray river—'the Mississippi of Australia'[26]—which formed part of the border between protectionist Victoria and (formally) free-trade New South Wales but flowed through South Australia before emptying into Lake Alexandrina near Adelaide. In the 1850s the massive demand for goods generated by the Victorian gold rush and the development of river transport via paddle steamers intensified intercolonial conflict over tariffs. Governments set up customs houses at river and land crossings; officials searched boats for dutiable goods; even travelers and their luggage on coaches—and, later, trains—would be examined, leading to calls to appoint female inspectors to deal with travelers of their own sex.[27] On the long, thinly populated land borders between South Australia and Victoria, and Queensland and the Northern Territory, it was the 1870s before there was any attempt to collect duties.[28]

Smuggling, both grand and trivial, was probably common, giving rise to a rich lore of variable plausibility about the lengths to which borderland residents went to avoid duties. Cattle and pigs were sent across the Murray

Figure 2.1 A depiction of some borderland scenes from Albury, a town on the Murray River's New South Wales side (*Australasian Sketcher*, May 13, 1876, p. 24. National Library of Australia). The Murray River separated that colony from Victoria

without attracting official notice, stills supposedly operated from caves along river banks, people walked across borders wearing several layers of clothes after a shopping trip; a young woman, ostensibly crossing a bridge to attend her weekly music lesson, was revealed to be smuggling groceries when on one occasion her violin case was accidentally knocked to the ground.[29] Given the vast length of the borders, smuggling must have been feasible, if not necessarily easy. Certainly, in 1879, on their way to rob the bank in Jerilderie, the Kelly Gang of bushrangers and their horses managed to swim across the Murray without attracting the attention of the large number of police searching for them. But 1879 was a drought year, which would have made a crossing easier.[30]

Some customs officials were zealous to the point of obnoxious. A Victorian farmer had to travel across the South Australian border to deliver some wheat to a mill. On the way home, he picked up some groceries at Bordertown, in South Australia:

> On meeting the Customs Officer at Lockhart Gate, he searched my groceries for dutiable goods and he charged me twelve shillings for having bacon, tea and sugar and a double duty on flour. I asked for a receipt. He retorted the only receipt you get is for me to allow you to proceed to Victoria.

Another customs officer, along the same border, noticed a selector (small farmer) with 40 head of cattle on his block. When pressed, the selector admitted that the cattle had indeed come from across the border, but he refused to pay a duty on them. The customs officer then began taking away the cattle, which quickly changed the farmer's mind. In this region, selectors also deeply resented the duties they had to pay on machinery and parts. South Australians even had to pay 20 per cent of the value of furniture, as estimated by the customs officer, that they carried with them to take up land in Victoria. There were complaints about border officials charging more than items were worth, and some travelers simply refused to pay and abandoned their goods on the spot.[31] One annoying official went so far as to lock the gate on the bridge at Corowa when he was off-duty, thereby closing off access entirely.[32] There were occasions when cross-border trade was entirely prohibited.[33]

The communities living in borderlands naturally came to resent tariff duties. While colonial governments agreed to bury the hatchet from time to time, each 'treaty' invariably collapsed amid recriminations. The price was paid by borderland communities, which were arguably 'natural' areas of economic and social interchange and political cooperation. Government railway construction in borderlands was also affected by intercolonial economic rivalry. The Murray river is much closer to Melbourne than Sydney, so that even allowing for each colony's unfortunate habit of adopting its own rail gauge, any link with the south was likely to strengthen economic relations between southern New South Wales and Melbourne. Unsurprisingly, the Victorian government invested more heavily in the borderland than its northern counterpart.[34] Meanwhile, borderland communities in southern New South Wales agitated for railway development, berated governments—especially in faraway Sydney—for their neglect of the region, and occasionally threatened to secede, most likely to frighten Sydney into giving them the railways they wanted. The Victorian railway network's arrival in the borderland (at Echuca in 1864 and Wodonga, across the river from Albury, in 1873) set off a new round of intercolonial tariff warfare, as the New South Wales government sought to prevent the southern colony from exercising a complete economic dominion over the region.[35]

The enforcement of Victoria's high duties was deeply damaging to a developing New South Wales town such as Albury, which had much to gain from being able to import Victorian manufactured goods easily, as well as by sending on to Melbourne produce from its rich pastoral and agricultural hinterland.[36] Indeed, the Murray valley and Riverina were oriented economically to Melbourne, and increasingly so as the railways developed. Melbourne was one of the British Empire's largest cities by the 1880s and a case-study in what James Belich has called 'explosive colonisation.' Its economic empire extended right up Eastern Australia into the Riverina, the silver mines of Broken Hill in outback New South Wales, and the massive sheep runs of western Queensland, as well as across the Tasman to New Zealand.[37] Victorians argued that the Murrumbidgee river rather than the Murray would have been a more rational border between the colonies, thereby awarding Victoria a large section of southern New South Wales with which it was economically integrated. During 1872—one period in which the colonies engaged in tit-for-tat tariff 'warfare'—there was agitation in the Riverina for either annexation by Victoria or a separate colony. When the colonies ended their dispute in mid-1873, there were celebrations in the streets of Albury but a feeling of neglect at the hands of Sydney remained.[38]

In the borderlands, customs duties and the method of their collection nurtured a powerful sense of grievance. People living along borders resented what they saw as an unfair burden of taxation, one not endured by those who made the laws in Sydney, Melbourne, Brisbane, or Adelaide. The late nineteenth century saw two contradictory developments. Partly as a response to the devastating economic depression of the 1890s, borderland fiscal warfare became worse, as governments increased tariff rates.[39] Yet at the same time, in purely physical terms, the most significant borders were becoming less daunting. Railways now converged, from both sides, on the main borders in southeastern Australia, and there was a flurry of bridge-building across the Murray.[40] The movement of people reinforced the sense that hard borders were irrational. In the last quarter of the nineteenth century, a combination of declining gold yields in Victoria and more favorable land laws in New South Wales had led large numbers of Victorians to settle across the border. The depression of the 1890s also prompted much movement of people, although the phenomenon of inland Australian migration remains under-researched.

Unsurprisingly, given this combination of circumstances and grievances, the borderlands became strongly pro-federationist, since it was always taken for granted that federation would include the free passage of goods across internal Australian borders. It was possible to conceive of a customs union without political unity, but not vice versa. Pro-federation leagues emerged in virtually every town in the Murray borderland and in 1893 the border town of Corowa was the venue for an important, if unofficial, federation conference. It did much to revive a flagging federal movement but there was disagreement over what to do about 'those cursed Border duties.' Some commercial interests and border representatives favored their immediate abolition, while other delegates saw intercolonial free trade as less worthy than the achievement of political union, and they worried that a customs union might delay rather than expedite it. Some protectionists were simply hostile to lowering tariffs.[41] Nonetheless, when the colonies voted on federation, border towns were among the strongest supporters. In 1898, the nine Riverina electorates of southern New South Wales voted four to one in favor of the bill whereas the overall New South Wales average was just 52 percent in favor.[42]

People

The cross-border movement of white people was relatively straightforward, provided they were willing and able to pay duties on the goods they carried with them, but the situation for the Chinese was different. While they had earlier been a presence in the pastoral industry as shepherds, thousands of Chinese men arrived in Australia during the 1850s to dig for gold. The imposition of poll taxes on Chinese migrants by colonial governments in order to stem the flow of these sojourners meant that intercolonial borders were places of racial exclusion. There was some cooperation between colonial governments on this issue, although enforcement remained difficult, not least because Chinese with British nationality were exempt.

The open border between Victoria and South Australia in the 1850s allowed Chinese people to land at Robe on the southeast coast of South Australia, and walk overland to the Victorian goldfields to avoid the tax. At Albury in the late 1850s, Chinese descended on the border in parties of up to two hundred, which proved overwhelming for the customs officer charged with collecting the taxes from them.[43] After the Pine Creek gold rush in the Northern Territory during the 1870s, Chinese in 'great numbers' sought to enter Queensland without paying the poll tax.[44] Further south, the development of Chinese communities along the Murray raised similar problems to the imposition of customs duties on goods. Chinese going about their ordinary business of market gardening, farming, timber-cutting, and general laboring 'passed back and forth across the river in organised work parties' and, if they could not prove that they had been naturalized or born in the colony, faced harsh taxation, fines, or imprisonment.[45]

The administration of border control with respect to the Chinese is yet to attract much attention from historians and its details and chronology remain obscure. Nonetheless, some patterns and processes can be discerned. Towards the end of the century, colonial governments did become more draconian in dealing with unlawful Chinese border crossing; the New South Wales poll tax was £100 by the 1890s. In 1898 three Chinese men 'trying to sneak across the Murray River into New South Wales at Barham' received two years' hard labor, and there were several cases of Chinese going to prison for attempting to pass into Queensland from the Northern Territory. Nonetheless, enforcement remained difficult, and the New South Wales collector of customs complained that his colony's long and sparsely populated land border with Queensland, and its river border with Victoria, posed great difficulty for authorities. Most Chinese who entered New South Wales across a border, he said, were being smuggled across the Murray.[46]

Colonial authorities believed that impersonation was rife—Chinese would use birth and naturalization certificates that did not belong to them—and in 1900 New South Wales authorities claimed to have recently impounded 300 such documents. At this time, there was an exchange between public servants in Victoria and New South Wales about the respective merits of using hand-prints and photographs for identification. New South Wales already used the former; Victoria believed its current method of photographic identification adequate, but there were concerns in Melbourne that even after federation on January 1, 1901 New South Wales might prevent the passage of Chinese whose documentation did not include a handprint.[47]

While there is no reason to imagine that white borderland settlers were any less racist than their fellows elsewhere, some did sympathize with the difficulties faced by the Chinese people living among them. One Murray borderland newspaper commented that 'here on the border the law is thought to press with undue severity on Chinese who have been many years in the district, and are reputable and law abiding men.' It reported that local magistrates were unwilling to enforce the two-year term of imprisonment set down by a law widely considered 'unreasonable and arbitrary.'[48] Another paper said that the place to enforce immigration law was at the coastal ports; it was absurd to punish the unfortunate Chinese because they crossed 'a defined or an imaginary boundary line between the colonies.'[49]

Conclusion

The borders between Australia's colonies were indeed 'defined' and 'imaginary' but they also had real consequences for the people living near them. Settler Australia had internal borderlands defined by colonial boundaries, which have not been taken seriously in the scholarship, and have not been widely acknowledged in Australian cultural memory. Indigenous Australia, equally, was a realm of borders and borderlands whose history is harder to trace using the techniques and frameworks provided by western historical scholarship, but which the Australian legal and political systems have been forced to take more seriously since the emergence of the land rights movement; for instance, in the formation of Aboriginal representative bodies, and in court and tribunal hearings over native title claims.

Over the last decade or so, many Australian historians have increasingly imagined a transnational future for their subject, in which the limitations of a purely national approach would be overcome by an emphasis on shared histories and the (literal and metaphorical) crossing of borders.[50] Yet this transnational turn, while illuminating the connections between far-flung parts of Britain's empire, has so far given little attention to the multiplicity of sovereignties, borders, and nations found within colonial Australia itself. It is arguably the historians of Aboriginal sovereignty such as Henry Reynolds[51] and Lisa Ford whose work has been most suggestive in its insistence on plurality and contingency, whereas so much orthodox national history has taken for granted the naturalness of a history concerned (in the words of the National Anthem, *Advance Australia Fair*) with a 'home ... girt by sea.' If historians undertake an Australian history of borderlands, they are likely to find a much richer story than they imagine—one with the potential to link indigenous and non-indigenous histories, settler and migrant experiences, the local with the national and global.

Indeed, such an approach has the potential to locate Australia in the context of a global history of borders and borderlands. The process by which Australian Aboriginal borders and borderlands have been ignored, or mapped, by settler culture clearly has parallels in other settler colonial (and postcolonial) situations, such as those in the Americas and in southern Africa.[52] Similarly, the evolution of borders as places for the assertion of sovereignty by governments, as well as the reduced salience of borderlands as new sovereignties overlaid old, would suggest the possibility of comparisons with other circumstances in which smaller political units have been absorbed into larger unions or federations. In short, by paying greater attention to borders and borderlands, Australian historians might find that there is a transnational history waiting to be written on their own doorstep.[53]

Notes

1. M. Quartly, 'Alfred Deakin's Speech to the Annual Conference of the ANA, Bendigo, 1898,' *New Federalist*, 2 (1998), 66–8.
2. A. Atkinson, '2005 Eldershaw Memorial Lecture: Tasmania and the Multiplicity of Nations,' *Papers & Proceedings: Tasmanian Historical Research Association*, 52 (2005), 197. See also See also Atkinson, 'Federation, Democracy and the Struggle against a Single Australia,' *Australian Historical Studies*, 44 (2013), 262–79, and A. Coote, 'Imagining a Colonial Nation: The Development of Popular Concepts of Sovereignty and Nation in New South Wales with Particular Reference to the Period between 1856 and 1860,' *Journal of Australian Colonial History*, 1 (1999), 1–37.
3. J. Hirst, *The Sentimental Nation: The Making of the Australian Commonwealth* (South Melbourne, 2000), ch. 1.
4. R. West-Pavlov, 'Fencing in the Frontier,' in R. West-Pavlov and J. Wawrzinek (eds), *Frontier Skirmishes: Literary and Cultural Debates in Australia after 1992* (Heidelberg, 2010), p. 81.
5. R. Ward, *The Australian Legend* (Melbourne, 1958).
6. P. Edmonds, *Urbanizing Frontiers: Indigenous Peoples and Settlers in 19th-Century Pacific Rim Cities* (Vancouver and Toronto, 2010), p. 5.
7. H. Lawson, 'The Men Who Made Australia,' in *Henry Lawson: Collected Verse, Volume Two: 1901–1909*, ed. Colin Roderick (Sydney, 1968 [1901]), pp. 7–9.
8. H. Lawson, 'Hungerford,' in *Prose Works of Henry Lawson* (Sydney and London, 1948 [1893]), pp. 21–3.
9. See, for instance, A. Coote, 'Out from the Legend's Shadow: Re-thinking National Feeling in Colonial Australia,' *Journal of Australian Colonial History*, 10 (2008), 103–22; S. Macintyre, 'The Writing of Australian History,' in D.H. Borchardt (ed.), *Australians: A Guide to Sources* (Broadway, 1987), pp. 1–29; Macintyre, *A History for a Nation: Ernest Scott and the Making of Australian History* (Carlton, 1994).

10. M. Baud and W. van Schendel, 'Towards a Comparative History of Borderlands,' *Journal of World History*, 8 (1997), 224.
11. K. Blackburn, 'Mapping Aboriginal Nations: The "Nation" Concept of Late Nineteenth Century Anthropologists in Australia,' *Aboriginal History*, 26 (2002), 132, 135. See also G. Knapman, 'Mapping an Ancestral Past: Discovering Charles Richards' Maps of Aboriginal South-eastern Australia,' *Australian Aboriginal Studies*, 1 (2011), 19–34, and N. Etherington, 'Putting Tribes on Maps,' in Etherington (ed.), *Mapping Colonial Conquest: Australia and Southern Africa* (Crawley, 2007), pp. 97–101.
12. P. Sutton, *Country: Aboriginal Boundaries and Land Ownership in Australia* (Canberra, 1995), p. 25.
13. E. Young, 'Aboriginal Frontiers and Boundaries,' in Sutton, *Country*, pp. 88–92. For an interpretation that suggests more definite borders, see S.L. Davis and J.R.V. Prescott, *Aboriginal Frontiers and Boundaries in Australia* (Carlton, 1992).
14. L. Ford, *Settler Sovereignty: Jurisdiction and Indigenous People in America and Australia, 1788–1836* (Cambridge, MA, 2010).
15. H. Reynolds, Review of L. Ford, *Settler Sovereignty*, in *Australian Book Review*, 320 (2010), 7; P. Sutton, *The Politics of Suffering: Indigenous Australia and the End of the Liberal Consensus* (Carlton, 2009), pp. 91–8.
16. Even after federation, each state retained control of Aboriginal affairs within its own territory. The federal government had jurisdiction in the Northern Territory from 1911. For instances of Aboriginal people crossing a state border in order to escape a harsh regime, see B. Attwood, *Rights for Aborigines* (Crows Nest, 2003), pp. 32, 42–53; F. Davis and P. Grimshaw, 'Living on the Margins at Cummeragunja Aboriginal Reserve,' in A. Mayne and S. Atkinson (eds), *Outside Country: Histories of Inland Australia* (Kent Town, 2011), pp. 287–309; 'Amazing Story of a Burst for Freedom,' *Australian Abo Call*, June 1, 1938, p. 2.
17. Commonwealth of Australia, *Aboriginal Welfare, Initial Conference of Commonwealth and State Aboriginal Authorities held at Canberra, 21st to 23rd April, 1937* (Canberra, 1937), p. 22.
18. S. Mullins, *Torres Strait: A History of Colonial Occupation and Culture Contact 1864–1897* (Rockhampton, 1994), ch. 7.
19. R. Ganter, 'The View from the North,' in M. Lyons and P. Russell (eds), *Australia's History: Themes and Debates* (Sydney, 2005), pp. 41–62.
20. H. Reynolds, *North of Capricorn: The Untold Story of Australia's North* (Crows Nest, 2003); R. Ganter, *Mixed Relations: Asian–Aboriginal Contact in North Australia* (Crawley, 2006).
21. H. Laracy, 'Rays, Marquis de (1832–1893),' *Australian Dictionary of Biography*, National Centre of Biography, Australian National University, http://adb.anu.edu.au/biography/rays-marquis-de-4453/text7255 (accessed Oct. 2, 2011).
22. There were some formal changes to Australia's internal borders during the twentieth century. Under the provisions of the Australian constitution, New South Wales transferred a couple of thousand square kilometers to the federal government in 1909 to create a national capital. The Northern Territory passed to the federal government in 1911 and between 1927 and 1931, an area called Central Australia was carved out of it and separately administered.
23. *Serviceton: A Frontier Town on No Man's Land 1887–1987*, compiled by the Serviceton Centenary Committee, Serviceton (Serviceton, 1987), pp. 35–40.
24. The traditional centers of northern new-state sentiment voted strongly in favor of leaving New South Wales but the inclusion of a large and heavily populated area of the Hunter Valley, not far from Sydney, ensured the defeat of the measure.

25. See, for instance, A. Bashford, 'Quarantine and the Imagining of the Australian Nation,' *Health*, 2 (1998), 387–402, and Bashford, 'At the Border: Contagion, Immigration, Nation,' *Australian Historical Studies*, 33 (2002), 344–58.

26. D. Day, *Smugglers and Sailors: The Customs History of Australia 1788–1901* (Canberra, 1992), p. 416.

27. B. Pennay, *Federation at the Border: A Thematic History and Survey of Places Related to Federation in the Albury and Corowa District* (Albury, 1997), p. 13.

28. *Serviceton*, p. 26; Day, *Smugglers*, pp. 431–2.

29. A. Morris, *Rich River* (Belmont, 1979), pp. 164–5.

30. I. Jones, *Ned Kelly: A Short Life* (Sydney, 2008), pp. 222–3, 226; J.J. Kenneally, *The Complete Inner History of the Kelly Gang and their Pursuers* (2nd ed., Melbourne, 1929), pp. 114–16.

31. *Serviceton*, pp. 27–8.

32. Pennay, *Federation*, p. 18.

33. Day, *Smugglers*, pp. 418, 427.

34. B. Pennay, 'The Murray,' in J. Hagan (ed.), *People and Politics in Regional New South Wales, Volume 1: 1856 to the 1950s* (Annandale, 2006), pp. 249–51.

35. Pennay, 'The Murray,' pp. 249–51, and *Federation*, p. 10.

36. W.A. Bayley, *Border City: History of Albury New South Wales* (Albury, 1976), ch. 7.

37. J. Belich, *Replenishing the Earth: The Settler Revolution and the Rise of the Anglo-World, 1783–1939* (Oxford, 2009), chs 6 and 11.

38. Bayley, *Border City*, p. 93.

39. Morris, *Rich River*, p. 164.

40. Pennay, *Federation*, p. 18.

41. *Official Report of the Federation Conference Held in the Court-House, Corowa, on Monday, 31st July, and Tuesday, 1st August, 1893* (Corowa, 1893), pp. 26–30.

42. Pennay, *Federation*, p. 21, and 'The Murray,' p. 255.

43. Pennay, *Federation*, p. 11.

44. L.A. Miller, *The Border and Beyond: Camooweal 1884–1984* (Camooweal, 1984), p. 100.

45. Pennay, *Federation*, p. 11.

46. *Brisbane Courier*, Oct. 1, 1898, p. 9; Day, *Smugglers*, pp. 434–5; Pennay, *Federation*, p. 11.

47. Chief Secretary's Inward Correspondence, Public Record Office of Victoria, VPRS 3992/P0000/825, 1900/L13090, and M12157.

48. *Wahgunyah and Rutherglen News*, Aug. 18, 1893, quoted in Pennay, *Federation*, p. 18.

49. *Albury Banner*, Aug. 25, 1893, quoted in Pennay, *Federation*, p. 19.

50. A. Curthoys, 'Does Australian History Have a Future?,' *Australian Historical Studies*, 33 (2002), 140–52.

51. H. Reynolds, *Aboriginal Sovereignty: Reflections on Race, State and Nation* (St Leonards, 1996).

52. Etherington, 'Putting Tribes on Maps.'

53. My thanks to Lyndon Megarrity, Phil Griffiths, Alan Atkinson, and Nicole and Margaret McLennan, and the editors of this volume for their guidance.

Part II
Borderlands, Territoriality, and Landscape

3
Environment, Territory, and Landscape Changes in Northern Mexico during the Era of Independence

Cynthia Radding

What do we mean by territoriality and how do we connect territory, landscape, and borderlands in the ways in which we build historical narratives? Thinking about the title of this volume in reference to the northern borderlands of Mexico brought me back to the material features of land forms, vegetational patterns, stream flows, and water management systems, which are based on the natural forces of geology and climate. Nevertheless, their changing morphology is demonstrably and even measurably affected by human technique and culture.

The present chapter explores the historical content of cultural landscapes in the arid plains and mountainous valleys of northwestern Mexico. It focuses on the period of accelerated change that marked the transition from the late colonial economies of the Spanish Bourbon administration to the rupture of the imperial order and the tenuous foundations of the Mexican Republic, from approximately 1780 to 1840. It addresses the historical debates that were prompted by the bicentennial commemorations observed in Spain and the Americas concerning the participation of indigenous peoples in the contentious movements for independence of early nineteenth-century Latin America. This essay brings an innovative perspective to those debates grounded in the environmental exigencies of the borderlands of northern Mexico. Recent scholarship has developed well-researched topics concerning land tenure and local autonomy, especially in reference to indigenous communities.[1] This chapter will build on that literature by training its lens on the colonial landscapes that were altered by mining enterprises, livestock grazing, and irrigated cropland to consider how they affected the political agendas of rural communities and the cultural meanings they ascribed to 'liberty' in local settings.

The arid lands of the Sonoran and Chihuahuan Deserts are seemingly sparse and 'natural,' yet they bear the imprint of human culture through millennial patterns of gathering, hunting, burning, and transplanting cacti, succulents, bushes, and trees. In ancient as well as recent historical periods both

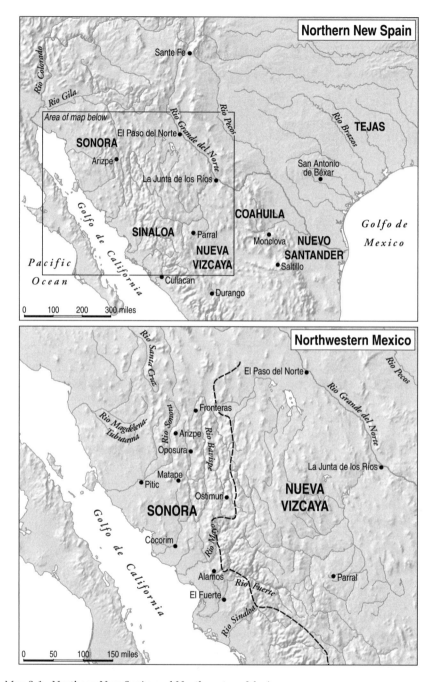

Map 3.1 Northern New Spain and Northwestern Mexico

indigenous and European peoples have 'gathered the desert,' harvesting the fruits, flowers, seeds, and vines of a wide diversity of plants. Through the processes of appropriating the resources of the desert, they propagate some species, propitiate new varieties, and alter the terrain.[2] These borderlands constitute cultivated landscapes, even if they do not produce the kinds of crops that would be familiar to the farming peoples of temperate climates.

Agrarian landscapes emerge from the highland valleys that border the deserts along both the eastern and western slopes of the Sierra Madre Occidental. This basin-and-range topography is carved by streams and rivers that, during seasonal floods, deposit layers of silt onto the floodplains of irregular sizes and shapes, changing their configuration along with the meandering streams. Village-dwelling peoples, who gathered local grasses and desert plants and experimented with maize and other Mesoamerican crops like gourds, squashes, beans, and amaranthus, developed techniques for damming the stream flow and planting living fence rows diagonally across the floodplains, thus turning silt into soil and engineering simple but durable irrigation canals. These agrarian landscapes characterize the piedmont and low ranges of central Sonora and western Chihuahua, attested to by archaeological remains from both the ancient and historical periods and by contemporary observations.[3]

The technologies that made possible the production of both desert and agrarian landscapes arose from different natural settings, to be sure, and befit distinct kinds of communities and modes of organizing labor. Long before European contact in the Americas, it is a well-founded hypothesis that ancient hunter-gatherers pursued certain species of megafauna into extinction.[4] Similarly, during the prolonged drought conditions of the thirteenth and fourteenth centuries, associated with the Little Ice Age in what today are the great deserts of northern Mexico and southwestern United States, the town-dwelling agriculturalists that were largely dependent on raising annual harvests of maize for consumption and trade were forced to disband, migrate, and resettle in smaller villages and seasonal encampments.[5]

Following the sixteenth-century Iberian invasions and the spread of colonial enterprises in northern Mexico during the seventeenth and eighteenth centuries, livestock ranching, European cultigens, and mining industries magnified the transformative impacts of human technologies in these arid plains and mountainous valleys through grazing, timber-felling, and increased consumption of both surface and underground water sources.[6] Furthermore, these technologies and the colonial economies that were grafted onto them gave rise to both creative and destructive processes. The introduction of European grains, wheat in particular, as winter crops may have complemented the indigenous summer growing season for maize and brought new staples into the Indians' diet. Iberian livestock management practices developed in Andalucía and transferred to northern New Spain were based on the transhumance of cattle from marshy lowlands (in the dry season) to higher ground or the scrub-lands (*agostadero* or *monte*) during the rainy season. These practices may have mitigated the impact of thousands of quadrupeds (domestic and feral) on the grasslands and brush of the Sonoran and Chihuahuan deserts.[7] Nevertheless, repeated references to droughts and crop failures in the colonial documents, as well as increasing occurrences of raiding by nomadic equestrian bands, point to falling water tables in riverine valleys and the disruption of hunting patterns in the *monte*. Based on calculations for San Luis Potosí in northeastern Mexico, mining consumed enormous quantities of timber for scaffolding in the excavation shafts and for fuel in the ore processing mills.[8] The environmental consequences of resource consumption to supply regional markets with foodstuffs and other commodities and to meet the demands of the imperial economy for bullion had reached critical levels by the turn of the nineteenth century, with severe impacts for indigenous communities.

Multiple Voices and Contested Histories from Central New Spain

Richly nuanced historiographies developed over the last two decades have overturned the epic narratives of political struggles for independence that were dedicated to demonstrating the heroism of their leaders or the tragic outcomes of their efforts. These narratives have been replaced by histories that recognize distinct voices in the processes of change that unraveled the Spanish imperium and led to the uneven formation of nation-states in Latin America. Regionally focused studies, based on careful analysis of primary sources, have shown that the socially and ethnically mixed sectors of indigenous pueblos and urban classes of artisans, muleteers, and day laborers—both men and women—joined in the political climate of the turn of the nineteenth century. Local plebeian groups and peasant villagers crafted their own interpretations of the principles that were codified by the Constitution of Cádiz and the derivative constitutional documents produced and circulated in the Spanish-American colonies.[9] The extraordinary summons of delegates from both the Iberian mainland and the American colonies for the Cortes de Cadiz (1810–12), precipitated by the Napoleonic invasion of Spain, established the bases for a constitutional monarchy, but failed to achieve balanced representation between the metropole and its overseas possessions.

The Indian pueblo, or the *república de indios*, had endured more than three centuries as a central figure of colonial juridical institutions and administrative practices. As such, it became a primary political and cultural space in which to debate the meaning of citizenship (*ciudadanía*) and community membership (*vecindad*) not only in New Spain but also in the Spanish viceroyalties of South America.[10] Locally and regionally based research has shown that these legally defined Indian pueblos were mixed ethnically and socially; yet, their political structures entwined with religious rituals upheld their corporate identity. Internal divisions between nobles and commoners and the segmentation of communities into 'head villages' and dependent hamlets often turned violent. Nevertheless these very conflicts provided the political spaces in which the pueblos reconstituted themselves and remained integral parts of the emerging nineteenth-century nations.[11] The corporate identity of the pueblos rested primarily on their claims to territory, including land and water rights, political autonomy in reference to their internal governance, and their symbolic representation through hybrid religious symbolism and the personal figure of the elected town council (*cabildo*) or the hereditary *cacique*.

The struggles over territory in the rural sectors of New Spain at the turn of the nineteenth century were rooted in the materiality of the agrarian landscapes and in the structural transformations of labor and commodity markets with varying repercussions in different regions of the viceroyalty. Population growth, the expansion of mining, and the commercialization of agricultural production during the preceding half-century, while supporting the splendor of Bourbon prosperity, had worked palpable hardships on the urban and rural working classes of the colony. In a recent synthesis, John Tutino compared the economic pressures and subsistence crises in four main regions of Mexico: the Bajío, the central plateau, the southern provinces dedicated to the commercialization of cochineal, and the semiarid plains of the Mesquital and San Luis Potosí, where haciendas had dispossessed subsistence grain producers and converted grasslands into maguey (agave) plantations. Tutino characterized the protests of peasants and rural workers as a stance for 'ecological autonomy' to redress the impoverishment of their living conditions and the loss of dignity as independent producers and the patriarchal heads of their households and families.[12]

If, indeed, the economic and ecological crises of the *ancien régime* were felt across New Spain, local conditions shaped the ways in which specific communities and regions experienced these critical years and interpreted their significance. Eric Van Young developed the theme of local perceptions articulated with the increasingly severe material conditions for the rural and urban poor. He has shown that the structural economic crises of New Spain, reaching the very roots of physical subsistence, were *perceived* differently according to 'local histories and contingencies.' Thus, he argues, the idea of *crisis* that gathered currency at the turn of the nineteenth century had different meanings for the colonial elites and the working classes and—among the latter—it shifted with the particular circumstances observed from region to region.[13] Elite merchant and landowning classes of New Spain had prospered under the liberalization of trade laws during the Bourbon administration, but they often chafed under the diminishing access to public office in the colonial bureaucracy and defended the autonomy of their town councils. After 1808, they focused on the monarchical crisis of Madrid and the events unfolding around the unprecedented convocation of the

Cortes of Cádiz. Indigenous villagers, day laborers, and mineworkers, by way of contrast, protested the loss of lands, the diminishment of their earning power, and the elevated cost of basic foodstuffs. Rural day laborers suffered a loss of 25 percent in their basic wage from 1774 to 1810. Multiple local tumults and uprisings reveal, in Van Young's analysis, that the political content of subaltern protests often assumed the language of religious symbolism in order to assert claims to local identities, material resources, and autonomy.[14]

In what ways did these local sensibilities with strong religious and ecological content translate into a discernible perception of political rights? Antonio Annino has argued that when the principles of Enlightenment Liberalism set forth in the Constitution of Cádiz, promulgated in 1812, crossed the Atlantic and were planted in the soil of New Spain, they were transposed into a re-vindication of local rights, centered in the communities and focused on territory and on the political autonomy of their internal institutions of governance. Reinforcing Van Young's emphasis on the local character of political identities, Annino underscored the corporate nature of 'the horizontal citizenship of commoner-residents (*vecinos-comuneros*)' deeply rooted in the juridical institutions and cultural traditions that had shaped the colonial pueblos. The political crisis that confronted the pueblos arose from the putative equality of all citizens announced by the Constitution of Cadiz. Faced with the formal disbandment of the *república de indios*, as this was contemplated by the delegates to Cadiz and by the first constitutional convention of Mexico in 1824, the pueblos reconstituted themselves politically as *ayuntamientos*—elected town councils—thus taking possession of a liberal political instrument in order to defend their lands and water and to ward off the territorial division of their colonial patrimony.[15] Annino's interpretation of the mobilization and political savviness of indigenous pueblos is convincing for central New Spain, in the cultural area of Mesoamerica. In this region, the density of peasant populations supported the 'indianization' of the *ayuntamientos* founded under the aegis of the final years of imperial rule and institutionalized by the early Mexican republic.[16] In the vast northern provinces, however, where rural communities could scarcely muster a thousand *vecinos*, the number required to form a town council, the colonial councils (called *cabildos*) established in the mission villages faced diminishing legal options for sustaining their corporate identity and defending their territorial autonomy.

Landscapes and Political Theatres of Northern New Spain

The concept of ecological autonomy serves us well to understand the changes that occurred in the northern Interior Provinces during this same period. Following the upheavals in the internal governance and economy of the Indian pueblos that ensued from the royal order to expel the Society of Jesus from all Spanish dominions in 1767, the Bourbon regime devoted significant resources to reforming and consolidating the military garrisons known as presidios placed along the northern frontier from the Californias to Texas. In addition to militarizing the borderlands, King Charles III and the visitors he sent to New Spain renovated the administration of the Interior Provinces through the establishment of the Intendancies and the military General Commandancy for the entire northern region. Some of the missions were secularized, becoming tithe-paying parishes, but others remained under the administration of the Franciscan missionary colleges *de Propaganda Fide* based in Querétaro and Zacatecas. These measures led to the growth of the Hispanic population in the region and the expansion of grain and cattle haciendas and ranches in the valleys with floodplains susceptible to irrigation and in the grasslands of the semiarid slopes and high plains. During this same period military authorities negotiated peace agreements with the Comanches in New Mexico and established peace encampments with several bands of Apaches in the environs of the presidios of Sonora and Nueva Vizcaya. The relative prosperity of the Hispanic *vecinos* during the last two decades of the eighteenth century, together with the resurgence of mining and royal financial support for the region through missionary stipends and the salaries assigned to the presidial garrisons, strengthened local and regional markets and contributed to population growth and the development of urban centers such as Alamos (Sinaloa), Chihuahua (Nueva Vizcaya), and Real de Catorce (San Luis Potosí).

The indigenous pueblos of northern New Spain felt the impact of these changes in markedly different ways, according to the regions in which they lived and the combination of demographic, ecological, economic, and cultural factors that circumscribed their world. In the western Interior Provinces, including Sinaloa, Ostimuri, Sonora, Durango, Chihuahua, New Mexico, and the Californias, the growth of the presidial troops opened new avenues for advancement for indigenous warriors outside the confines of the missions. Indigenous militias, especially the Ópatas and Pimas, were recruited for presidial service with an assigned salary and the right to a parcel of land, not as commoners in their own villages, but as *vecinos-soldados*. The transfer of a substantial part of the male labor force from the mission villages to the presidios undoubtedly had repercussions for the missions' agrarian economy. It may have increased the tasks that fell to women and deepened class divisions among the Indians between those who paid their neighbors to work their land and those who did not have access to wage labor outside the missions. Politically, the prestige accruing to the indigenous militias and the elevation of certain leaders to the office of 'captain general of the Ópata or Pima nation' created parallel and competing lines of authority within the pueblos, between the presidial troops and captains general and the traditional cabildo.[17]

The eastern Interior Provinces, including Nuevo León, Coahuila, Texas, and Nuevo Santander, presented a geography of arid plains and low ranges, but with fewer extended river valleys that lent themselves to irrigated agriculture than in the foothills and terraced slopes of the Sierra Madre Occidental. In these eastern provinces of New Spain's northern borderlands, Jesuits and Franciscans labored to establish indigenous villages among the nomadic bands of hunter-gatherers that had survived the harsh consequences of wars, epidemics, and forced labor during more than two centuries.[18] Interspersed among the itinerant settlements of local indigenous bands, Tlaxcaltecan colonies founded under viceregal auspices, those most well known being in San Estéban de Nueva Tlaxcala facing the colonial villa of Saltillo and in Santa María de Parras near the Nasas River and the Laguna de San Pedro, represented Mesoamerican outposts of peasant agrarian communities. They endured as social and cultural enclaves, although surrounded by Spanish haciendas—most notoriously the entailed network of estates known as the *mayorazgo* of San Miguel de Aguayo.[19] Santa María de Parras, founded as a Jesuit mission in the late sixteenth century for diverse bands of Zacatecan and Coahuiltecan Indians, was secularized at mid-seventeenth century, although the Jesuits maintained a presence there through their College of San Ignacio. Eighteenth-century parish records reveal a mixed population of Hispanic and Tlaxcaltecan vecinos and locally based '*indios laguneros.*' Its local economy had prospered through vineyards and wine-making, and its commercial circuits were linked to the mining region of Mazapil and the Laguna de San Pedro.[20]

The eighteenth-century growth of the colonial economy and the expansion of the Hispanic sector in both the western and eastern corridors of the Interior Provinces were hailed as prosperity in the official reports and raised fiscal revenues for the viceroyalty. These same developments had somber consequences for numerous indigenous pueblos that lost control over land and water. The social and environmental processes of change that followed the accelerated commercialization of the late colonial economy led to uneven outcomes for specific communities, as we have suggested above for the indigenous militias of Sonora in the northwestern provinces and the Tlaxcaltecan colonies of northeastern New Spain. The following section examines in greater detail the transformations in the material landscape and the changes in the social and ethnic identities of the communities of the Sonora River valley, as these were crystallized by the establishment of the Intendency of Sonora in the Ópata pueblo of Arizpe.

Piedmont Communities, Landscapes, and Territories between the Desert and the Sierra

Arizpe, 'place of the large cave,' became the site of a consolidated community of Ópata villages and *rancherías* at the confluence of several tributaries that form the Sonora River. The territory surrounding Arizpe constituted a basin-and-range topography carved out of sandstone and granite by the streams and arroyos that flowed in the rainy season southwest from the escarpment of the Sierra Madre Occidental to the Sonoran Desert and the Gulf of California. Indigenous towns, organized in chiefdoms, produced considerable harvests of maize, beans, and squash in the river valleys, and they adapted their techniques for cultivation and irrigation to the rhythms of hunting and gathering in the *monte* beyond the watered valleys.[21] Ópata chiefdoms first confronted Spaniards through the violent incursions of the Francisco de Coronado expedition (1540–42) and, a century later, of Captain and Alcalde Mayor Pedro de Perea (1637–45). Perea's attempts to force Indian laborers into his mining and ranching enterprises provoked open rebellion and cost him his life. By the mid-seventeenth century three villages—Arizpe, Chinapa, and Bacoachi—had reconstituted themselves in the Jesuit Mission of Nuestra Señora de la Asunción de Arizpe. With a consolidated population and its command of fertile cropland, Arizpe became an important anchor for the entire Jesuit mission system in Sonora.[22]

The population figures for the Ópatas of Arizpe and its sister villages (Fig. 3.1) are, at best, estimates drawn from periodic Jesuit reports, listing families and 'souls,' that is, persons under the missionaries' jurisdiction and possibly excluding children not yet of age for religious instruction. Jesuit visitor Juan Antonio Balthasar (1744) observed irrigated mission lands planted in wheat, maize, vineyards, sugar cane, and orchards as well as livestock, including cattle, mules, and horses. Thus, during the first century of its mission history, the Ópata villagers of Arizpe maintained an agrarian landscape with floodplain fields and *milpas* (family plots) dedicated to maize and cultigens from Eurasia, such as wheat, fruit trees, grapes, and sugarcane. The irrigation works that Balthasar observed depended on indigenous techniques for managing stream flow during the summer rains and diverting the nutrient-laden silt over the floodplain to create layered *milpas* for planting. Villagers organized communal labor to build and maintain earthen dams and to dig canals leading water to the crops planted in the flood plain in addition to seasonal labor for gathering wild plants, fishing, and hunting. The missions depended on the Indians' skills and labor to produce foodstuffs to sustain the community and

Date	Source	Families	Individuals
1678	Juan Ortiz Zapata		1004
1730	Cristóbal de Cañas	180	
1744	Carlos de Rojas	260	803
	Juan Antonio Balthasar		
1778	Pedro de Corbalán	287	933

Figure 3.1 Population estimates for the Arizpe Mission[23]

surpluses to trade with the mining centers (*reales de minas*) and the presidios of Sonora and Nueva Vizcaya.[24]

The Jesuit missions of Arizpe and the Sonora valley did not, however, develop in isolation. Spanish *vecinos* sought watered lands for grain haciendas and bred livestock to graze in the open rangeland, thus invading the Indians' *monte* for hunting and gathering. These same *vecinos* depended on indigenous hydraulic techniques and recruited seasonal laborers, paid in kind, to plant and irrigate their fields. Furthermore, the Sonoran piedmont and mountainous slopes yielded silver mines, and Ópata pueblos were burdened with forced labor drafts under the system of *repartimiento* (minimally paid work crews for specific periods of time) sent to excavate the ore and process the silver. Indigenous laborers from Arizpe, Chinapa, and Bacoachi went regularly to the *reales de minas* in San Juan Bautista, Bacanuche, and Basochuca. Indigenous governors, so designated as the elected heads of the village cabildos, were charged with sending the requisite number of laborers to the mines.[25]

At mid-eighteenth century the Ópatas of the upper Sonora River valley remained as a recognizable cultural enclave within a network of Hispanic settlements and missions, and their villages were surrounded by mining camps and private ranches. Following the expulsion of the Jesuits (1767) and on the eve of the formal establishment of the Commandancy of the Interior Provinces (1779) and the Intendancy of Arizpe (1786), the Bourbon reforms weighed heavily on the communal patrimony and the ecological autonomy of the Ópata communities. The formal structures of mission governance and religious traditions were maintained by the indigenous cabildo and the Franciscans of the Province of Jalisco; nevertheless, the social transformations of this period inverted the demographic ratio of *vecinos* to Indians and accelerated the expansion of commercial agriculture and private land ownership. A 1778 census ordered by Intendant-Governor Pedro de Corbalán listed 390 persons (adults) for the pueblo of Arizpe, divided among Ópatas and 'Spanish settlers of all racial classes' (*españoles de todas las castas*); the census enumerated 120 adobe houses in Arizpe, 132 in Chinapa, and 35 in Bacoachi. The communal croplands for Arizpe were reduced to 50 hectares (14 *fanegas*), although mission herds still numbered in the thousands.[26]

Who passed as 'Indian'? And, by what criteria did individuals or whole families 'count' as Ópata or Spanish? These distinctions arose not so much from racial hue and phenotype as from the effective participation of indigenous or Hispanized peasants in the corporate labor and cultural traditions of the community. It is important to take into account that the growing number of Hispanic *vecinos*—understood not to live by the communal labor of the mission—was augmented by indigenous commoners, both men and women, who passed into the category of *vecinos* through military service, as noted above, and by marriage.[27] In 1781, the General Commandancy established two presidios in the former mission pueblos of Bacoachi and San Miguel de Bavispe, in the neighboring valley to the east, manned principally by Ópata troops. These indigenous soldiers received a plot of land in the area of the floodplain that was assigned to the presidios; thus, their usufruct claims to land and water passed from their communal rights in the pueblo to their service in the military garrison.[28]

The documentary basis for this brief historical profile of Arizpe and its surrounding villages is complemented by the cartographic record for this region, which helps us to visualize patterns of change in the territorial configuration of the province and in the cultural landscapes of its communities. In 1780, in anticipation of the location in Arizpe of the headquarters for both the Intendancy and the military Commandancy of the Western Interior Provinces, royal engineer Manuel Agustín Mascaró produced a place map (*plano*) of the Valley of Arizpe.[29] Mascaró reproduced to scale, as faithfully as possible, the cultural geography of the town of Arizpe and the topography of the valley; in addition, he projected the growth of its urban nucleus and the expansion of its agricultural fields. Mascaró's map illustrates in admirable detail the verbal description of Arizpe that was recorded in Pedro de Corbalán's 1778 report. Following the convention of the times, the map is oriented to the east, such that the stream bed of the Sonora River flows horizontally across the page, from north to south, between the low ranges that define its boundaries. Mascaró plotted the ongoing production of an agrarian landscape in the lines showing the principal irrigation canals (*acequias*) that flowed through a series of contiguous fields in both the northeastern and southern portions of the valley of Arizpe, surrounding the narrow core of village dwellings and extending into the floodplain. The 120 adobe houses reported by Corbalán are shown in this map on the plateau (*mesa* or *serranía*) on the west bank of the river. Some of the buildings indicated by black rectangles in the southeastern portion of the valley may project new construction that would have served the public functions of the Commandancy, thus turning the mission into a Spanish villa or city. (Arizpe received the title of city and, in 1779, was designated the official seat of the new Bishopric of Sonora, but it never achieved the urban growth that its title implied.)

Turning to the expansion of cultivated land in Arizpe, Mascaró's map designated the southern portion of the valley as the area where Hispanic *vecinos* would augment their plantings—in lands that a half-century earlier belonged to the mission. Here the floodplain widened, allowing for a larger network of distribution canals to irrigate new cropland. In support of this project, two years later, Mascaró presented a detailed engineering design for building a dam across the river and opening a new *acequia*. His carefully labeled illustration elaborated on the basic technique for building earthen dams that indigenous peoples had used for centuries to raise the water level to allow it to flow into the *milpas* they cleared for planting in the floodplain. Mascaró's plan seemed intended to create a small reservoir from which to irrigate a larger expanse of agricultural fields downstream, anticipating the growth of Hispanic *vecinos* and the larger urban footprint of Arizpe. We do not know whether the dam and *acequias* were built; however, the design points to an innovative project that was rooted in native technologies and labor for the production of humanly crafted spaces.[30]

Military and civilian authorities accelerated the division of communal lands in small family plots (*hijuelas*) throughout the province during the two decades following Mascaró's visit to Arizpe. Royal orders that had been issued by Visitor-General José de Gálvez (1767–69) and reiterated by the military command in 1785 and 1794, were carried out in a series of measured allotments called *suertes*.[31] The 1794 allotment reserved eight *suertes* for the commons of each pueblo and awarded two or three suertes each to indigenous caciques, generals, governors, and cabildo officers, exempting them as well from the obligatory communal labor. The division of traditional mission lands throughout the Sonoran piedmont had lasting effects on the Ópata, Eudeve, and Nebome villagers of the Sonora, Oposura, Bavispe, and Mátape valleys. Land allotment deepened the class divisions between the indigenous elite, composed by the cabildo and the militia captains, and the commoners—small peasant producers and hacienda workers.[32]

At the close of the eighteenth century the indigenous peoples of Sonora confronted a markedly changed landscape. They faced the fragmentation of their human and territorial patrimony and the weakening of their internal structures of governance. Furthermore, the demographic ascendancy of Hispanic *vecinos* in the pueblos, coupled with the seasonal or permanent migration of indigenous peasants seeking a livelihood in the presidios and in the labor markets that coalesced around the mining camps and haciendas, rendered their communal identity ambivalent. Archival documents for this period, although abundant, focus principally on the military campaigns directed against nomadic bands of Apaches in the Sierra and of Seris on the desert coast; however, they do reveal indigenous responses among the highland villagers to the transformations of their communal institutions and agrarian landscapes. Ópata militia captains, for example, petitioned the General Commandant and the Intendant directly to demand payment of their full salaries or to call for either the removal or the retention of the friars who served in their mission parishes. In 1789, a tumult occurred during the harvest season in the pueblo of Bacerác, involving the village governor, the captain general of the Ópata troops of Bavispe, and the Spanish lieutenant governor. The ensuing investigation revealed layers of conflicts and resentments over the political and social divisions that had severed the pueblo.[33] These local disturbances signal key moments in which communities defended what they considered to be their political and religious foundations, and obliged the authorities to acknowledge and even renegotiate the colonial pact that tied the pueblos to the monarch and to his delegates in the region.[34] Antonio Annino and Eric Van Young have interpreted similar disturbances for central and western Mexico through the language of indigenous petitions and using the lenses of political and cultural history.[35]

Territory and Community in the Transition to Independence

The murmurs of discontent grew louder through the imperial crises of the opening decades of the nineteenth century. In the northern provinces, and particularly in Sonora, the sense of crisis arose from recurring droughts and crop failures, the resurgence of Apache raiding following the failure of the Bourbon administration to maintain the peace encampments, and in the dispersion of presidial troops out of the province to combat the popular Insurgency in central New Spain. A significant number of Ópata troops, for example, were sent to the neighboring province of Sinaloa to the south, to defeat the rebel band led by José María González de Hermosillo.[36] These environmental, economic, and military disjunctures were compounded in the region because of the structural transformations in landholding and in the material landscapes of the indigenous pueblos set in motion by the Bourbon administration.

At the close of the Independence wars, when insurgent and royalist leaders were beginning to negotiate a treaty in central New Spain under the terms of the conservative pact of 'Three Guarantees' heralded by Agustín de Iturbide, the Ópata troops stationed in Bavispe rose up in arms against the Commandancy that they had served so faithfully. Demanding not only the payment of their salaries in arrears, but also the right to elect their own captains general, the Ópatas raised a claim for local autonomy within the communities they had re-created through the presidial system. The militias of Bavispe, led by Juan Dórame, joined the indigenous troops from highland Sonoran pueblos in Arivechi, Pónida, Sahuaripa, and Tonichi in 1820 to confront the remaining royalist forces in the Commandancy General. This revolt was suppressed when their munitions were depleted. Indigenous warriors rebelled again in 1824, now within the framework of the Mexican Republic and the Estado de Occidente, against the federal military commander of Sonora, who had removed their captain general from office. In both instances, the Indians' protests centered on their right to elect their military leaders, thus defending long-standing practices of local political autonomy and—by inference—the common territories that marked their communities of origin.[37]

The merchant and landed classes of northwestern Mexico took control of municipal councils (*ayuntamientos*) and the state legislature under the juridical institutions of the Estado de Occidente (1824–30) and the State of Sonora (separated from Sinaloa in 1831), whose first capital was Arizpe. Taking a leaf from the book of their Bourbon predecessors, the new political elite defined citizenship in terms of individual land ownership and further accelerated the division and redistribution of lands that had remained in communal usufruct in the former mission pueblos. In a series of legislative initiatives carried out between 1828 and 1835, Sonoran governors and legislators undermined the corporate representation of the pueblos through their cabildos, effectively absorbing them into the newly created municipalities. While these same laws allotted small plots of land to indigenous villagers, they placed increasing extensions of potential cropland and grazing *monte* into the category of untitled *realengos* or *baldíos*, thus making them accessible for private land claims that were formalized through the protocols of bidding, measurement, and entitlement. Labeled as progressive in the liberal discourse of the nascent republic, these actions provoked new indigenous uprisings in 1832–33, bringing together different ethnic groups, most notably the Ópatas and Yaquis of northern and southern Sonora. In a formidable armed movement led by Juan Banderas and Dolores Gutiérrez, rebel forces defended their principles of 'ecological autonomy' focused on territorial rights to land and water and on the integrity of their local government.[38] Notwithstanding the military defeat of this rebellion and the execution of its leaders, indigenous leaders continued to defend their local autonomy and their territory by legal actions and by force of arms. Over the course of the nineteenth century they repeatedly challenged the conflictive unfolding of the nation-state in the former Western Interior Provinces of Spain's North American borderlands.

Conclusions

The indigenous communities of the Sonoran piedmont lost significant political spaces during the formative years of the federal republic of Mexico, in contrast to research findings summarized by Antonio Annino and Leticia Reina for the central and southern regions of Mexico. In northwestern Mexico the indigenous pueblos articulated their own meanings of citizenship in these key transitional years from colony to republic, thus demonstrating the maturity of their corporate political culture as it had been forged in the colonial missions and nourished by their deep roots in the ecological practices of their physical surroundings. Notwithstanding the clarity of their concept of citizen-commoner, with its dual connotations of community membership (*vecindad*) and territorial integrity (*el común*), the Sonoran pueblos no longer controlled the demographic and economic resources sufficient to take possession of local institutions of government under the *ayuntamientos*. I have argued here that the truncated power and diminished territories of the Sonoran communities in the early nineteenth century can be traced directly to the ecological transformations of their landed base. In order to understand these apparent political weaknesses and to analyze the accelerated rhythm and multiple directions of change during the unraveling of the Viceroyalty of New Spain and the shaky construction of the Mexican Republic, it is necessary to place the events of this period in a framework of *longue durée* processes of both transformation and continuity. It is equally important to connect the episodic trail of historical events with the landscapes that are, themselves, an ineffaceable part of the historical record.

The historical narratives and perceptions of regionality that emerge from this tumultuous period of change in northern New Spain illustrate the radical sense of locality that scholars like Eric Van Young, Peter Guardino, Ethelia Ruiz Medrano, and John Tutino have researched so deeply for other regions of western, central, and southern Mexico.[39] Their scholarship illustrates a growing body of published research that is framed by both historical and anthropological methods and perspectives. It has taught us, among many things, the religious cast of political culture and the spatial dimension of historical events. Decision-making and actions taken by men and women of the popular classes evinced a sense of place that was rooted in religious symbols, social networks, and in the landscapes of subsistence and of cultural belonging.

These same material and symbolic constructions of place took on enhanced meanings in the northern borderlands of both the Spanish imperium and the formative Mexican nation-state. The militarization of what Bourbon authorities in the Iberian metropole and the viceregal court considered to be the frontier of Spain's effective dominion in North America increased the layers of colonial authority in these borderlands of seminomadic peoples. At the same time the increased population that claimed Hispanic heritage and asserted private claims to property in agriculture, livestock ranching, mining, and commerce deepened the social complexity of these northern provinces. Their growing numbers and economic resources, together with the indigenous villagers who aspired to the status of *vecinos*, placed additional pressures on the commoners who defended traditional ways of holding and managing cropland in the riverine valleys and the desert-like expanses of the *monte* for hunting, gathering, and grazing. The political and environmental histories of northwestern Mexico are entwined in this brief account focused on the community of Arizpe and its surrounding territories in the Province of Sonora. The changing landscapes of this narrative point to physical changes in the configuration of cropland, grazing lands, and uncultivated desert, to social changes in the access to land, to political changes in local governance, and to cultural changes in the meaning of community.

The borderlands of northern New Spain had signified for colonial authorities an externally conceived moving frontier that was intended to distinguish the territories effectively held under Spanish dominion from the vast spaces occupied by indigenous confederations of Comanche captaincies and Apache bands and from French imperial claims to Louisiana. At the turn of the nineteenth century and across the divide of Mexican Independence, the external construct of imperial borderlands had turned inward, constituting a web of internal borderlands in which authority was disputed at both local and regional levels.

In this new spatial and political configuration of borderlands, social divisions were codified ethnically and opposing concepts of citizenship were manifested in open conflict in relation to the institutions of state power and to the material resources and symbolic meanings of culturally produced landscapes.

Notes

1. See, for example, J. Tutino, *From Insurrection to Revolution in Mexico: Social Bases of Agrarian Violence, 1750–1940* (Princeton, 1986); L. Reina and E. Servín (eds), *Crisis, Reforma y Revolución: México: historias de fin de siglo* (2nd ed., México, 2002 [2001]); P. Guardino, *The Time of Liberty: Popular Political Culture in Oaxaca, 1750–1850* (Durham, NC, 2005); E. Van Young, *The Other Rebellion: Popular Violence, Ideology, and the Struggle for Mexican Independence, 1810–1825* (Stanford, 2001); A. Escobar Ohmstede, R. Falcón and R. Buve (eds), *Pueblos, comunidades y municipios frente a los proyectos modernizadores en América Latina, siglo XIX* (Amsterdam, 2002); M.T. Ducey, *A Nation of Villages: Riot and Rebellion in the Huasteca, 1750–1850* (Tucson, 2004).
2. G. Nabhan, *Gathering the Desert* (Tucson, 1985).
3. W. Doolittle, 'Misreading between the Lines: Evidence and Interpretation of Ancient Settlements in Eastern Sonora, Mexico,' in P.H. Herlihy, K. Mathewson and C.S. Revels (eds), *Ethno- and Historical Geographic Studies in Latin America: Essays Honoring William V. Davidson* (Baton Rouge, 2008), pp. 283–98; S.K. Fish, 'Hohokam Impacts on Sonoran Desert Environment,' in D.L. Lentz (ed.), *Imperfect Balance: Landscape Transformations in the Precolumbian Americas* (New York, 2000), pp. 251–80; T. Sheridan, *Where the Dove Calls: The Political Ecology of a Peasant Corporate Community in Northwestern Mexico* (Tucson, 1988).
4. P.S. Martin, *Twilight of the Mammoths: Ice Age Extinctions and the Rewilding of America* (Berkeley, 2005).
5. C.L. Riley, *Becoming Aztlan: Mesoamerican Influence in the Greater Southwest, AD 1200–1500* (Salt Lake City, 2005).
6. R.C. West, *The Mining Community in Northern New Spain* (Berkeley, 1949); West, *Sonora: Its Geographical Personality* (Austin, 1993).
7. T. Sheridan, *Landscapes of Fraud: Mission Tumacácori, the Baca Float, and the Betrayal of the O'odham* (Tucson, 2006), pp. 38–42; A. Sluyter, *Colonialism and Landscape: Postcolonial Theory and Applications* (Lanham, 2002).
8. D. Studnicki-Gizbert and D. Schecter, 'The Environmental Dynamics of a Colonial Fuel-Rush: Silver Mining and Deforestation in New Spain, 1522 to 1810,' *Environmental History*, 15 (2010), 94–119.
9. A. Escobar Ohmstede, R. Falcón and R. Buve (eds), *Pueblos, comunidades y municipios frente a los proyectos modernizadores en América Latina, siglo XIX*; A. Annino, 'El primer constitucionalismo mexicano, 1810–1830,' in M. Carmagnani, A. Hernandez Chávez and R. Romano (eds), *Para una historia de América* v. 3, *Los nudos* (México, 1999), pp. 140–89; Guardino, *The Time of Liberty*.
10. M. Irurosqui, 'El bautismo de la violencia: indígenas patriotas en la revolución de 1870 en Bolivia,' in J. Salmón and G. Delgado (eds), *Identidad, ciudadanía y participación popular desde la colonia al siglo XX* (La Paz, 2003), pp. 115–52; B. Larson, *Trials of Nation Making : Liberalism, Race, and Ethnicity in the Andes, 1810–1910* (Cambridge, 2002); C.F. Walker, *Smoldering Ashes: Cuzco and the Creation of Republican Peru, 1780–1840* (Durham, NC, 1999).

11. E. Van Young, 'Agrarian Rebellion and Defense of Community: Meaning and Collective Violence in Late Colonial and Independence-Era Mexico,' *Journal of Social History* 27 (1993), 245–70; F. Mallon, *Peasant and Nation: The Making of Postcolonial Mexico and Peru* (Berkeley, 1995); M.D. O'Hara, *A Flock Divided: Race, Religion, and Politics in Mexico, 1749–1857* (Durham, NC, 2010).

12. J. Tutino, 'Globalizaciones, autonomías y revoluciones: poder y participación popular en la historia de México,' in L. Reina and E. Servín (eds), *Crisis, Reforma y Revolución: México: historias de fin de siglo* (2nd ed., México, 2002 [2001]), pp. 25–86.

13. Van Young, 'Agrarian Rebellion'; Van Young, *The Other Rebellion*; Van Young, 'De tempestades y teteras: crisis imperial y conflicto local en México a principios del siglo XIX,' in Reina and Servín (eds), *Crisis, Reforma y Revolución*, pp. 161–208, at 173–4.

14. Van Young, 'De tempestades y teteras,' pp. 164–5, 179–86.

15. A. Annino, 'El Jano bifronte: los pueblos y los orígenes del liberalismo en México,' in Reina and Servín (eds), *Crisis, Reforma y Revolución*, pp. 209–51; Annino, 'El primer constitucionalism mexicano,' pp. 140–89.

16. L. Reina (ed.), *La reindicanización de América, siglo XIX* (Mexico, 1997).

17. C. Radding, *Wandering Peoples: Colonialism, Ethnic Spaces, and Ecological Frontiers (Northwestern Mexico, 1700–1850)* (Durham, NC, 1997), pp. 288–92; Biblioteca Nacional Fondo Franciscano (BNFF) 35/767 (1790), f. 3–11.

18. J. Cuello, 'The Persistence of Indian Slavery and Encomienda in the Northeast of Colonial Mexico, 1577–1723,' *Journal of Social History*, 21 (1988), 683–700; S.M. Deeds, *Defiance and Deference in Mexico's Colonial North: Indians under Spanish Rule in Nueva Vizcaya* (Austin, 2003); P. Osante, *Orígenes del Nuevo Santander, 1748–1772* (Mexico and Cd. Victoria, 2003); C. Sheridan, *Anónimos y desterrados. La contienda por el 'sitio que llaman de Quauyla,' siglos XVI–XVIII* (México, 2000).

19. E. Butzer, Elisabeth, *Historia social de una comunidad tlaxcalteca: San Miguel de Aguayo (Bustamante, NL), 1686–1820* (Tlaxcala, 2001); J.G. Martinez Serna, 'Vineyards in the Desert: The Jesuits and the Rise and Decline of an Indian Town in New Spain's Northeastern Borderlands,' PhD dissertation, Southern Methodist University, Dallas, 2009; M. Vargas-Lobsinger, *Formación y decadencia de una fortuna : los mayorazgos de San Miguel de Aguayo y de San Pedro del Álamo, 1583–1823* (México, 1992), p. 48.

20. 'Libro de Fábrica 1774,' Archivo Parroquial Histórico del Templo de San Ignacio de Loyola, in University of Texas Nettie Lee Benson Latin American Collection (UTNLBLAC), exp. 514, folios s/n.

21. Doolittle, 'Misreading between the Lines,' 300–3. While the term *chiefdom* implies a wide variety of meanings, archaeological and ethno-historical evidence for this region confirms the existence of supra-village political entities based on allegiance to outstanding male leaders prior to and at the time of Spanish contact, in the mid-sixteenth century.

22. Don Pedro de Perea, Gouernador y Capitan a guerra de la Prouincia de Sinaloa y la Nueua Andaluzia (Mexico, 1637) del Catálogo Colectivo de Impresos Latinamericanos hasta 1851 (CCILA), http://cbsrdb.ucr.edu/cgi-bin/starfinder/3618/lastc.txt #6994 (accessed June 22, 2010).

23. Jesuit Juan Ortiz Zapta, *Relación* 1678, Archivo General de la Nación (AGN), *Misiones* 26; Carlos de Rojas, N.S. de la Assumpción de Arizpe, Bancroft Library M-M 176, v. 1–77; F. Molina Molina (ed.), *Estado de la Provincia de Sonora, 1730* (Hermosillo, 1979); E.J. Burrus and F. Zubillaga (eds), *El noroeste de México: Documentos sobre las misiones jesuíticas, 1600–1769* (México, 1986), pp. 189–190; BNFF 34/733, 736 (1778). The final population figure of 933 is calculated from the house-to-person ratio given

in the source for the village of Arizpe and applied to the villages of Chinapa and Bacoachi, for which only the number of houses is given in the 1778 census.

24. Radding, *Wandering Peoples*, pp. 48–54, 70–91.
25. I. del Río, 'Repartimientos de indios en Sonora y Sinaloa,' in *Memoria del VII Simposio de Historia de Sonora* (Hermosillo, 1982), pp. 7–22; West, *Sonora*.
26. BNFF 34/733, 736, 'Estado que manifiesta el número de poblaciones correspondientes a esta jurisdicción, distancias y rumbos de la capital, bienes que poseen sus habitantes, y lo demás que se expresa en las casillas' (1778); Radding, *Wandering Peoples*, 92–3. A *fanega* signified the land needed to plant a measure of grain, usually calculated for wheat in 3.5 hectares.
27. Informe del Obispo Antonio de los Reyes al Virrey Matías de Gálvez, 1784, en BNFF 34/759; Informe del Asesor Pedro Galindo Navarro, 1785, en Archivo de la Mitra, Hermosillo, Archivo Diocesano, I.
28. The distribution of these plots is represented in the local maps of the presidios drawn up during the military inspection of the northern frontier carried out by the Marqués de Rubí and the royal engineer Nicolás de la Fora (1766–68): British Library, London: Presidio de Fronteras, Sonora, BMK19931.
29. 'Plano general de la mission y pueblo de Arispe, que Su Magestad en sus reales instrucciones destina para la capital de las Provincias Internas de Nueva España. Relación de los edificios, tierras, caminos, etc. del territorio representado y de los proyectos de construcción de nuevos edificios'—Museo Naval (Madrid) MNM881.
30. M. Agustín Mascaró, 'Plano, perfil y elevación de la nueva Presa de Arispe y parte de su nueva acequia' (Arispe, 24 Apr. 1782, in MNM R-3697, P-C-13–1); H. Lefebvre, *The Production of Space*, trans. D. Nicholson-Smith (Oxford, 1991).
31. Comandante General de la Provincias Internas Pedro de Nava, 1794, Archivo General de Indias (AGI), Guadalajara, 586. A *suerte* ('lot') corresponded to approximately 10.5 hectares or 26.5 acres.
32. Radding, *Wandering Peoples*, pp. 180–93.
33. BNFF 35/767, 1789–1790, 12ff.
34. C. Radding, *Entre el desierto y la sierra: Las naciones o'odham y tegüima de Sonora, 1530–1840* (México, 1995); C. Radding, 'Sonora-Arizona: The *común*, Local Governance, and Defiance in Colonial Sonora,' in J.F. de la Teja and R. Frank (eds), *Choice, Persuasion, and Coercion: Social Control on Spain's North American Frontiers* (Albuquerque, 2005), pp. 179–99.
35. Van Young, 'De tempestades y teteras'; Annino, 'El Janos bifronte.'
36. J.M. Medina Bustos, 'La crisis de la monarquía hispánica en la Intendencia de Arizpe (1808–1812),' in Z. Márquez Terrazas (ed.), *Coloquio Camino Real: Bicentenario de la Independencia de México* (Chihuahua, 2009), pp. 15–47.
37. C. Radding, *Landscapes of Power and Identity: Comparative Histories in the Sonoran Desert and the Forests of Amazonia from Colony to Republic* (Durham, NC, 2005), pp. 388–9; R.W.H. Hardy, *Travels in the Interior of Mexico, in Baja California, and around the Sea of Cortés* (Glorieta, 1977 [1829]), pp. 164–5.
38. E. Hu-Dehart, *Adaptación y resistencia en el Yaquimi. Los yaquis durante la Colonia*, trans. Z. Marcela (México, 1995); A. Figueroa, J. Montané Martí and E. Villalpando, *Los que hablan fuerte, desarrollo de la sociedad Yaqui* (Hermosillo, 1985).
39. Van Young, *The Other Rebellion*; Guardino, *Time of Liberty*; Tutino, 'Globalizaciones, autonomías y revoluciones'; J. Tutino, *From Insurrection to Revolution in Mexico*; E. Ruiz Medrano, *Mexico's Indigenous Communities: Their Lands and Histories, 1500 to 2010* (Boulder, 2010).

4

'We Are Comfortable Riding the Waves': Landscape and the Formation of a Border State in Eighteenth-Century Island Southeast Asia

Timothy P. Barnard

In 1820 a British civil servant named Robert Ibbetson traveled to eastern Sumatra. He was not impressed with what he saw. In his report to the British authorities Ibbetson described the region as 'a number of petty principalities lying along the seashore and bordered inland by various tribes, while poverty, misrule and piracy contend for mastery and serve to nullify the natural advantages of the country and its numerous resources.'[1] Within this contempt for the form of rule among the various communities in the region Ibbetson did point out an important factor influencing British interest in the area, as well as the difficulties in controlling it: the natural landscape. 'Eastern Sumatra,' stretching from Palembang to modern-day Medan, is a geographic region within which the polity known as Siak (after its major river) emerged in the eighteenth century to dominate the entire area. While boundaries fluctuated continually, eastward Siak included the coastal areas bordering the Melaka Straits as well as the offshore seas and islands; to the west it ended at an elevation of 100 m above sea level, which in such a swampy low-lying region was some 200 km inland. It was a region where authority did not seem to follow either European, or Malay, understandings of power. It was a borderland region in which the natural landscape influenced state formation. It may have appeared to a European observer to suffer from poverty, misrule, and piracy, but its natural advantages and resources led to trade that supported the creation of a state located between the waters that united the Malay World and the highlands of Minangkabau gold, coffee, and rice in the interior of Sumatra.[2]

Four years after Ibbetson wandered through the various principalities of eastern Sumatra under Siak rule, British and Dutch authorities created a boundary in this region. The signing of the Anglo–Dutch Treaty of 1824, which was negotiated in Europe to settle post-Napoleonic era differences between the United Kingdom and the Netherlands with regard to their overseas territories, created a border that still exists today between Malaysia and Indonesia. The

84

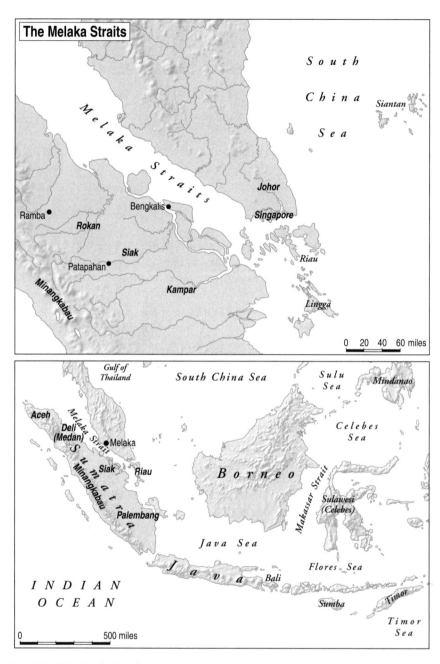

Map 4.1 The Melaka Straits

British carved out a sphere of influence that included the Malay Peninsula, while the Dutch gained nominal control over Sumatra and the islands south of Singapore. While the treaty did divide a region that was environmentally united around the trade that flowed through the Melaka Straits, this new boundary did not create the tremendous political and economic problems often found when Western powers drew arbitrary lines in maps in other parts of the world. This was because each 'petty principality' within Siak was able to develop its own form of rule based around control of various diverse eco-niches. They could act together, or on their own.[3]

The presence of these divisions within a larger state reflected the flexible nature of boundaries in the precolonial world, and also is reflected in the academic literature. Most studies of Siak have focused on its hybrid state structures in which Malay and Minangkabau traditions were blended to create a unique state apparatus that ruled over a swampy frontier between the sea and the mountains.[4] Traditionally, the area was considered Malay, in which all of the peoples were loyal to a centralized Malay sultan. Obtaining the loyalty of downstream and coastal groups was one of the keys to gaining economic and social control over regional trade. With the power that trade provided, rulers would enter the frontier of eastern Sumatra where they attacked and subjugated the local chiefs. The defeated ruler usually was allowed to maintain his position, with additional honorific titles and favors, as long as he remained loyal to the Malay sultan. This system was recreated on a smaller scale at the regional level as the local ruler gained the loyalty of smaller groups, particularly the *orang asli* ('original people,' a term for people who did not follow a world religion and were slightly nomadic). Thus, there was a layering of authority that stretched from local rulers up to the regional centers. Relations between *orang asli* communities and outside rulers were maintained through intermediaries, usually Malays or Minangkabau who had married an *orang asli* woman and had taken on honorific titles. This culminated in a centralized Malay ruler, who was perceived as controlling the entire region. Each of these groups could be allied with the center, or a rival, leading to a constant shifting of loyalties and power.[5]

In the environmentally diverse region of islands, bays, ocean, and forests along the Melaka Straits, however, the presentation of titles, charisma of a ruler, and the use of force had a limited effect. It was the geographical characteristics of this region, its landscape, which determined the internal boundaries, acting as the main influence on how polities in the region were formed. This landscape of coasts and seas, as well as the sinews of connecting rivers, came to influence the development of society in eastern Sumatra. This chapter will focus on how this landscape influenced the development of boundaries and borders for polities in this region in the era prior to that of high imperialism, when Western powers could arbitrarily draw a line through the Melaka Straits. The first section will break this region into zones of highlands, swamps, and seas, while the second will discuss how the rulers of the region used their knowledge of these zones to create a polity and expanded its boundaries until it was one in which mountain dwellers transformed into a people who were 'comfortable riding the waves,' as described in the polity's main historical-literary text, the *Hikayat Siak*.[6] In the precolonial context, there were borders. These borderlands, however, consisted of eco-niches, resulting in a flexible form of rule to accommodate the diverse communities among them.

An Ecological Borderland

Siak arose in the late seventeenth and early eighteenth centuries out of a cultural and ecological frontier between Malay and Minangkabau worlds that allowed it to flourish in a harsh region. Because three major rivers cross the lowlands, communities were able to develop in eco-niches where they subsisted on trade, limited agriculture, and the collection of forest products. Each of these communities was basically self-supporting, but never isolated. These communities were located between the foothills of the Bukit Barisan mountain range, which runs down the western side of Sumatra, and the Melaka Straits, and were connected by the wide, slow-moving rivers that flow into the Straits. This region of dense tropical rainforest slowly shifted into lowland swampy mangrove forest before meeting the cosmopolitan trading world of the Straits. In the Straits, communities were dotted up and down the east Sumatran coast, fanning outwards into the islands of the South China Sea. The forest, swamps, and numerous inlets allowed for the development of niches where settlements of both indigenous inhabitants and migrants thrived. Similar eco-niches were also present in the seas and shallows of the Melaka Straits and South China Sea. The natural products of the ocean and jungle, such as rattan, sago, and fish, provided the backbone for trade while also allowing communities to function on a relatively independent basis.[7]

Although some 150 km inland, the upstream regions of eastern Sumatra are still less than 20 m above sea level. The flat land of the coasts slowly blends into rolling hills, which begin to flow through the countryside some 50 to 75 km upstream, and in combination with the forest cover often made walking a difficult if not impossible task, as locals themselves noted.[8] The climate in this region is hot and humid, with daily temperatures usually above 30° Celsius. Extreme amounts of precipitation, which fall during the monsoonal rainy season usually lasting from November to April, compound the effects of high humidity. Even during the driest months, it is common for over 100 ml of rainfall to be recorded at stations spread throughout the region. The differences in rain, however, did make for a distinct growing season for plants and animals, in contrast to the controlled planting and harvesting of wet rice in the Minangkabau highlands of western Sumatra. These harsh conditions made it difficult for communities with large populations to develop.[9]

While the precolonial Melaka Straits has been described as 'population poor' and an 'empty center' of Southeast Asia, small settlements upstream did exist.[10] The limiting factor for the growth of larger communities lay in its soils, which are very poor and highly susceptible to erosion since they have been formed from highly acidic parent materials, mainly sandstone in this region, which has been leeched of any nutrients that the soil may have contained.[11] Despite the complexity and diversity of flora in the lowland tropical rainforest that grow out of these soils, the land is not suited for agricultural production.[12] Thus, small trading communities, which transferred goods between the Malay coast and the Minangkabau interior, mainly inhabited much of central eastern Sumatra and were limited to dry fields near streams and rivers. Beyond these communities was a seemingly endless forest that provided items for export.

The Minangkabau who settled in the region were drawn into a vital and complex trade system that had involved local populations for hundreds of years. The trade ranged from common items such as timber to exotic resins and animal products. An example of their unique, valuable, and exotic nature was the bezoar stone, called *guliga* in the Malay World. A bezoar stone is a hard stone-like substance found in the stomachs of forest animals, and during the early modern era such stones were known in Aceh as *pedro de porco siacca*, emphasizing the center of bezoar stone collection and distribution in the Straits region. The value of the stones was such that they were 'valued at ten times [their] weight in gold,' and those from Southeast Asia were considered to be particularly potent medicinally. Shavings from a bezoar stone, eaten directly or placed in a glass of wine, were believed to 'have the virtue of cleansing the stomach, creating an appetite, and sweetening the blood.'[13] Faith in the efficacy of bezoar stones was even more prevalent in Southeast Asian society, where they acted as talismans and became important items in family heirlooms and the regalia of rulers.[14]

The exotic nature of the collection and attributes of products such as bezoar stones made central-eastern Sumatra an area of desired trade but also one of uncertainty and mystery. It was difficult to access valuable forest products, and thus full cooperation with those possessing specialized knowledge, usually *orang asli* groups, was required. Since these items were important spiritually as well as economically, their collection remained closely guarded secrets that allowed the communities that gathered and controlled them to retain a certain amount of power that could be used to their advantage. In addition to bezoar stones there were other trade goods from eastern Sumatra that reflected the economic and spiritual importance of the forest product trade. The region was particularly renowned for resins such as dammar, agul wood, and camphor. For example, dammar, a form of tree resin that includes benzoin, was usually found in diseased *dipterocarp* trees. Since dammar forms in diseased trees, the search for older trees, more likely to possess the resin, was the work of specialists. White specks of resin in the bark alerted the specialist, who then oversaw the tapping. Softer resins, which were recently tapped, were used as torches, while 'fossil resins,' usually found in the ground near the tree, were utilized for caulking boats or making buttons.[15]

Not all forest products, however, were as exotic as dammar or bezoar stones. The tropical lowland rainforest was also a productive supplier of relatively common items that actually made up 'the principal exports' of the region.[16] These products were traded in high volume on a daily basis in settlements throughout the lowlands. Among these products was rattan, including numerous species of solid-stemmed climbing palms from the subfamily Calamoideae, most of which cling to dipterocarp trees with thorny grapples. The collector harvested the rattan by cutting at its base and then pulling down the plant from the crown of the tree. As this was done, the thorns would be stripped from the stem. The collector then would roll up the rattan and carry it to a cleared location near a stream, where it could be dried near a fire. The main use for rattan was as the basic material in a variety of objects used in indigenous households, including baskets, buildings, fences, furniture, and tools.[17]

Although *orang asli* groups were similar in that they harvested forest products and served the rulers in certain traditional functions, they considered themselves distinct through their origin tales and area of settlement. Among the best known of the upstream *orang asli* groups was the Petalangan, who lived between the Siak and Kampar rivers in an area of lowland tropical rainforest.[18] The collection of the honey and wax of the *Sialang* tree constituted one of the main economic activities of the Petalangan, and was imbued with great spiritual significance. Once harvested, the Petalangan were supposed to present the honey and wax to the ruler or his representative, under the auspices of *serahan* (submission) trade, in exchange for iron, salt, or cloth.[19] These rules were rarely followed, however. Many traders married *orang asli* women in order to gain access to forest products,[20] and although the ruler did receive a portion of these products, much of it entered the trading system surreptitiously, with benefit to the local communities.

In the late seventeenth century, the ability of the *orang asli* to gather valuable forest products mixed with the economic experience of the migrant Minangkabau who began to settle in upstream communities. The collection and distribution of forest products became linked, allowing for new possibilities for the exploitation of the forest, and bringing into question any loyalty the *orang asli* supposedly showed toward distant Malay or Minangkabau rulers. Through intermarriage and growing trade relations between these groups, upstream central Sumatra became an increasingly hybrid region, neither Minangkabau nor Malay. Each community had a leader with an honorific title who negotiated for better relations with other upstream communities, in addition to those located in the Minangkabau highlands and along the Melaka Straits. By the end of the seventeenth century such interaction resulted in a fluid region that contained a variety of locally valuable trade products and communities that were acting independently to trade and benefit from their presence. The borders that the eco-niche created were made flexible through personal and trade relations within the larger region.

While the upstream areas of eastern Sumatra were the location of communities of Minangkabau migrants and indigenous peoples, they were connected to the downstream regions through many wide and slow-moving rivers that flowed between the Bukit Barisan range and the Melaka Straits.[21] From the Musi to the Rokan these rivers were not only links to various regions of eastern Sumatra, but also provided numerous advantages to the communities who lived along their banks. In William Marsden's words,

> the distance of the range of hills not only affords a larger scope for the course of the rivers before they disemboque, presents a greater surface for the receptacle of rain and vapours, and enables them to unite a greater number of subsidiary streams, but also renders the flux more steady and uniform by the extent of level space ...[22]

This was particularly the case in comparison with the fast and steep rivers of the western coast of Sumatra. On the wide stretch of eastern Sumatra three rivers are of particular interest for the development of the border region in the immediate precolonial era: the Rokan, Siak, and Kampar. Each of these rivers has its own natural advantages that help determine the nature and extent of external influences and economic trade that flows up and down them.

Both the Rokan and Kampar rivers originate in the Minangkabau highlands of the Bukit Barisan range. They are sufficiently deep for the navigation of ocean-going ships over 100 km inland, while their estuaries are quite wide. In his eighteenth-century survey *The History of Sumatra*, Marsden remarked that the mouth of the Rokan was so broad that it should not be thought of as part of a river but 'considered as an inlet of the sea.'[23] Although both would seem to have a tactical trade advantage over the Siak River, which only originates in the foothills of the Bukit Barisan and thus has no direct access to Minangkabau regions, the presence of several obstacles at the mouths and in the estuaries of the Kampar and Rokan rivers hampers entry. Anderson described the area at the mouth of the Rokan, which could also apply to the Kampar estuary, as 'the most dangerous part of the coast' due to the presence of sandbanks, mudflats, and islets.[24] In addition, John Anderson noted in 1823 that the Rokan River was almost dry at low tide, and the rapidity of the tides was estimated at 7 to 10 km an hour while they rose and fell up to a height of 10 m, strong enough to cut the anchor rope of his boat.[25]

The difficulties these shallow areas and tides created for travelers in the estuaries of the Kampar and Rokan rivers, however, paled in comparison to the tidal bore present on both rivers. A tidal bore, called a *beno* along the east coast of Sumatra, is a massive wave caused by the force of a river flowing downstream colliding with a rapidly rising tide. The presence of the *beno* on the Kampar was among the earliest descriptions of the river, when Tome Pires wrote in his early sixteenth-century account *Suma Oriental* that its power was such that it 'overthrows and breaks up anything it finds.'[26] The danger of these tidal bores was so great that most local vessels would not venture in Rokan and Kampar in the seventeenth and eighteenth centuries without a pilot, while some Europeans avoided the rivers completely.[27] Once past these estuarine difficulties, the Rokan and Kampar rivers were excellent locations for trade and habitation, despite further problems due to the rapidity of the stream flow. The rivers were wide, and in the case of the Kampar River, its depth reached some 8 m while the dark brown water was 'without flavor and quite suitable for drinking.'[28] Numerous smaller rivers and streams, similar to those found all along the eastern Sumatran coast, fed both rivers, and they split into branches some 100 km inland. In the early nineteenth century the banks of the Rokan and Kampar rivers and their tributaries were studded with villages 'in a state of tolerable civilization.'[29]

In contrast to the difficulties the Kampar and Rokan rivers posed for inhabitants and traders, the Siak River was ideal. Although not as wide as its neighbors, the Siak River, as John Anderson wrote, 'ranks first in all other respects, as being the deepest, most free from obstructions, and the channel of conveyance for the most valuable and extensive commodities and commerce.'[30] Various surveys of the river supported such claims. In the early nineteenth century, the Englishman Francis Lynch reported that its depth was quite consistent for more than 100 km, thus permitting large vessels to travel as far as the trading center of Pekanbaru, while the tide was less than that on the Kampar and Rokan rivers. Anderson agreed with Lynch when he came to the conclusion that the Siak River was, 'without exception, the finest river I ever saw.'[31] Anderson, however, was not the first visitor to sing the praises of the Siak River. Its ideal conditions had led the sixteenth-century Portuguese visitor Jorge Botelho to claim that the Siak River supported 'a land of plenty.'[32]

Some 150 km upstream, the Siak River splits into two branches, the Tapong Kiri and the Tapong Kanan. This was the nineteenth-century boundary between upstream and downstream since large vessels could not travel farther up the Siak River. Limitations on shipping in this section of the river were not due to the depth, which continued to be 8 to 9 m, but to the rapidity of the current, as the branches of the Siak began their descent from the foothills of the Bukit Barisan range. During the rainy season flooding made the current so strong that ropes were required to pull boats upstream.[33] Of these two branches, the Tapong Kiri was more popular since it led to footpaths that acted as transfer routes for trade goods between the Minangkabau highlands and the coast.[34]

battle since his rowers 'were from the river reaches, and his minwinmen from the Minangkabau interior.[49] Following Raja Kecik's death in 1746 the rulers of Siak would have to expand their influence beyond a swampy border region. They would need to traverse the sea.

The key figure who initiated the move beyond the vast Sumatran littoral was a son of Raja Kecik, known as Raja Alam. Following the death of Raja Kecik, there was a tense conflict between two of his sons, Raja Mahmud and Raja Alam. As the loser in this conflict, Raja Alam fled to new territories where he could find supporters. This new environment was centered on the island of Siantan. Located in the South China Sea in a region of scattered islands spread between the Malay Peninsula and Borneo, Siantan was an important source of manpower for many of the warring communities in the region during the eighteenth century. From the perspective of the European trading companies, Siantan was simply a pirates' lair. It was home to numerous Orang Laut ('sea peoples,' an *orang asli* group), as well as Malay, Bugis, and Chinese migrants. Raja Alam used his base in Siantan to gather followers and material. Although it took him several years, by 1761 he gained control over Siak with the help of these various groups as well as the VOC (Vereenigde Oost-Indische Compagnie—United East India Trading Company). This was the beginning of a period in which rule over Siak rotated between various descendants of Raja Alam and Raja Mahmud.[50]

While Europeans interpreted the constantly shifting leadership of Siak—usually appearing in sources as a series of bloodless coups—as a sign of unpredictability, its apparent instability reflected the ability of its rulers to balance the needs of the various communities along the border region of eastern Sumatra. Siak was neither an inland state nor a maritime state; rather it was a borderland from which exiled princes could flee either inland or seaward, gain resources and loyalty, and then return triumphant. The cycle repeated itself five times between the 1750s and 1790s. The numerous participants and the fluctuations in their fortunes can become confusing; nevertheless, the key factors at the core of these historical cycles were the environment and the means of gaining access to its sources. While the eastern Sumatran littoral held numerous valuable forest products, it was access to these resources combined with a mastery of the sea that led to the greatest successes for Siak rulers as they moved beyond the 'river reaches' that had limited Raja Kecik. While Raja Alam provided a glimpse at this possibility, the best example of this phenomenon was Raja Ismail, the son of Raja Mahmud.

Raja Ismail began his exile from Siak in 1761 when Raja Alam came to power. Raja Ismail also fled to Siantan, much like his uncle had over a decade earlier, and from this base Raja Ismail gained the support of the Orang Laut and began raiding ships. Eventually, the followers of Raja Ismail became famous throughout the Malay World for their brutal attacks, which led Dutch observers to categorize Raja Ismail as a 'notorious pirate.'[51] The martial skills of the followers of Raja Ismail were such that local rulers often offered them refuge for fear that the Siak prince would direct his anger towards their states. It was precisely these unsettled conditions that allowed a princely raider such as Raja Ismail to carve out a sphere of influence—a polity—on the sea.

Key to its rise was the ability of its rulers to command the respect of not only the peoples of the forest, but also the sea peoples off the coast of Sumatra. Raids and other examples of increased power during this period could only be accomplished with the support of *orang asli* groups. Raja Ismail and his brothers received the support of one of the Orang Laut groups soon after they arrived in Siantan. Pirate raids were a vital component of the Malay states that would rise to dominate the Straits. Raiding was a way to direct trade to ports where Orang Laut patrons lived. It allowed for a 'policing' of the Straits. While it was an activity that was labeled as piracy by even early Chinese visitors to the region, it was a common practice in the Malay World.[52] Such raiding assured that trade goods, and their interlinked prestige, flowed through an allied port. If raids were not necessary, the Orang Laut could serve the ruler, and enter trade networks, by supplying produce from the sea, such as *trepang* (sea slugs) or turtle shells, which were highly desired in the marketplace. The ability of a ruler to gain some control over Orang Laut raiding as well as its produce, and to direct it for his purposes, was the cornerstone of power and trade in this border region.

This Orang Laut assistance for Raja Ismail was secured through his connection with the spiritual authority of Raja Kecik. In addition, the riches Raja Ismail obtained in raiding attracted followers. His power grew to the point that he became known as *raja di laut* (king of the sea). With these loyal followers at his side, Raja Ismail terrorized local as well as international shipping—from the various East India Companies—in the region. By basing a state on the sea, Raja Ismail was attempting to develop a form of rule that exploited the potential of a different 'territory' beyond the numerous eco-niches of eastern Sumatra. He was extending power into the South China Sea. Although most Europeans dismissed him as a pirate, Raja Ismail's concept of power looked beyond the land to consider a more lasting base for a Siak polity.[53]

Eventually, rule over Siak took into consideration all of the environmental niches in this border territory. Raja Ismail ruled the sea and coasts, while his cousin (Raja Muhammad Ali) ruled the inland regions of Sumatra. While they would compete to rule over the entire polity, and there was often tremendous tension between the various factions, the key was to place numerous princes and other followers loyal to the ruling family in each eco-niche. This is perhaps best seen in a series of letters sent in 1775. One letter between the rulers of two communities—Ramba and Patapahan—describes how Raja Ismail had contacted upstream leaders to ensure all that all trade that used the neighboring Rokan

River to circumvent the Siak capital would be under his control.[54] The presence of Raja Ismail supplemented the authority of the Siak sultan; it directed the violence against the enemies of Siak and allowed for a closer administration of its vast sea and land frontiers. As a result of these policies, the economic power of Siak during late eighteenth century was so great that many neighboring communities were jealous, laying the ground for nineteenth-century memories of Siak leaders as men who 'lusted after the riches of this world.'[55]

By the 1790s Siak was the dominant authority in the Melaka Straits. The lack of any competitors allowed a borderland region in eastern Sumatra to expand and encompass the territories of neighboring Malay polities. Trade now flowed through Siak ports, generating wealth for most of the nobility as well as the various communities. This success was grounded in the development of a system of rule that promoted the sharing of power and was suited to the eastern Sumatran environment, which was supported by a fairly ruthless maritime policy. Trade was either funneled through Siak ports on the Sumatran coast, or it was destroyed.

Siak rulers had created a polity based on a ruling lineage that initially controlled valuable inland trade along vast rivers in Eastern Sumatra. At that time, they were seen as rulers of the rivers. Over the century, however, the scope of their power expanded as they gained the support of a variety of people who controlled the important international sea-lanes of the Straits of Melaka and the South China Sea. The continual mixing of different groups with their hybrid form of government and environment had generated a sense of 'being Siak' far more compelling than any attractions offered by some distant Malay or Minangkabau locale. They now mastered a vast and diverse border region in which a simple description of its inhabitants was no longer viable. This period of transformation is perhaps best understood through the example of a Siak noble being offered the possibility of an alliance with Minangkabau princes and the south Sumatran ruler of Palembang. The Siak noble refused to consider the proposition because 'We are descendants of the ruler of Siak, not Palembang. We are comfortable riding the waves as sailors, not like those from the hinterland.'[56]

Conclusion

The people of Siak originated from a diverse mixture of inland Sumatra residents who adapted their way of life to the rivers of the swampy lowlands of the eastern littoral of that vast island. Over a century, they mastered this region of dense jungles and mangrove forests as well as the open ocean that served as a link to the outside world. Each of these zones in the lowlands of eastern Sumatra allowed communities to provide unique resources for international trade, while also allowing for relative isolation from larger forces. As they originated in the Minangkabau highlands, the rivers that ran through eastern

Sumatra connected them to a series of coastal communities along the Melaka Straits, and to a larger world of international trade. Prior to European colonial rule, the ability to rule this diverse region lay in the ability to gain the loyalty of leaders of these various eco-niches.

While James C. Scott has argued that upland groups in Mainland Southeast Asia actively opposed appropriation as 'the art of not being governed,'[57] in the Melaka Straits the various communities negotiated their place within a larger geographic space. They did not 'flee' from attempts at control from larger states in the Melaka Straits, as Scott has argued for their northerly neighbors. These borderland communities had resources that allowed for numerous rulers, and systems, to emerge. If a ruler could gain control over multiple centers, such as occurred when Raja Ismail of Siak came to control sea-lanes in the Straits as well as access to forest goods in the Sumatran mainland, he could expand the state to encompass numerous other borderlands. Their nature as borderlands allowed them to exist as relatively self-contained entities, which larger state structures tried to attract, often for socioeconomic reasons. It required a leader to develop a variety of tools, ranging from mythology and intermarriage to the ability to distribute trade goods throughout the region. And this was often supported through force. Any such violence, however, was not directed within but along trade routes to secure goods and services, which would then be funneled back to supporters to gain access to the products of their environment that would lead to greater trade and further riches and prosperity. The environment ultimately influenced, but did not determine, the flexible nature of the polity in early modern Southeast Asia, particularly in borderlands between large eco-logical zones, calling for a greater 'give-and-take' between constituents if a ruler hoped to incorporate them into his polity.

The economic and governing strategies Siak leaders developed over time and through conflict allowed them to rule a borderland. Its statecraft may have been based in violence as well as access to exotic jungle and sea products found in a harsh environment—to the extent, indeed, that it often appeared to be marked by 'misrule' and 'piracy.' Yet, the capacity of Siak rulers to oversee and control everything from the production of honey in the interior to sago on the coasts, to the transport of such goods over the channels, oceans, and rivers that were the vital transport networks in the region, can fairly be described as masterful. The loosely united federation of ports and polities beholden to Siak resulted in the development of a powerful state that controlled the Melaka Straits region in the eighteenth century. The result was a precolonial transna-tional boundary that made sense environmentally. The rulers and residents of Siak were 'comfortable' because of their flexible adaptation to their environ-ment, exploiting its products and idiosyncrasies, allowing them to ride the waves of diverse eco-niches until they became the ultimate power in the region on the eve of colonial rule.

Notes

1. British Library, Straits Settlements Records, Ibbetson report, Sept. 30, 1820, f. 75–6v; H.M.J. Maier, *In the Center of Authority: The Malay Hikayat Merong Mahawangsa* (Ithaca, NY, 1988).
2. T.P. Barnard, *Multiple Centres of Authority: Society and Environment in Siak and Eastern Sumatra, 1684–1827* (Leiden, 2003).
3. N. Tarling, *Anglo–Dutch Rivalry in the Malay World, 1780–1824* (Cambridge, 1962); T. Winichakul, *Siam Mapped: The History of a Geo-body of a Nation* (Honolulu, 1994).
4. Barnard, *Multiple Centres of Authority*.
5. N.N. Dodge, 'The Malay–Aborigine Nexus under Malay Rule,' *Bijdragen tot de Taal-, Land- en Volkenkunde*, 137 (1981), 4–7; T.N. Harper, 'The Politics of the Forest in Colonial Malaya, *Modern Asian Studies*, 31 (1997), 5–7.
6. T. Said, *Hikayat Siak* (Kuala Lumpur, 1992).
7. Barnard, *Multiple Centres of Authority*.
8. Said, *Hikayat Siak*, p. 146; D.J. Goudie (ed.), *Syair Perang Siak: A Court Poem Presenting the State Policy of a Minangkabau/Malay Family in Exile* (Kuala Lumpur, 1989), pp. 84–5.
9. The region has been described as 'Ever-Humid Semi-Hot' in *FAO-UNESCO, Soil Map of the World: Volume IX, Southeast Asia* (Paris, 1979); BAPPEDA (*Badan Perencana Pembangunan Daerah* [Regional Body for Planning and Development]), *Riau dalam Angka: 1984* (BAPPEDA, 1985), pp. 30–1.
10. C.A. Trocki, 'Chinese Pioneering in Eighteenth-Century Southeast Asia,' in A. Reid (ed.), *The Last Stand of Asian Autonomies: Responses to Modernity in the Diverse States of Southeast Asia and Korea, 1750–1900* (Basingstoke, 1997), p. 86.
11. For a discussion of tropical soils and how they are formed, see R. Dudal, F. Moormann, and J. Riquier, 'Soils of Humid Tropical Asia,' in *Natural Resources of Humid Tropical Asia* (Paris, 1974), p. 168; *FAO-UNESCO, Soil Map of the World*, vol. IX, p. 49.
12. D.J. Mabberley, *Tropical Rain Forest Ecology* (New York, 1992); A.J. Whitten, et al., *The Ecology of Sumatra* (Yogyakarta, 1984), pp. 249–339.
13. Anonymous, 'Acheen in 1704,' *Calcutta Monthly Journal and General Register* (Calcutta, 1837), p. 175.
14. C. Lockyer, *An Account of Trade in India* (n.p., 1711), p. 49; I.H. Burkill, *A Dictionary of the Economic Products of the Malay Peninsula* (2nd ed., 2 vols, Kuala Lumpur, 1966), I, p. 325; P. Borschberg, 'The Euro–Asian Trade in Bezoar Stones (Approx. 1500 to 1700),' in M. North (ed.), *Artistic and Cultural Exchanges between Europe and Asia, 1400–1900: Rethinking Markets, Workshops and Collections* (Farnham, 2010), pp. 29–44; H.A. Hijmans van Anrooij, 'Nota omtrent Het Rijk van Siak,' *Tijdscrift voor Indische Taal-, Land- en Volkenkunde*, 30 (1885), 276–7; D. Beeckman, *A Voyage to and from the Island of Borneo in the East Indies* (London, 1718), p. 151; B.W. Andaya, *To Live as Brothers: Southeast Sumatra in the Seventeenth and Eighteenth Centuries* (Honolulu, 1993), pp. 7, 34.
15. For an example of the use of specialized knowledge in the exploitation of forest resources, such as the 'camphor language' of the Sakai, see F.N. Nieuwenhuizen, 'Het Rijk Siak Sri Indrapoera,' *Tijdschrift voor Indische Taal-, Land- en Volkenkunde*, 7 (1858), 345–8. Burkill, *Dictionary*, I, pp. 768–76.
16. *Nationaal Archief*, the Netherlands [Hereafter: NA], VOC 3554: 'Notitie van de goederen, die van de overwal hier to koop gebragt en van hier derwaarts vervoeren worden,' Diverse Papieren, Reported to the Council, Jan. 12, 1779.
17. Burkill, *Dictionary*, II, pp. 1908, 1902–18; Beeckman, *Voyage*, pp. 147–8. Another important product from the region was timber, which was famous throughout

Southeast Asia since it made for excellent masts: T.P. Barnard, 'The Timber Trade in Pre-Modern Siak,' *Indonesia*, 65 (1998), 87–96.

18. J.W. Ijzerman, *Dwars door Sumatra. Tocht van Padang naar Siak* (Haarlem, 1895), pp. 472–3, 530–6; A. Turner, 'Cultural Survival, Identity, and the Performing Arts of Kampar's Suku Petalangan,' *Bijdragen tot de Taal-, Land- en Volkenkunde*, 153 (1997), 672–98; Hijmans van Anrooij, 'Nota omtrent het Rijk van Siak,' 295.

19. Hijmans van Anrooij, 'Nota omtrent het Rijk van Siak,' 338; Nieuwenhuizen, 'Het Rijk van Siak Sri Indrapoera,' 344–8.

20. J.A. van Rijn van Alkemade, 'Beschrijving eener Reis van Bengkalis Langs de Rokan-Rivier naar Rantau Binoewang,' *Bijdragen tot de Taal-, Land- en Volkenkunde*, 32 (1884), 23.

21. Although John Miksic counts ten major river systems, the rivers north of the Rokan are much steeper closer to the coast. In addition, rivers north of the Rokan originate in Batak cultural territory and not Minangkabau cultural territory, like those discussed here. See J. Anderson, *Mission to the East Coast of Sumatra in 1823* (Oxford, 1971), p. 8; J. Miksic, 'Traditional Sumatran Trade,' *Bulletin de l'Ecole francaise d'extreme-orient*, 74 (1985), 427.

22. W. Marsden, *The History of Sumatra* (Oxford, 1986), p. 15.

23. Marsden, *History of Sumatra*, p. 357; van Rijn van Alkemade, 'Beschrijving eener Reis van Bengkalis,' 25.

24. Anderson, *Mission*, p. 334. For a similar description of the Kampar estuary see J. Faes, 'Het Rijk van Pelalawan,' *Tijdschrift voor Indische Taal-, Land- en Volkenkunde*, 27 (1882), 492.

25. Anderson, *Mission*, p. 332; Marsden, *History of Sumatra*, 357; Van Rijn van Alkemade, 'Beschrijving eener Reis van Bengkalis,' 24.

26. T. Pires, *The Suma Oriental of Tome Pires* (London, 1944), p. 151.

27. Faes, 'Het Rijk van Pelalawan,' 492; Anderson, *Mission*, p. 157.

28. Faes, 'Het Rijk van Pelalawan,' 493: 'zonder smaak en zeer goed voor drinkwater.'

29. J. Phipps, *A Practical Treatise on the China and Eastern Trade* (Calcutta, 1835); Faes, 'Het Rijk van Pelalawan,' 492.

30. Anderson, *Mission*, p. 200; Marsden, *History of Sumatra*, p. 357.

31. Anderson, *Mission*, p. 164.

32. R.B. Smith, *The First Age of the Portuguese Embassies, Navigations, and Peregrinations to the Kingdoms and Islands of Southeast Asia, 1509–1521* (Bethesda, 1968), p. 86.

33. R. Everwijn, 'Verslag van een Onderzoekingsreis in het Rijk van Siak,' *Natuurkundig Tijdschrift van Nederlandsch-Indie*, 29 (1867), 295–6.

34. NA, VOC 1677: Malacca to Batavia, March 13, 1703, f. 51–2; Everwijn, 'Verslag van een Onderzoekingsreis in het Rijk van Siak,' 296–8, NA.

35. The size and shape of these communities, however, often was not on the model or scale of those in Europe or China and led to misunderstandings of their function and location. See J.N. Miksic, 'Urbanization and Social Change: The Case of Sumatra,' *Archipel*, 37 (1989), 3–29. *FAO-UNESCO Soil Map of the World*, vol. IX, p. 43; U. Scholz, *The Natural Regions of Sumatra and Their Agricultural Production Pattern* (Bogor, 1983), pp. 15, 141–2.

36. There are many references to the control of trade as the basis for the Malay social system. For one of the most influential articles see B. Bronson, 'Exchange at the Upstream and Downstream End: Notes toward a Functional Model of the Coastal State in Southeast Asia,' in K.L. Hutterer (ed.), *Economic Exchange and Social Interaction in Southeast Asia: Perspectives from Prehistory, History, and Ethnography* (Ann Arbor, 1977), pp. 39–52.

37. Anderson, *Mission*, pp. 173, 348.
38. Dudal et al., 'Soils of Humid Tropical Asia,' p. 170; *FAO-UNESCO Soil Map of the World*, vol. IX, pp. 52–3.
39. E.J.H. Corner, *The Freshwater Swamp-forest of South Johore and Singapore* (Singapore, 1978); Whitten, et al., *Ecology of Sumatra*, pp. 219–48.
40. D. Sopher, *The Sea Nomads: A Study of the Maritime Boat People of Southeast Asia* (Singapore, 1977), pp. 3–8; Whitten et al., *Ecology of Sumatra*, pp. 89–151.
41. C.E. Wurtzburg, *Raffles of the Eastern Isles* (Singapore, 1984), p. 70; J.R. Logan, 'Sago,' *Journal of the Indian Archipelago and Eastern Asia*, 3 (1849), 296.
42. Logan, 'Sago,' 300.
43. Logan, 'Sago,' 297–301; Burkill, *Dictionary*, II, pp. 1484–8.
44. Hijmans van Anrooij, 'Nota omtrent het Rijk van Siak,' 302–5, 348–50; Nieuwenhuizen, 'Het Rijk van Siak Sri Indrapoera,' 426–7.
45. For discussions of the relationship between the ruler and his followers from an economic standpoint see A. Reid, *Southeast Asia in the Age of Commerce, 1450–1680. Volume II: Expansion and Crisis* (New Haven 1993), pp. 202–66.
46. For examples of people in eastern Sumatra protesting their treatment as 'slaves' see NA, VOC 1911 (2nd part): Malacca to Batavia, Sept. 28, 1718, f. 99; and NA, VOC 2700: Malacca to Batavia, March 29, 1747, f. 424.
47. Drakard, *Kingdom of Words*; L.Y. Andaya, *The Kingdom of Johor, 1641–1727* (Oxford, 1975).
48. Barnard, *Multiple Centres of Authority*; Andaya, *Kingdom of Johor*.
49. Raja Ali Haji ibn Ahmad, *The Precious Gift: The Tuhfat al-Nafis* (Oxford, 1982), p. 83. For more on the context in which this text was recorded, please refer to the excellent introduction by Barbara Watson Andaya and Virginia Matheson of the English translation published by Oxford University Press.
50. Barnard, *Multiple Centres of Authority*, 79–126.
51. For a contemporary description of Raja Ismail as 'one of the greatest pirates' in the sea, see J.C.M. Radermacher, 'Beschrijving van het Eiland van Sumatra, in Zoo Verre Hetzelve tot Nu Toe Bekend Is,' *Verhandelingen van het Bataviaasch Genootschap van Kunsten en Wetenschapen*, 3 (1781), 89.
52. O.W. Wolters, *Early Indonesian Commerce: A Study of the Origins of Srivijaya* (Ithaca, NY, 1967), p. 187; T.P. Barnard, 'Celates, Rayat Laut, Pirates: The Orang Laut and Their Decline in History,' *Journal of the Malaysian Branch of the Royal Asiatic Society*, 80 (2007), 33–49.
53. T.P. Barnard, 'Texts, Raja Ismail, and Violence: Siak and the Transformation of Malay Identity in the Eighteenth Century,' *Journal of Southeast Asian Studies*, 32 (2001), 331–42.
54. NA, VOC 3467: Letters to/from rulers of eastern Sumatra, and other notes: Letter from King of Ramba to Bendahara of Patapahan (received in Melaka June 13, 1775); Malacca Resolutions, Dec. 4, 1775, 373–376.
55. Ali Haji, *The Precious Gift*, p. 173.
56. 'Kita ini anak Raja Siak, bukan Raja Palembang, sudah kami ini biasa bermain gelombang, anak laut, bukannya anak ulu.' Said, *Hikayat Siak*, pp. 247, 209–74.
57. J.C. Scott, *The Art of Not Being Governed: An Anarchist History of Upland Southeast Asia* (New Haven, 2009).

5

From Constituting Communities to Dividing Districts: The Formalization of a Cultural Border between Mombasa and Its Hinterland

Daren Ray

The popular image of ignorant diplomats carving up blank maps of Africa into European colonies elides the extensive negotiations over space, power, and identity that played out in the towns and countrysides of Kenya until colonial rule ended in 1963.[1] In order to capture how Mombasa's residents participated in cementing the division of their communities, this chapter combines ethnographic consultations that I conducted with local residents in 2010 and 2011 with archival evidence written and collected by European officials in the early twentieth century, including ethnographic data collected by colonial anthropologists and petitions from Mombasa residents.[2] Reading archival documents through an ethnographic lens emphasizes that Western concepts do not often align with African perspectives; but this multidisciplinary methodology also shows that local residents and British colonial officials in Mombasa were aware of their differences and often attempted to translate their motivations to one another as they pursued goals that inevitably clashed. In addition, personally visiting the sites mentioned in the archives and consulting with residents who live there provides a textured view of strategies that indigenous communities have used to claim territory but which left no written artifacts.

After describing the ways in which residents anchored their social communities at key sites in the physical landscapes of Mombasa before the establishment of the British East Africa Protectorate in 1895, I describe how residents at Jomvu persuaded British officials in 1914 to realign an internal administrative boundary to accommodate their Muslim identities. While residents previously assembled around places of remembrance and healing in Mombasa's landscapes, the imperial strategy of dividing physical geography into discrete territories aimed to stifle interaction and 'contamination' among communities that officials considered culturally distinct. Officials directed residents to erect survey stones and posts that divided Mombasa's residents into adjacent districts and attempted to enforce laws that dictated where they could live and

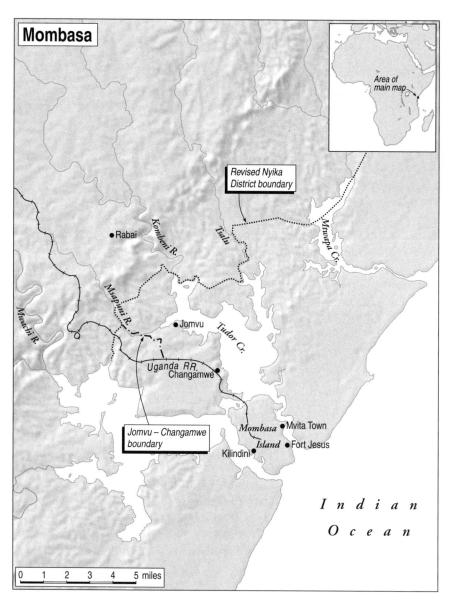

Map 5.1 Mombasa

work. Since British officials relied on their limited understandings about local cultural identities to demarcate the internal boundaries of the Protectorate, their rigid colonial policies transformed how Mombasa's residents constituted their communities. The experiences of Jomvu's residents demonstrate how local initiatives contributed to the formalization of cultural borders—as well as ethnic and religious identities—in colonial contexts. Instead of a shifting mosaic of composite communities, Mombasa became an ethnic borderland.

Local Landscapes: A Shifting Mosaic of Composite Communities

Mombasa Island is nearly engulfed by land: the rectangular island is surrounded on three sides by wide saltwater creeks that separate it from the Kenyan mainland in East Africa, so only its eastern shore faces out into the Indian Ocean. Fishing boats, canoes, and ferries have long plied the watery boundary around Mombasa, and at low tide, one can cross to the mainland by foot near the island's westernmost point of Makupa.[3] The erection of bridges spanning the creeks in the early twentieth century has further eased communication, transportation, and trade between the island and its hinterland. And the tributaries of the island's saltwater creeks extend about 12 miles into the mainland through a broad plain until they lap at the base of a steep escarpment which, together with the arid plains that lie beyond it, local residents call the *nyika*.[4]

As is common throughout East Africa, the variations in Mombasa's physical landscape correspond well with the dominant ethnic communities that its residents articulated in the twentieth century: the Swahili and Mijikenda. Swahili is a derivation of the Arabic word for coast, and Swahili communities stretch along the East African shore from southern Somalia to northern Mozambique. The Mijikenda reside mostly within the *nyika* hinterland of Mombasa, with communities stretching from the Kenyan city of Malindi in the north to Northern Tanzania in the south. Their name means 'nine towns' and refers to nine mostly abandoned hill-top forest settlements known as *kayas* with which the various Mijikenda communities are affiliated.[5] But before hinterland communities articulated a shared ethnic identity as Mijikenda in the twentieth century, coastal residents and foreign visitors referred to hinterland residents as Wanyika, a generic appellation with negative connotations that hinterland residents rejected at the time in favor of their local *kaya* identities. Swahili and Mijikenda communities distinguish themselves through distinct styles of dancing, music, and clothing, as well as cuisine, religion, and language.

Members of the two ethnic groups, however, were at least as entangled in one another's lives as the landscapes they occupied: just as the tidal creeks extend into the *nyika* and Mombasa Island physically joins the hinterland at low tide, individuals from both areas have frequently crossed into and joined one another's communities. In the nineteenth century, for example, the broad plain that might have served as a natural boundary between the coast and the *nyika* became instead a place of extensive cooperation. Foreign Omani and local Swahili investors joined their resources with the efforts of young men from the hinterland and imported slaves to establish plantations that supplied grain to the slave, ivory, and spice trades of East Africa.[6] Beyond the plain, German missionary J.L. Krapf often commented in the nineteenth century on what he saw as the intrusion of Swahili individuals who built their distinctive square homes and lived among the Wanyika.[7] He also met with delegations from Wanyika communities in Mombasa.

Today, Mijikenda women often marry into Swahili families; but, the reverse is discouraged by Islamic customs on the coast which emphasize marriage within the faith community. Mijikenda men can join Swahili communities through conversion to Islam, though their efforts are often contested.[8] And, Swahili individuals rarely reciprocate these conversions by joining Mijikenda communities.[9] Mombasa's communities have innovated a number of strategies to normalize such movements while assiduously maintaining their distinctive identities. Despite noticeable differences in their languages, the Swahili and Mijikenda ethnic groups share a common linguistic heritage stretching back more than 1500 years that informs the ways in which they constitute their communities.[10]

Long before affiliating in the colonial era as Swahili and Mijikenda, the residents of Mombasa Island and its hinterland organized their dispersed settlements into a mosaic of composite communities. The residents of the Mombasa region organize their settlements into towns (*miji*) on the coast and locations (*malalo*) in the hinterland, a pattern which has persisted since at least the sixteenth century.[11] Residents of coastal towns like those on Mombasa Island divided their settlements into neighborhoods composed of 'stone houses' made of lime and coral for those residents who could afford them and less durable homes for the poor. In several coastal towns, residents also built walls enclosing the section of town built in stone, leaving many of the dwellings of poorer coastal residents outside. These walls have disappeared with urban consolidation but are preserved in settlements such as Gedi that were abandoned prior to the eighteenth century.[12]

Residents in hinterland locations form their settlements into clusters of homesteads from several clans. They tend to arrange their mud-and-wattle, cement, or grass homes around a shared yard, all of which are surrounded by the fields, gardens, and groves that they maintain and cultivate. They mark off their homesteads in each location by planting a variety of durable trees and sisal plants. Because of the frequent dislocation and opportunities for expansion, most clans have lands in more than one location, resulting in a mosaic pattern of settlement: the Mwamkamba clan, for example, claims land in at least three locations within the Rabai territory, and in each of these places they are surrounded by other clans.[13] There are also some villages (*vijiji*) where anyone may build stand-alone homes regardless of their clan or ethnic affiliation.[14] Mijikenda communities no longer reside permanently in the hill-top *kayas* that they claim as their ancestral homes; the hills are difficult to ascend and far enough from water sources, fields, and markets that they would be inconvenient residences today.

East Africans often abandoned their settlements and founded new ones because of war or drought. Oromo raiders forced people to abandon their settlements during the seventeenth century throughout the Kenyan littoral and hinterland, sparing only those located on islands like Mombasa and Lamu to the north and the protected forts (*kayas*) that hinterland residents built in hill-top forests.[15] In addition, severe droughts prompted residents to disband entire settlements in the hinterland and relocate about once a generation as crops failed and water sources dried up.[16] Also, large numbers of slaves escaped coastal plantations established in the nineteenth century to join the retinues of ambitious young men from the hinterland who were claiming stretches of the coastal plain for themselves once the raids began to diminish.[17] Untold numbers of these runaways may have also joined hinterland communities that offered refuge and kinship; others founded independent enclaves.[18]

The basic unit of social organization in these communities was a lineage consisting of two or three generations but also including slaves and clients treated as subordinate kin. Both hinterland and coastal residents considered land to be the property of clans known locally as *mbari*; but, in practice, land and houses claimed by a settler would pass on in a much narrower line to his direct descendants or brothers. Only if no male heir was available within this range would the property pass outside the lineage of the original owner to another man of the same clan.

Among the Mijikenda, these land-owning clans still predominate and are conceptually grouped into larger clans—also called *mbari*.[19] In turn, these larger clans compose confederations focused on ancestral settlements known as *kayas*. And the confederations are the communities linked to the nine *kayas* that constituted the Mijikenda ethnic group in the mid-twentieth century.[20] Most residents of Mombasa Island followed similar practices of corporate land ownership until the nineteenth century when they increasingly relied on Islamic laws of inheritance to establish personal property rights over land. As early as the seventeenth century, Mombasa Island's residents assembled their land-owning clans into *taifas*, an Arabic word translated in colonial times as 'tribe' but now as 'nation.'[21] Since both the hinterland *kaya* confederations and coastal *taifas* are composites of clans, I will refer to both as clan confederations.[22]

The residents of Mombasa's hinterland and island regard their large clans and clan confederations as direct descendants of communities that originated in specific settlements. Thus, among hinterland residents, the names associated with their clan confederations usually reference physical settlements as well as the original social groups that inhabited them. For example, the large clans of the Rabai clan confederation (Mwezi and Kiza) maintain distinct land plots (*lwandas*) in a *kaya* called Mudzi Muvya despite abandoning it as a residence.[23] This association of corporate identity and territory is also true of the confederations in Mombasa Island, with the caveat that many of their names refer to locations elsewhere on the coast to signal their origins as refugees from coastal settlements to the north.

From the seventeenth to the mid-nineteenth century, the confederations of Mombasa Island lived in two settlements. Mvita was located on the north side of the island near the Portuguese-built fort that served as a capital for the various foreigners who have asserted dominion over Mombasa since the sixteenth century.[24] It was originally the home of nine clan confederations, including refugee groups from elsewhere on the coast and Jomvu. The Wajomvu (people of Jomvu) claim that they left Mombasa Island and settled along a tidal creek leading westward into the hinterland in order to escape Portuguese privations while continuing their participation in coastal commerce.[25] The second town, Kilindini, was on the south of the island. It was the home of the Kilindini and Tandana clan confederations. They also allied with the Changamwe clan confederation across the western creek on the mainland which borders the more distant Jomvu.[26]

Residents of Mombasa Island reflected their division into two settlements by forming two political confederations called the Three Tribes (Thalatha Taifa) of Kilindini town and Nine Tribes (Tissia Taifa) of Mvita town. But the Three Tribes relocated in 1837 to Mvita alongside the Nine Tribes after allies of Seyyid Said, Sultan of Oman and Zanzibar, razed Kilindini.[27] Mombasa residents further united their *taifas* into the Twelve Tribes (Thenashara Taifa) in the early twentieth century.[28] Hinterland residents similarly emphasized the composite nature of their ethnic identity when they consolidated nine clan confederations into the Mijikenda ('nine towns') in the mid-twentieth century.

Investors from Arabia and India, enslaved laborers, and migrants from the African interior flocked to Mombasa Island after Omani Sultan Seyyid Said secured his claims to coastal East Africa in the 1830s and began integrating its towns into his commercial empire.[29] This inward migration continued to increase as British officials invested in the island's infrastructure after they established the British East Africa Protectorate in 1895.[30] Since both the Omani and British courts gave preference to written records over oral testimony, Mombasa's indigenous residents began to rely more consistently on Islamic principles of individual ownership. The original residents attempted to retain previously established rights of joint stewardship over specific neighborhoods, mosques, wells, and cemeteries through their membership in a clan. But, their collective claims attenuated when residents and city planners dealt with the flood of immigrants seeking work by rapidly dividing the island into over

50 neighborhoods.[31] While the new neighborhoods were not claimed exclusively by clans, many new immigrants continued to affiliate with one of Mombasa Island's two clan confederations (the Three Tribes or the Nine Tribes) until the mid-twentieth century.[32]

Mosques and Kayas: Spatial Anchors for Assembling Communities

In addition to managing and adjudicating land claims, clan affines support one another in times of sickness and death, and these commitments to communal well-being are honored by the *kaya* elders of clan confederations as well. Even as residents relocated to new settlements, or returned to previously abandoned ones, they maintained relationships with their clans. And despite regular dislocations, some sites remained more-or-less permanently occupied or maintained. These sites have served as spatial anchors for the residents of Mombasa Island and the hinterland to reassemble their communities as they have been forced to resettle time and again. The most prominent of sites in the physical landscape of Mombasa are mosques, gravesites, *kayas*, and water sources. These landmarks unified residents not only because they were prominent in the physical landscape, but also because residents associated them with the original settlers that they honored as their ancestors and the 'owners of the land.' They made such places the focus of regular rituals of remembrance and healing that helped mitigate the trauma of frequent relocations by providing safe havens and promising protection.

Although urban consolidation has obscured the natural features of Mombasa Island, mosques and walled cemeteries continue to memorialize its now defunct clan confederations. Residents often organized their neighborhoods around the mosques where they worshipped, socialized, and fetched water for ritual cleansing and daily use. Since some mosques have been in place for over three centuries, they are among the most permanent landmarks on the island.[33] Under the Wakf Ordinance of 1900 British officials gazetted many mosques to protect them from desecration. British officials listed them as protected monuments in the *Official Gazette of the East Africa and Uganda Protectorates* and appointed trustees as part of their obligations to the Sultan of Zanzibar. While the Sultan had granted administration over his coastal territories to the British Foreign Office in 1895, he retained nominal sovereignty and required the colonial government to uphold Islamic law, which forbids the desecration of mosques.[34]

Mosques are accessible to all Muslims, but residents associate the buildings with the historical communities that established them. The British ordinances were too late to preserve the 'Mosque of the Three Tribes' which was demolished along with the town of Kilindini in the 1830s, but prominent members of the Three Tribes went on to establish half of the new Swahili mosques built in Mombasa during the nineteenth century.[35] Mijikenda communities to the south of Mombasa also began constructing mosques in the nineteenth century, when they began converting to Islam in great numbers.[36]

Cemeteries are another set of enduring landmarks in urban Mombasa, and the only places in which the island's residents continue to express their affiliations to clan confederations. Today, cemeteries (or sections of them) are reserved for the descendants of particular clan confederations; the historical divisions among Mombasa's residents are thus enacted as families visit the graves of their forefathers on Islamic holidays such as Id al-Fitr.[37] In an effort to protect these cemeteries, Swahili families, even in relatively isolated Jomvu, have begun to erect walls and gates around them. In order to bury a family member in some of these cemeteries, its steward must first affirm that the deceased belongs to the appropriate lineage or clan.[38]

In addition, the Twelve Tribes celebrate their corporate identity during 'Swahili New Year' at the grave of Shehe Mvita, the eponymous founder of the original home of the Nine Tribes whom they claim was executed by the Portuguese in the sixteenth century.[39] This celebration, as practiced in the nineteenth century, linked the mosques of Mombasa with Shehe Mvita's gravesite by driving a bull through the main street; celebrants would stop at each mosque along the route as the *muezzin* made a call to prayer to invite the whole community to participate.[40] Afterwards, a butcher slaughtered the bull according to Islamic requirements and offered a free stew to all residents in the city, particularly the poor. After the feast, all the leftovers and the inedible portions of the animals were tied in a bag and entrusted to a Swahili fisherman to cast into the deep ocean, thus symbolizing the cleansing of the island.[41]

Mijikenda communities also commemorate their ancestors at gravesites during New Year rituals. Early in the morning, women cast ashes onto the exterior walls of their homes. Then, the eldest man in the homestead offers a meal of ashes mixed with water to the ancestors in a temporary shrine on a path leading to a shared yard. After prayers and singing, the adult residents clean the gravesites of the ancestors buried on the edges of the yard by removing any plant growth and sweeping the dust immediately around the grave. They describe these ceremonies as a way to cleanse and protect their homesteads from the spirits of the deceased.[42]

Like the practices of Friday sermons performed in mosques and prayers at grave sites, the *kayas* are associated with rituals of communal protection and healing, which the residents in the hinterland use to symbolize, unify, and assemble their clan confederations. For example, just outside the hill-top settlement of Kaya Mudzi Muvya in Rabai is a circular path around 30 m long that encloses a seemingly unremarkable patch of shrubs and trees. In times of sickness or danger, *kaya* elders walk around this path several times in a metaphoric encirclement of the territory occupied by the Rabai clan confederation. When finished, they build a small fire at an intersection with the main path to the *kaya*.[43] On the fire, they make medicines with herbs collected from within and around the *kaya*. Then they deposit these herbs in four water sources on the far corners of Rabai: a pond, a river, a dam, and an abandoned mining pit that serves as a reservoir. Until recently, elders also traveled to the various locations of Rabai and summoned residents with an antelope horn to receive medicine from the *kaya*.

Such rituals emphasize how *kaya* elders, who represent the original home of the various clans, maintain the community of dispersed residents by distributing healing herbs. The water sources in which these herbs are placed are not boundaries; instead, they are anchors that secure the affiliation of people who were frequently dislocated and forced to begin anew. The hill-top *kayas* are no longer large enough for everyone to live within them and far enough from water sources, fields, and markets that they would be inconvenient residences today. But rituals focused on the *kayas* symbolically assemble the now-dispersed communities affiliated with them. Thus Rabai's constituent communities can continue associating with the *kayas* from which their ancestors came and rely on the resources of the *kayas* for continued health. In addition, immigrants and newly organized clans can associate with a *kaya* by participating in these and similar rituals. The physical landscape, including its built and natural features, provides effective anchors around which the residents of Mombasa Island and its hinterland have organized and imagined their communities for centuries.

Formalizing a Cultural Border at Jomvu

At the end of the nineteenth century, Mombasa's residents added another strategy for constituting their communities: the association of communal identities with formally demarcated territory. This extension of the logic which associates communities with places was prompted by the introduction of a survey system for marking claims to land that British and Indian surveyors developed in South Asia.[44] British administrators introduced this system throughout Africa at the turn of the twentieth century to help them govern the unfamiliar terrain of their newly acquired protectorates and colonies.[45] They increased their efforts in Mombasa after passing the 1908 Land Titles Ordinances, which required landowners to register their titles with proof of ownership and a survey. The ordinance also provided a legal basis for the colonial government to secure any land without a registered title.[46] Though it classified all unregistered land as government property, colonial officials prepared detailed surveys in order to avoid potential lawsuits when they reserved specific pieces of land for government projects.

These survey techniques referenced a global grid of imagined meridians (that is, latitudes and longitudes), which required a corps of specialists to lay out and maintain. In addition to introducing a framework for conceptualizing the Mombasa landscape that was novel to its residents, this system relied on a new set of landmarks: cement and wooden survey beacons that were recorded on a written index kept at central administrative offices. They used these beacons to specify the locations of other colonial landmarks, such as railways, telegraph poles, roads, schools, and health centers. Administrators then used them as reference points for dividing the people of the Protectorate into discrete, manageable districts. Officials relied heavily on local knowledge to select where they would place these landmarks and tried to represent both the local names of natural features and the complexities of the social landscape in their maps.[47]

Although European diplomats negotiated many of Africa's intercolonial borders with little knowledge of local landscapes, District Commissioners and surveyors within the British East Africa Protectorate routinely consulted with local residents, whom they regarded as experts on identifying recognized boundaries between established communities. The following excerpt from the official notification setting out boundaries for the new Nyika Reserve and Rabai District in 1916 demonstrates how administrators demarcated their districts using a mix of landmarks erected by the surveyors and others established and identified by local residents:

> thence by a straight line in a south-easterly direction to the most westerly corner of *shamba* [farm] No. 7 of Group No. VII ... thence by a general eastern limits [*sic*] of this village [Ziani] to a point marked by a *beacon* ... thence by a straight line in a westerly direction through *Laiti hill* to the *baobab tree* which demarcates the northern end of the boundary between the *lands of Changamwe and Jomvu*: thence by a straight line in a south-westerly direction to a point on the *Sampuni* [Msapuni] *river* on line between the above mentioned *baobab tree* and the *telegraph post* at mile 9/5 on the *Uganda Railway* ... thence by the generally northerly and westerly boundaries of *Land Office Portion* Nos. 1031, 1013, [and nine more portions] ... thence by the *low tide mark* along the coast in a generally southerly direction to the mouth of the Cha Simba or Pemba river ...[48]

While natural features of the landscape are prominent in these boundary descriptions, so too are the communities established by local residents. For example, the boundary between Changamwe and Jomvu, belonging to the rival clan confederations of Mombasa, was marked in 1896 by the erection of red posts after British Consul Arthur Hardinge settled a dispute between the two communities.[49]

Despite their attention to local perceptions, the demarcation of the settlements of the Nyika District within restrictive borders grossly simplified how Mombasa's residents imagined the complex mosaic of their communities.

When British officials first assumed control of the Protectorate in 1895 they decided to administer the ten-mile wide strip of coast that the Sultan of Zanzibar had ceded to them separately from the remainder of the East African Protectorate.[50] Officials were compelled to incorporate preexisting Islamic courts into their legal system on the coastal strip; but in the hinterland and other interior territories, they presided over native courts with the assistance of elders who helped interpret native custom as assessors.[51] After 1909, the policy of separation became more explicit under the leadership of Provincial Commissioner Charles Hobley. As he wrote to a subordinate, 'It is generally advisable to keep the administration of Swahili governed by Mohamedan Law and Nyika tribes who come under tribal law quite separate.'[52]

Although colonial officials developed distinct policies for the 'Mohamedan' coast and the 'native' interior, the application of these policies was complicated by practical considerations that prevented a formal demarcation of the ten-mile strip: if measured from the low-water mark as stipulated in the treaty, Swahili settlements such as Jomvu would have been included in the strip because they nestled against the saltwater creek that extended into the hinterland but then retreated each day with the tide.[53] But this interpretation of the treaty language also would have included Rabai and a number of other non-Muslim communities that the British did not want to 'contaminate' with Islamic law, as if it was a disease whose growth could be arrested by carefully regulating the movement of people.[54] Thus Jomvu was originally included in the Rabai subdistrict in 1908 when it was provisionally marked out as a 'Nyika Reserve.'[55] Although Jomvu was considered 'Mohamedan,' it was closer to the district station at Rabai than Mombasa. Furthermore, Jomvu's leaders refused to be lumped together with the neighboring and rival Changamwe community under the legal jurisdiction of a single Islamic official.[56]

By 1910, District Commissioners had formed several councils of elders in local communities to perform administrative work.[57] Since Jomvu had previously refused to be in the same jurisdiction as Changamwe, British officials administered Jomvu through a council of elders led by a 'headman' as they did in other hinterland communities. This arrangement subtly transformed earlier institutions for settling claims that were also shared by hinterland neighbors. Previously, conflicts had been resolved by an ad hoc council of clan elders with knowledge of the land in dispute.[58] In severe cases, complaints could be brought to the elders of a *kaya*, or in the case of Mombasa's clan confederations, Islamic judges resident in Mombasa. Under the British system, District Commissioners appointed a permanent council of elders and gave them the responsibility to collect hut-taxes, enforce government policies, and resolve disputes over land and marriage.

In a memorandum dated May 2, 1913, British Assistant Commissioner J.M. Pearson expressed the desire to centralize district authority at Rabai as an independent district, instead of remaining a subdistrict of Mombasa. At the time he noted that 'no Arabs live in Jomvu ... so it should not be difficult but for the contrary influence of the Tissia Taifa [Nine Tribes].'[59] Apparently, Pearson felt that Islamic law need apply only to the Arab subjects of the Zanzibari Sultan. Yet, the Swahili Nine Tribes based in Mombasa were actively asserting themselves as communities of Arab descent in order to gain access to rights that the British government reserved for non-native Arabs—not coincidentally the right to register individual land titles. In Jomvu, prominent individuals from the Nine Tribes had sponsored a rival to the headman appointed by the British, which was the interference that prompted Pearson's complaint.

In September 1913, the conflict between the British-backed council at Jomvu and the Nine Tribes escalated when Jomvu's council decided to evict Mfaki bin Salim from his home in Maunguja, one of three settlements that composed Jomvu. Mfaki bin Salim had refused to acknowledge the authority of the council when they tried to reprimand him for immoral behavior, allegedly related to gambling and prostitution.[60] Instead of complying, he rushed to Mombasa, hired a European solicitor, and contested his eviction on the grounds that the people of Jomvu ought to be subject to Islamic law, since they were Muslims. He also sought help from some prominent members of the Nine Tribes, who secured the assent of Provincial Commissioner Charles Hobley on his behalf to halt the eviction order. Part of his petition reads,

3rd Recently the Government Mzee [elder] of Maunguja with a so called council have taken upon themselves to decide cases and have caused much dissatisfaction among the rest of the people of Jomvu and the powers exercised by this Mzee going even to the extent of expulsion from the Village are entirely illegal and *even if such Powers are in accord with Wanyika custom* they cannot be legally exercised over Mohamedans [Muslims] such as the Petitioners. The Wajomvu are similar to the Wachangamwe and in Changamwe there is a duly appointed Mudir [a kind of Islamic judge] who decides cases according to the Mohamedan Law.

4th The Petitioners *very strongly object to any return to Paganism and pagan customs* such as is implied in a grant of new Powers to the Mzee and a council.[61]

In this petition, Mfaki bin Salim expressly associated the administrative councils of elders with paganism while seeking to place Jomvu within the Islamic jurisdictions of Mombasa. He even went so far as to compare Jomvu favorably with Changamwe, the neighboring and rival clan confederation to the south whose jurisdiction Jomvu's elders had previously refused to share.[62]

Pearson argued against Mfaki bin Salim's characterization of councils as pagan, noting that Jomvu's elders enforced Islamic law inasmuch as they understood it.[63] Communities on the East African coast, such as those in Mombasa, had been incorporating Islamic law into their customs for centuries. As early as 1331, the North African scholar and traveler Ibn Battuta noted that Mombasa followed the Shafi school of Islamic law.[64] However, trained judges were few and far between, so most communities made do with their limited knowledge of Islamic law and much more extensive communal sensibilities of justice. But Pearson's claim that councils could administer Islamic law was undercut by the same policy that distinguished the system of councils from a coastal judicial system of trained Islamic judges. Ultimately, the eviction order was revoked.

Soon after this incident, Pearson submitted a new recommendation for the formation of the Nyika Reserve and Rabai District that recognized the people of Jomvu as 'Muslims' and expressly separated it from the Nyika on the grounds of cultural difference:

> By this boundary it is noted that Jomvu is cut out from the Rabai District. This has been done because Rabai is to be a Nyika District with tribal organization of local councils to be supervised from the District Office, whereas the Wa Jomvu are Mohammedans alleged Arabs from Shirazi with no innate tribal organization, whose interests are with the Coast and not in the Nyika country. ... *As it is they form an alien community with different customs,* different religion, different standards of morals and conduct, different history, different temperament, from any other component part of the district.[65]

Pearson's recommendation that Jomvu be moved to the Mombasa district settled the border of the coast at Jomvu—but nowhere else. Other points of the border were left unmarked, or followed the boundaries of the several plantations that Arab and Swahili landowners had successfully registered and titled in Mombasa, again showing that Mombasa's residents had significant influence in setting the internal (and incipient ethnic) borders of the Protectorate.[66] The residents of Jomvu who opposed the council had effectively seized the policy enacted by the British and used it to formalize cultural distinctions from their neighboring communities that would more closely align them with politics at the coast. For Mfaki bin Salim, the immediate result was freedom from the authority of elders in return for lax oversight from distant Islamic judges.

For others in Jomvu, the new boundaries represented an added burden. In an effort to restrict the movement of itinerant traders across district boundaries, British officials established a pass system that required visitors and migrants to check in with the District Commissioner before conducting any business within his domain and inform government-appointed tax collectors before relocating.[67] When women who had routinely traveled from Jomvu to Rabai to sell chickens and clay pots complained about the pass system, officials waived the requirements for them but still insisted that anyone seeking to transport and trade livestock secure a pass first.[68] After officials strengthened the pass system with the Stock Traders Licence Ordinance of 1918, the livestock trade between districts collapsed, causing a steep rise in stock prices in Mombasa District.[69] Officials also closed the borders during epidemics and epizootics to avoid further contamination among people and livestock.[70] This strategy of controlling mobility stands in stark contrast to local methods of assembling communities around settlements protected from war and disease with walls, forests, and medicines. Instead of integrating people around an ancestral home, survey methods carved up geographical space into distinct bounded territories.

The realignment of the district border to 'cut out' Jomvu according to cultural criteria suggests how the contrasting legal regimes that British administrators formalized at the coast and hinterland provided residents of the Mombasa region with a new strategy of differentiation. By accepting the arguments offered by the Nine Tribes, they formalized the distinction between territories controlled by 'pagan' councils and those subject to Islamic courts. Districts were no longer arranged simply for proximity to district stations but based on perceived cultural difference. This new strategy undercut generations of complementary interactions among their clan confederations and reframed rivalries among neighbors as contests among cultural groups. By suppressing hundreds of similar local community boundaries within the borders of only a few manageable districts, British administrators helped create competing ethnicized enclaves—territories of undifferentiated space where all residents were assumed to be essentially the same.[71]

Conclusion

Colonial borders shaped not only African communities but also the ways in which scholars have examined their historical and contemporary experiences. While interstate borders in Africa have generally been considered more artificial, arbitrary, and porous than national borders in Europe and America, they became a permanent fixture in African landscapes once the member states of the newly formed Organization of African Unity agreed to retain colonial-era boundaries in 1964. Most scholars similarly remained within these state boundaries as they explored the experiences of colonial and postcolonial Africans. But for those scholars who happened to situate their studies near a border, it was only a matter of time before, as David Coplan writes, 'the festive bustle and *demi-monde* enterprise they encountered upon every crossing or visit to a border post led them to realize that this is where some of the most revealing and important forms of social and economic transactions were taking place.'[72]

The first scholars to explicitly apply borderland analysis to African communities focused on movements of people and goods across interstate borders.[73] These scholars sought to shift attention away from European negotiations and their scramble for African resources to the ways Africans used interstate borders to disrupt and facilitate local political, social, and economic strategies.[74] For instance, Donna Flynn's article '"We Are the Border": Identity, Exchange, and the State along the Bénin–Nigeria Border' exemplifies the emphasis in African borderland studies on borderlander communities whose experiences revolved around economic transactions and personal mobility across interstate borders.[75] In addition, scholars have explored border dynamics for insights into the limits of state power. So, while Eric Allina-Pisano examined how colonial officials and local chiefs negotiated over labor movements across the borders of Mozambique and Southern Rhodesia, he emphasized that the contours of colonial power could be discerned at internal boundaries as well as intercolonial borders—an observation that also applies to boundary negotiations in the British East Africa Protectorate.[76]

Although borderland scholarship in Africa has always explored the human efforts that initiated and reinforced borders, most studies have examined dynamics across borders that had already been fixed in place. In this chapter, I have focused instead on the ways colonized communities contributed their perceptions of local landscapes to the legal and administrative jurisdictions imposed upon them. The correspondences among the spatial and social organizations of Mombasa's island and hinterland landscapes over the past several centuries suggest that, contrary to the perspective of British administrators, the residents of the Mombasa region did not have a different history. Yet, as residents introduced their vision of physical and social landscapes into colonial policies, they also helped to collapse their mosaic of composite communities into a borderland. Jomvu's realignment into an Islamic legal jurisdiction shows how some individuals seized upon colonial policies and institutions to persuade British officials to accept and reinforce the local distinctions that they favored. Indigenous residents thus participated in establishing a crude border that separated Mombasa Island from its hinterland.

Although British officials often misapprehended the relationships among Mombasa's communities as they attempted to align district borders with cultural groups, the movement of Jomvu from Rabai District to Mombasa District demonstrated that British officials consulted closely with local communities because they wanted people to recognize and respect these boundaries. Thus, the imaginary lines across geographical space that British surveyors carved into posts and printed in official gazettes roughly corresponded with local perceptions of physical and social landscapes. As British officials endorsed specific claims of difference by incorporating them into colonial administrative structures, however, these distinctions assumed a degree of cultural homogeneity that local communities had never previously articulated. And colonial officials further differentiated the experiences of these communities by sponsoring distinct institutions on either side of the border that impaired free mobility and affiliation among coastal and hinterland communities.

In particular, the British policy of administering coastal towns through Islamic law and Muslim judges while relying on native elders to adjudicate disputes in hinterland communities through 'tribal custom' introduced a new possibility for constituting communities by coalescing what had been dozens of clan confederations into the essentialized categories of Muslims and 'pagans.'[77] From the 1920s until Kenya's independence in 1963, hinterland residents learned to employ their new government-recognized cultural identities to seek benefits from the colonial government as members of the Mijikenda ethnic group, while the indigenous residents of Mombasa Island found common cause with other Muslims in coastal East Africa as Swahili.[78]

Notes

1. Simon Katzenellenbogen carefully disputed the popular notion that the Berlin Conference sparked the scramble for Africa in '"It Didn't Happen in Berlin": Politics, Economics and Ignorance in the Setting of Africa's Colonial Boundaries,' in P. Nugent and A.I. Asiwaju (eds), *African Boundaries: Barriers, Conduits and Opportunities* (London, 1996), pp. 21–31.
2. This chapter draws on dissertation research conducted as a Dissertation Proposal Development Fellow with the Social Science Research Council in 2008 and a Fulbright–Hays Doctoral Dissertation Research Abroad Fellow in 2009–10. I also received support from the Corcoran Department of History of the University of Virginia to conduct follow-up research for this chapter in the summer of 2011.
3. Personal Conversation with Muhammad Shalo, May 21, 2011. See C.W. Hobley, 'Safari journal,' Dec. 23, 1891, Kenya National Archives (KNA)/MSS/10/50.
4. The ridge stretches from Mwangea Hill near Malindi to the Shimoni Hills south of Mombasa—a distance of about 120 miles. *Nyika* means 'wilderness' and properly refers only to the scrub brush plains to the west of the ridge; but, by association with historical Wanyika communities, the place name also refers to the ridge.
5. From south to north the nine Mijikenda communities are Digo, Duruma, Rabai, Ribe, Chonyi, Kauma, Jibana, Kambe, and Giriama. A local alternative to the name Mijikenda is Makayakenda, making the relationship between the ethnic group and the nine *kaya* clan confederations explicit. See T. Spear, *The Kaya Complex: A History of the Mijikenda Peoples of the Kenya Coast to 1900* (Nairobi, 1978).
6. F. Cooper, *Plantation Slavery on the East Coast of Africa* (Portsmouth, NH, 1997); C. Brantley, *The Giriama and Colonial Resistance in Kenya, 1800–1920* (Berkeley, 1981).
7. J.L. Krapf, *Travels, Researches, and Missionary Labours, during an Eighteen Years' Residence in Eastern Africa* (2nd ed., London, 1968).
8. D. Parkin, 'Swahili Mijikenda: Facing Both Ways in Kenya,' *Africa*, 59 (1989), 161–75. It should be noted that Digo Mijikenda are predominantly Muslim; many individuals from other Mijikenda communities have recently converted to Islam as well: see D.C. Sperling, 'The Growth of Islam Among the Mijikenda of the Kenya Coast, 1826–1930,' PhD dissertation, University of London, 1988.
9. Janet McIntosh explores the intricacies of interethnic relationships among Swahili and Mijikenda communities near Malindi, Kenya in *The Edge of Islam: Power, Personhood, and Ethnoreligious Boundaries on the Kenya Coast* (Durham, NC, 2009).
10. D. Nurse and T. Hinnebusch, *Swahili and Sabaki: A Linguistic History* (Berkeley, 1993).
11. R. Helm, 'Re-evaluating Traditional Histories on the Coast of Kenya: An Archaeological Perspective,' in A.M. Reid and P.J. Lane (eds), *African Historical Archaeologies* (New York, 2004), pp. 59–89. Mombasa's residents have lived in dispersed settlements in 'locations' since the sixteenth century; the practice of living in dense 'stone towns' and forts is a much older practice.

12. J. Kirkman, *Gedi: The Palace* (The Hague, 1963).

13. Personal Communication with William Tsaka, 25 May 2010; walking tour and interview with Abdalla Mwamganga, Aug. 9, 2010.

14. Interview with Ndeje Pekeshe, May 24, 2011.

15. F.J. Berg, 'The Swahili Community of Mombasa, 1500–1900,' *Journal of African History*, 9 (1968), 45–7; Brantley, *Giriama and Colonial Resistance*, pp. 23–4. Rabai traditions also emphasize the depredations of Maasai raiders in the nineteenth century (interview with Gona Dzoka, Aug. 16, 2010).

16. To the consternation of British officials, this strategy of frequently relocating endured into the colonial era: 'Individuals break up their villages and move off by night without giving anyone a Qoheri [kwa heri, 'goodbye'] and tax rolls and lists of all kinds are confused. [W]azee instructed if anyone wants to move he is to inform his Headman first. ... This is a most necessary order but there seems little hope of its ever being carried out' (R.F. Palethorpe, 'Diary of Tour, Nov[ember] 2nd–26th 1916,' KNA/PC/Coast/1/12/264, p. 4).

17. Brantley, *Giriama and Colonial Resistance*, pp. 22–3.

18. For an overview of runaway slave communities in the Mombasa hinterland, see F. Morton, *Children of Ham: Freed Slaves and Fugitive Slaves on the Kenya Coast, 1873 to 1907* (Boulder, 1990).

19. Rabai residents claim they have 26 *mbari*, though my consultants only identified around 20. The list I compiled matches about half of the 19 *mbari* identified by Alice Werner in the early twentieth century. Her informants grouped these *mbari* into four larger clans (A. Werner, *The Bantu Coast Tribes of the East Africa Protectorate* [London, 1915]).

20. Alternatively, Spear has referred to the Mijikenda as a 'super-tribe' since it was composed of many smaller 'tribes': T.T. Spear, 'Neo-Traditionalism and the Limits of Invention in British Colonial Africa,' *Journal of African History*, 44 (2003), 20.

21. The nineteenth-century Mombasa poet Muyaka referred to these Mombasa clan confederations as *miji*, but in current usage *miji* means 'cities' and is currently an inexact reference for the several *miji* which lived within the city of Mombasa: M.H. Abdulaziz, *Muyaka, 19th Century Swahili Popular Poetry* (Nairobi, 1979).

22. F.J. Berg seems to be the first historian to have described the *taifa* as a confederation: Berg, 'Swahili Community,' 41. I extend his term to *kaya* confederations in part because Daniel Begarero, a *kaya* elder, defined *kayas* as *miji* (see preceding note): interview with Daniel Begarero, Apr. 26, 2010.

23. Author's tour of Mudzi Muvya, July 19, 2010.

24. M.N. Pearson, *Port Cities and Intruders: The Swahili Coast, India, Portugal in the Early Modern Era* (Baltimore, 1998).

25. Interview with Jomvu elders, May 21, 2011.

26. Berg, 'Swahili Community,' 47.

27. Berg, 'Swahili Community,' 52.

28. The most comprehensive treatment of the relationship between these two confederations is Berg, 'Swahili Community,' but for the politics surrounding their unification as the Twelve Tribes see H. Kindy, *Life and Politics in Mombasa* (Nairobi, 1972).

29. Berg, 'Swahili Community,' 52; A. Sheriff, *Slaves, Spices, & Ivory in Zanzibar: Integration of an East African Commercial Empire into the World Economy, 1770–1873* (London, 1987).

30. H.J. de Blij, *Mombasa: An African City* (Evanston, 1968), pp. 28–44. The British Foreign Office proclaimed the establishment of the East Africa Protectorate on July 1, 1895 when the trustees of the Imperial British East Africa Company (IBEAC) relinquished their concession over coastal territories they had purchased from the Sultan

of Zanzibar in 1887. When the Foreign Office assumed the concession from IBEAC to counter German presence in Tanganyika, they also assumed the terms of the original agreement. The stipulations of the concession are available in P.L. McDermott, *British East Africa; or, Ibea; a History of the Formation and Work of the Imperial British East Africa Company* (London, 1893).

31. F.J. Berg and B.J. Walter, 'Mosques, Population, and Urban Development in Mombasa,' *Hadith*, 1 (1968), 47–100; de Blij, *Mombasa: An African City*.

32. For example, new immigrants from Pate in northern Kenya affiliated with the WaPate clan confederation and by extension the Nine Tribes. Berg ('Swahili Community') noted that almost all the residents he interviewed in the 1960s recalled the *taifa* to which they belonged but not the clans or lineages which formerly composed them. Only one of my local consultants knew his clan when asked, but many recalled their *taifa*.

33. For example, Mandhry Mosque was built in 1570; Berg and Walter, 'Mosques, Population, and Urban Development,' 57.

34. Even the ruins of mosques were preserved and the rubble left on site: 'Wakf Commissioner to Provincial Commissioner, Coast,' Jan. 14, 1915, KNA/PC/Coast/2/3/1.

35. Berg, 'Swahili Community,' 63.

36. Sperling, 'Growth of Islam.'

37. I refer here to private family visits, not the festive processions to visit the graves of local saints that accompany other celebrations along the coast.

38. Author's tour of Old Town Mombasa with Muhammad Ahmed Mossin, Mombasa, June 24, 2010.

39. Portuguese accounts suggest Shehe Mvita was killed in a conflict with a Mosseguejo pastoral community while conducting a raid on the Swahili town of Malindi: M.A. Hinawy, *Al Akida and Fort Jesus Mombasa* (Nairobi, 1970), p. 5; Berg, 'Swahili Community,' 45.

40. Interview with Hassan Muhammad, Aug. 24, 2010.

41. P.J.L. Frankl, 'Siku Ya Mwaka: New Year's Day in Swahili-Land (with Special Reference to Mombasa),' *Journal of Religion in Africa*, 23 (1993), 125–35; Berg, 'Swahili Community.' The Twelve Tribes is now defunct; a much more limited ceremony is sponsored each year by Mombasa's Wamiji Foundation, a Swahili cultural and civic organization. They no longer organize a grand procession through the main road and attendance at the morning ceremony has dwindled to fewer than 100 people. I attended and recorded the celebration on July 18, 2010. Participants told me that discarding the remains of the feast in the ocean protects the island from disease as well as violence. The recording is on deposit as 'E020 Siku ya Kibunzi,' Ray Research Deposit, Audio-Visual Department, Fort Jesus Museum, Mombasa. A similar ceremony in which the leftovers of a feast are cast into one of Mombasa's tidal creeks is performed by *kaya* elders in Rabai.

42. I recorded the early morning ceremonies of a New Year celebration on Oct. 24, 2010 and interviewed some of the participants on May 27, 2011. The recording is on deposit as 'E027 Rabai New Year Mwaka Muvya,' Ray Research Deposit, Audio-Visual Department, Fort Jesus Museum, Mombasa.

43. This ritual was described to me by William Tsaka of the National Museums of Kenya. He works closely with the *kaya* elders on community development projects; ashes were apparent on the path when I visited the site on July 19, 2010.

44. K. Raj, 'Circulation and the Emergence of Modern Mapping: Great Britain and Early Colonial India, 1764–1820,' in C. Markovits, J. Pouchepadass, and S. Subrahmanyam (eds), *Society and Circulation: Mobile People and Itinerant Cultures in South Asia, 1750–1950* (Delhi, 2003), pp. 23–54.

45. T.H. Holdich, 'African Boundaries, and the Application of Indian Systems of Geographical Survey to Africa,' *Proceedings of the Royal Geographical Society and Monthly Record of Geography*, 13 (1 Oct. 1891), 596–607.

46. J. Willis, *Mombasa, the Swahili, and the Making of the Mijikenda* (New York, 1993), p. 122.

47. Correspondence among colonial administrators is littered with discussions of boundaries, vandalized beacons, and the lack of qualified personnel to conduct surveys, as well as officials' reliance on local knowledge to make sense of the landscape. For example, Assistant District Commissioner R.G. Fanant wrote, 'the only satisfactory means of defining the boundary will be to make a tour round the respective area with natives having local knowledge, erecting beacons at intervals which can be numbered; and the names of the rivers and hills which may be used' (R.G. Fanant, Assistant District Commissioner, Voi, to C.W. Hobley, Provincial Commissioner,' July 26, 1915, KNA/PC/Coast/1/11/101).

48. 'Government Notice No. 73 Schedule 1,' *Official Gazette of the East Africa Protectorate and Colony*, March 15, 1916, pp. 97–8 (emphasis added), quoted in 'Boundaries of the Nyika Reserve and Rabai District,' KNA/DC/KFI/3/2.

49. 'Asst. District Commissioner J.M. Pearson to Provincial Commissioner C.W. Hobley,' May 2, 1913, KNA/PC/COAST/1/3/62; 'Rabai Sub-District,' n.d., KNA/DC/KFI/3/2.

50. The territories to the west of the Sultan's coastal strip were acquired through purchase and conquest, both by the East Africa Protectorate and its predecessor, the Imperial British East Africa Company; for details see E. Hertslet, *The Map of Africa by Treaty* (3 vols, 3rd ed., London, 1967), and McDermott, *British East Africa*.

51. The principle of separation was codified in 'Native Courts Regulation of August 12, 1897,' which directed that District Commissioners' court assessors in the 'Mohammedan Coast Region' should be a *kadhi* while an elder should be selected in 'non-Mohammedan' regions. 'Circular, Chief Native Court, Mombasa,' Dec. 18, 1900, KNA/AG/11/16.

52. 'Provincial Commissioner C.W. Hobley to E.V. Hammond, District Commissioner of Shimoni,' Oct. 10, 1913, KNA/CC/12/15.

53. Willis, *Making of the Mijikenda*, p. 118; colonial officials used 'Mohamedan' as a pejorative adjective for 'Muslim.'

54. Willis, *Making of the Mijikenda*, pp. 117–44.

55. *Official Gazette of the East Africa Protectorate*, May 1, 1908, p. 271.

56. The official over Changamwe was a lower-level Islamic judge known as a *mudir* who was under the supervision of the District Commissioner of Mombasa ('Rabai Sub-District,' n.d., KNA/DC/KFI/3/2).

57. These councils were formalized as Local Native Councils in 1924: D.G. Schilling, 'Local Native Councils and the Politics of Education in Kenya, 1925–1939,' *International Journal of African Historical Studies*, 9 (1976), 220.

58. For example, see 'Notes on Duruma Kambi,' 1917, KNA/DC/MSA/5/1.

59. 'Asst. District Commissioner J.M. Pearson to Provincial Commissioner C.W. Hobley,' May 2, 1913, KNA/PC/Coast/1/3/62.

60. I have not identified Mfaki bin Salim in any other records, though J.M. Pearson reports the following about his co-petitioners: 'The objectors to the council are the young men of locations, who would equally resent any system of order and restraint. ... These young men appear to be an idle and dissolute lot and are believed to be living on the proceeds of prostitution. They have no *shambas* [farms] nor any visible means of subsistence' ('District Commissioner J.M. Pearson to Provincial Commissioner C.W. Hobley,' Sept. 8, 1913, KNA/PC/Coast/1/3/62).

61. 'Petition to Provincial Commissioner, Mombasa,' n.d., KNA/PC/Coast/1/3/62.
62. Jomvu elders reasserted the higher status of their community over Changamwe in a group interview conducted by the author on May 21, 2011.
63. 'District Commissioner Hemsted to Provincial Commissioner C.W. Hobley,' Sept. 8, 1913, KNA/PC/Coast/1/3/62.
64. I. Battuta, *Ibn Battuta: Travels in Asia and Africa*, trans. H.A.R. Gibb (London, 1929).
65. 'Rabai District Boundaries,' Sept. 1, 1913, KNA/DC/KFI/3/2.
66. Willis, *Making of the Mijikenda*, p. 132.
67. R.F. Palethorpe, 'Diary of Tour, Nov[ember] 2nd–26th 1916,' KNA/PC/Coast/1/12/264, p. 4.
68. 'P.L. Deacon, District Commissioner to R.W. Lambert, Assistant District Commission, Rabai,' Sept. 2, 1918, KNA/DC/MSA/5/1.
69. 'W.S. Marchant, Asst. District Commissioner of Mombasa to C.W. Hobley, Provincial Commissioner,' Sept. 1, 1919, DC/MSA/5/1.
70. For example, A.N. Bailward, Asst. District Commissioner of Mombasa, noted 'Reports received that small pox was very bad on borders of District and it was arranged with Ag. District Commissioner, Shimoni, that people should not cross the border from one District to the other until the epidemic had abated' ('Safari Diary,' July 22, 1920, KNA/DC/MSA/5/1).
71. For a broader narrative of the process of demarcation and its consequences, see Willis, *Making of the Mijikenda*, ch. 5.
72. D. Coplan, 'Introduction: From Empiricism to Theory in African Border Studies,' *Journal of Borderland Studies*, 25 (2010), 1–5.
73. The early Africanists who conducted borderland studies were historians and anthropologists focusing primarily on West Africa and Southern Africa. For examples, see D.J. Thom, *The Niger–Nigeria Boundary, 1890–1906: A Study of Ethnic Frontiers and a Colonial Boundary* (Athens, NY, 1975); A.I. Asiwaju, *Partitioned Africans* (London, 1984); A.I. Asiwaju and P.O. Adeniyi, *Borderlands in Africa: A Multidisciplinary and Comparative Focus on Nigeria and West Africa* (Lagos, 1989); P. Nugent and A.I. Asiwaju (eds), *African Boundaries*; P. Nugent, *Smugglers, Secessionists & Loyal Citizens on the Ghana–Toga Frontier: The Life of the Borderlands Since 1914* (Athens, NY, 2002); W.F.S. Miles, *Hausaland Divided: Colonialism and Independence in Nigeria and Niger* (Ithaca, NY, 1994); M.L. Fleisher, *Kuria Cattle Raiders: Violence and Vigilantism on the Tanzania/Kenya Frontier* (Ann Arbor, 2000). Recently researchers from several disciplines have established the African Borderlands Research Network (http://www.aborne.org) to keep track of ongoing borderlands research throughout Africa. An extensive bibliography of African Borderland studies is available at http://www.aborne.org/bibliography.html.
74. For an example of fine-grained scholarship on European negotiations over colonial borders see A.C. McEwan, *International Boundaries of East Africa* (Oxford, 1971).
75. D.K. Flynn, '"We Are the Border": Identity, Exchange, and the State along the Bénin–Nigeria Border,' *American Ethnologist*, 24 (1997), 311–30.
76. E. Allina-Pisano, 'Borderlands, Boundaries, and the Contours of Colonial Rule: African Labor in Manica District, Mozambique, c.1904–1908,' *International Journal of African Historical Studies*, 36 (2003), 81.
77. For a history of Islamic legal institutions in the East Africa Protectorate and Kenya, its successor, see K. Mwangi, 'The Application and Development of Sharia in Kenya: 1895–1990,' in M. Bakari and S. Yahya (eds), *Islam in Kenya* (Nairobi, 1995), pp. 252–9.
78. R.M. Mambo, 'Nascent Political Activities among the Mijikenda of Kenya's Coast during the Colonial Era,' *Transafrican Journal of History*, 16 (1987), 92–120; A.I. Salim, *The Swahili-Speaking Peoples of Kenya's Coast, 1895–1965* (Nairobi, 1973).

Part III
Borderlands and State Action

6

Not by Force Alone: Public Health and the Establishment of Russian Rule in the Russo–Polish Borderland, 1762–85

Oksana Mykhed

The Russo–Polish borderland of the 1760s was subject to longstanding competition for land and resources between the Russian Empire and the Polish-Lithuanian Commonwealth. It was an example of a borderland peculiar to Europe before 1700, one that Daniel Power and Naomi Standen have described as a buffer zone between two states whose populations experienced intense interaction, while also being influenced by the core areas of their respective states.[1] The Russo–Polish competition for control led to long-running disagreement over border demarcation.[2]

In 1770, local Russian authorities of the Kyiv province in left-bank Ukraine described the unsatisfactory state of the border in reports that aimed to attract the attention of central imperial officials to border insecurity, and at the same time petition them for improvements to be made to this state of affairs. The reports suggested that the border meant little to local inhabitants. Indeed, in some areas on the right bank of the Dnieper River, Russian and Polish villages were located in such close proximity that peasants could easily exchange goods, use common grasslands, rivers, and forests, and go fishing together.[3] According to the Russian administrators, local peasants were not even aware that a legal border existed between the two states. Polish subjects captured by border guards during cross-border pilgrimages to Kyiv, visits to relatives, or fishing activities were not able to tell whether they lived in the Polish-Lithuanian Commonwealth or the Russian Empire; nor were they able to discern where the Empire ended and the Commonwealth began.[4]

The disappointing state of border security improved dramatically in the last decades of the eighteenth century, when the Russian Empire created a definitive border with the Commonwealth. The introduction of a vicegerency in the Kyiv province of the Russian Empire in 1782 secured this territory for the empire. The creation of the new province increased the empire's presence in the borderland, and blocked Polish conquest and colonization in the central

Map 6.1   The Kyiv Vicegerency of the Russian Empire and the Kyiv Palatinate of Poland-Lithuania, 1782

Dnieper area. By changing the landscape of the borderland with the intensive construction of outposts, barriers, customs, quarantine facilities, and hospitals in the 1770s and 1780s, the empire reduced uncontrolled population migration, peasant flight to the Commonwealth, bandit attacks, smuggling, and the spread of the plague in left-bank Ukraine. These improvements made the empire's further expansion into the Polish palatinates acceptable for the population of these territories, and may have contributed to the second and third Partitions of Poland in 1793 and 1795 respectively.

This essay will explore the methods and tactics used by the Russian Empire to establish control over its borderland with the Polish-Lithuanian Commonwealth. First, it will demonstrate that despite the signing of the Eternal Peace Treaty of 1686, which defined the status of Russian and Polish territories in left- and right-bank Ukraine, this region lacked a clear-cut border. Second, the essay will argue that the empire's struggle with the outbreak of bubonic plague in the Russian Kyiv province in left-bank Ukraine motivated the closure of the border in the 1770s, and led to a complete rebuilding of the border's infrastructure. Finally, this study will demonstrate how the transformation of the border into a complex state mechanism replete with an expanding state infrastructure subordinated left-bank Ukraine and the borderland to the empire. This transformation was achieved not only by force, but primarily through the advancements in health care and migration control that allowed the empire to influence the loyalties of the borderland population.

This study contributes to the history of frontiers, borderlands, and empires in eighteenth-century Eastern Europe. The existing literature on this subject concentrates mainly on the ideological and political competitions among the empires for land, resources, and the stateless population.[5] By exploring more physical and material spheres of rivalry such as public health and migration control, this research reveals that during the early modern period the policies of improvement in the police, border infrastructure, and health care were important for the empires struggling to incorporate borderlands and attract their population and elites.

The Borderland in Russian and Polish Diplomatic Affairs

When Catherine II ascended the Russian throne in 1762, she continued the Petrine reforms of the empire and planned to unify all imperial provinces and standardize their administration.[6] This plan included profound reforms of both the government and the administrative structure of the provinces. The work on the realization of this plan began with the collection of information about the provinces and negotiations with the nobility governing them. Following the abolition of the traditional rule of hetmans in left-bank Ukraine in 1764, Catherine introduced the new office of the governor general subordinated to her, akin to the post of Jacob Sievers in the Novgorod province. Piotr

Rumiantsev, who became a governor general of left-bank Ukraine, initiated important reforms of administration and legislation, and cooperated with the local nobility to develop a specific plan of reforms in the region from 1765 to 1768.[7]

Catherine's reign was marked by the exploration of diverse regions of the empire, the redrawing of political and topographic maps, the creation of well-defined provincial boundaries and careful and elegant city plans. In the 1760s, Russian imperial cartographers produced atlases and maps, which not only depicted parts of the empire but aimed to provide detailed information about the population, natural resources, trade routes, and administrative divisions of the empire. This new mapping of the empire stretched the imagination of officials, and increased their knowledge of each imperial province.[8] Mapping the Russo–Polish borderland, however, was a difficult task. The expedition of Major General Aleksandr Bibikov, organized to explore the borderland in March 1765, demonstrated that Russian officials and cartographers had only a very limited knowledge of the region. Bibikov's expedition confirmed that no border demarcation existed, Russian garrisons and border guards were ineffective in policing the border, and the Commonwealth lacked state border guards in the region.[9]

Early in her reign, Catherine II considered the resolution of conflicts with the Polish-Lithuanian Commonwealth to be one of the most important aspects of Russian foreign policy. The Russo–Polish border demarcation had remained an unresolved dispute in relations between the two states since the seventeenth century. Thus Catherine's first instructions to Russian diplomats in negotiations with the Commonwealth in 1764 addressed the question of borders.[10] Following the election of a new Polish king, Stanislaw August Poniatowski, Russian diplomats strove to confirm the earlier Russo–Polish territorial and political agreements based on the Eternal Peace Treaty of 1686.[11]

The original treaty had used the Dnieper River to divide Ukraine into a left-bank section, which together with Kyiv belonged to the Russian Empire. The right-bank section had been placed under Commonwealth sovereignty.[12] Yet despite this straightforward division, the treaty introduced a significant Russian presence in right-bank Ukraine. Not only the whole city of Kyiv, which earlier in the century had belonged to the Polish Kyiv palatinate (*Wojewodztwo Kijowskie*), but also the territories surrounding the city and the Dnieper River became a zone of Russian control. Taking into account these territorial concessions and the dispute over their status, the Polish representatives had been unwilling to concede a part of the Kyiv palatinate and its major city to the empire, and only signed the treaty under Russian military and diplomatic pressure.

The Eternal Peace Treaty of 1686 required the establishment of a five-mile depopulated buffer zone in the borderland between the two states. This provision was a temporary measure designed to postpone more precise border demarcation. However, the arrangement never worked in practice because the two states failed to implement it.[13] The region which was supposed to remain depopulated experienced a wave of colonization by newcomers from the Polish and Russian states in the early eighteenth century. Both states failed to stop the colonization of the buffer zone in right-bank Ukraine, and the Polish magnates—concerned to protect their landownership rights—blocked the ratification of the treaty by the Diet until 1710, and later opposed any attempts at border demarcation. These problems set the stage for further Russo–Polish disputes concerning the borderland and the city of Kyiv. The landowners in the Polish Kyiv palatinate refused to recognize the treaty. For the Russian state, however, the treaty became an important diplomatic tool used for more than a century in negotiations with the Polish-Lithuanian Commonwealth.

The election of Stanislaw Poniatowski, a noble with close ties to Catherine II and the Russian Empire, provided new opportunities to define and demarcate the Russo–Polish border.[14] During Poniatowski's election and coronation diets, both Russian and Polish diplomats confirmed their adherence to the treaty of 1686 as the basis for the development of further agreements between the two states. However, as noted above, the treaty did not provide any details on the demarcation of borders between the Commonwealth and the Russian Empire. Russian and Polish diplomats agreed to review the conditions of the borderland and properly demarcate the border upon the confirmation of the treaty.[15]

Nikolai Repnin, the Russian ambassador to Poland, received numerous recommendations from Catherine II and Nikita Panin, Russian minister of foreign affairs, to pursue border demarcation as a method of solving serious problems which undermined the economy in the empire's western provinces. These problems were: the mass flight of Russian peasantry westward and their settlement in the buffer zone of the Russo–Polish borderland, smuggling, and the exposure of the western provinces of the empire to frequent epidemics of bubonic plague and other contagious diseases. Beginning in the early eighteenth century, outbreaks of the plague reoccurred periodically in the borderland due to its proximity to major river and land trade routes to the Crimean peninsula and the Ottoman Empire.[16] In fact, just before the establishment of Rumiantsev's government in left-bank Ukraine several severe outbreaks of the plague occurred in the borderland. These outbreaks depopulated whole villages and towns, and caused the demise of up to 80 per cent of their inhabitants.[17] The Russian intention to demarcate the border ran against the determination of the Polish nobility to attract peasantry to their estates and recolonize the lands of right-bank Ukraine. Hence, despite the persistent efforts of Repnin, in 1764 the Polish Diet, under the sway of powerful magnate families, postponed consideration of the demarcation. Repnin's attempts to resolve the issue at the diets of 1766 and 1767 proved to be equally fruitless, thus leaving the borderland in its ambiguous and unruly condition.[18]

The only visible sign of the zigzagging and vanishing traces of the border between the Russian and Polish states in the Russian Kyiv province and the Polish Kyiv palatinate was a line of outposts built by the Russian Empire after the conclusion of the Eternal Peace Treaty in 1686. Inspections by Russian authorities revealed that some outposts had collapsed due to lack of maintenance and neglect. Sporadic renewals of the outposts and reconstructions of the line did not allow for the complete closure of the border between the states. The Kyiv or Vasylkiv line of outposts tracked the Dnieper River and the triangle marked by the rivers Irpin and Stuhna near Kyiv. This line was about 393 miles long. The distance between each outpost varied between approximately 1.3 and 4 miles. In the 1760s, each outpost had two to four Cossacks guarding and patrolling the surrounding territory. Typical outposts were small huts located in unpopulated forests or near the major roads.[19]

In the 1760s, the imperial authorities in left-bank Ukraine were unable to guarantee an appropriate defense of the border due to a lack of funding and an insufficient supply of border guards and patrols. Usually, some of the outposts were left unprotected, the guards being concentrated on particularly dangerous parts of the border where attacks by the Haidamaks (free militiamen who often pillaged noble estates in right-bank Ukraine)[20] or epidemics of bubonic plague were more likely. Russian imperial representatives in left-bank Ukraine were well aware of the inadequate defense of the border. However, local administration lacked the necessary human resources for the establishment of permanent border guards, militia, and patrols along the whole line of outposts.[21] In addition, guards and patrols did not possess steady supplies of gun powder, food, and forage. Underfunded by the empire, guards even abandoned the outposts or lived too far from them to fulfill their duties effectively.[22]

The mass recolonization and development of new private Russian and Polish settlements in the buffer zone displaced some of the outposts and erased parts of the borderline. Surrounded by forests and unpopulated territories, the newly established towns boasted private garrisons as an alternative to the official outposts. Many places such as Bila Tserkva, Vasylkiv, and Korsun included large fortresses which facilitated recolonization. Fastiv, Motovylivka, and Didiv possessed smaller forts which hosted unofficial border outposts. Originally founded by the Polish and Lithuanian nobility as defensive complexes against Tartar invasions from the steppe,[23] these towns attracted a diverse population from both the Polish-Lithuanian Commonwealth and the Russian Empire. This population only occasionally interacted and cooperated with Polish and Russian authorities. Upon their settlement in the borderland, they enjoyed the status of free colonists or paid moderate fees to the Polish nobles which granted them freedom from enserfment and the high taxes typically experienced by peasants in the Russian provinces.[24]

Unregulated Migration and the Bubonic Plague

Piotr Rumiantsev, governor general of left-bank Ukraine between 1765 and 1789, highlighted the problem of the chaotic population movement and migration of peasants from the Russian Empire to the Commonwealth in his reports to Catherine II. Rumiantsev observed that mass emigration from left-bank Ukraine resulted from the introduction of military conscription, numerous taxes, and serfdom in the Russian-controlled region. In order to stimulate the recolonization of their lands in the Kyiv palatinate, the Polish nobility granted new settlers land and freedom from serfdom and military conscription for 20 years, and sometimes even longer periods.[25] Thus, the unprotected border allowed the Commonwealth to outcompete the empire in the contest for the borderland population. Because their serfs fled to the Polish lands, Ukrainian and Russian landowners were unable to pay taxes established by the empire. The 'unlimited freedom and willfulness' of the peasants leaving the estates and migrating to the Polish territory of the borderland deprived the Russian provincial government of essential human resources and tax revenues.[26] In order to curtail this mass peasant flight, Rumiantsev proposed to divide all inhabitants of left-bank Ukraine into social groups, introduce peasant indentureship to land, and assign peasants to certain landlords. These reforms, however, could produce positive results only after the establishment of a well-protected border.[27]

In the 1760s, many peasants from left-bank Ukraine fled to the Polish-Lithuanian Commonwealth where they were eagerly welcomed by local landlords. The porous border allowed whole families and village communities to migrate to the Commonwealth with all their household items, carriages, horses, and livestock.[28] Russian authorities could not use force to catch runaway peasants after they crossed the border because Polish nobles interpreted these military expeditions as illegal interventions, and any such complaints to Polish diplomats and the Diet in Warsaw threatened to further strain Russo–Polish relations. Desperate about the loss of peasants, Russian landowners illegally sent small groups of armed men to search for fugitives. However, the runaways outnumbered these small groups, meaning that Russian soldiers could not force peasants to return even if they found them. These conditions emboldened runaway peasants to attack, rob, and expel particularly aggressive and insistent Russian patrols.[29]

In order to abate mass peasant flight, repatriate runaways to the empire, and improve relations with the Commonwealth, Russian officials turned to nonviolent methods of returning fugitives to left-bank Ukraine. Russian officers traveling to the borderland secretly distributed propaganda letters urging peasants to return, promising them not only a pardon for their crimes in the Russian Empire but even certain tax incentives. Russian authorities ordered border guards to be very tolerant towards any peasants returning to left-bank Ukraine, to protect them from bandits and their former Polish landlords, and even to accompany peasant families on their journeys back into the Russian Empire.[30] These measures proved ineffective, and few peasants returned. The Kyiv and Smolensk provinces of the empire continued to experience large-scale peasant flight and resettlement in the borderland or the inner Polish palatinates.[31]

Harboring runaway peasants, deserters from the army, and free settlers, the Russo–Polish borderland experienced high levels of banditry and smuggling. Lured by opportunities for quick economic advancement, and facing few deterrents, some of the newcomers resorted to robbing merchants or diplomats who traveled through these lands, thus provoking international scandals.[32] Both Russian and Polish officials received numerous complaints from foreign and domestic travelers, but they were not able to apprehend the robbers and recover any stolen goods. The local population was itself frequently involved in disposing of stolen property, harboring criminals, and otherwise aiding and abetting the banditry. Unable to curtail the brigandage in the borderland and close the border, Russian and Polish authorities provided large military escorts to guard prominent travelers during their journeys. These efforts, however, did not change the troublesome situation in the borderland and robberies continued.[33]

Uncontrolled migration to the borderland and the spread of brigandage culminated in 1768–69 when two social cataclysms occurred almost simultaneously in right-bank Ukraine. The Confederation of Bar, a rebellion organized by Polish nobles, was followed by a peasant and Haidamak revolt, known as the Kolii Uprising. The Confederation was initiated by the nobility of right-bank Ukraine dissatisfied with Russian attempts to create a pro-Russian party from the non-Catholic population of the Commonwealth at the Diet of 1767. However, the majority of the peasant population who lived in right-bank Ukraine welcomed Russian participation in the domestic politics of the Commonwealth. On the other hand, the noble rebels interpreted all Russian initiatives as interferences in the affairs of the borderland. The noble movement swiftly developed into a civil war that paralyzed the economy and agriculture, and destroyed peasant households, which in turn led to a mass revolt of the local population against their landlords. In 1768–69 the Russian policy of creating a pro-Russian party among the Commonwealth's subjects greatly undermined noble–peasant relations in the eastern Polish palatinates and sharply polarized the attitudes of the borderland's elite and peasants toward the empire.[34]

The social unrest and the subsequent Russian efforts to suppress it helped turn the borderland into a territory of devastation, famine, and refugees. Many towns and villages were abandoned or depopulated due to outbreaks of the plague, which did great damage in this troubled context. Typically, following the appearance of disease, the local population fled their homes to seek refuge in forests or the steppe. Abandoned houses attracted occasional travelers or robbers who got infected and spread the pestilence. In theory, the majority of Polish towns possessed police units and medical professionals who blocked population movements during the outbreaks of plague, sent sick individuals to quarantine facilities, and burned all infected houses. However, these anti-plague precautions were ineffectual in many towns because they had not recovered from the devastation caused by the uprisings, and their town councils and regular police units had been recruited to defend the nobility in the calamities of 1768–69.[35]

There was no treatment for the plague in the eighteenth-century Russian Empire or the Polish-Lithuanian Commonwealth. In the 1760s the most effective method of combating the disease was separating sick and healthy individuals from each other, and relocating all travelers or suspicious persons to hospitals and disinfecting their personal belongings. Borders and other human or natural barriers played a key role in this process. It was very important to observe the situation in neighboring areas carefully, and promptly to close all roads and borders before the plague reached a new place. Therefore, protection from the plague was an essential component of border security. However, in 1768–70, the Russian imperial authorities in Kyiv frequently disregarded the role the plague played in border security and underestimated the need for vigilance.

In order to protect the Russian Kyiv province from the plague, the governor general dispatched small detachments of soldiers to the Polish territory of the borderland to investigate the approaching disease. In theory, sanitary cordons against the plague included a network of quarantine facilities and main outposts that blocked major roads. Outpost guards and militia closed most border crossings and allowed incoming travelers to enter the western provinces of the empire only after quarantining them for three to six weeks and disinfecting their belongings. Clothes and the majority of household items were treated with smoke or herbal powders. Packages with food, tobacco, salt, and spices were sprinkled with vinegar or vodka, which were believed to sanitize and disinfect.[36]

While effective in theory, this system was ineffectual when applied to the porous Kyiv outpost line. The line did not encompass all border crossings, and except for Vasylkiv all its quarantine facilities were designed as temporary institutions constructed for short periods of time and dismantled when the fear of the plague in the borderland subsided. The Kyiv line consisted of only two or three quarantine surgeons, some of them without experience in treating plague patients.[37] Because physicians were in high demand to attend to Russian wounded soldiers returning to Kyiv from the Russo–Turkish war, only outpost guards without any medical training examined travelers at border crossings. In order to avoid prolonged border inspections and quarantines, travelers often bribed guards to bypass examinations. Due to low salaries and lack of accountability, outpost guards were keen to accept such inducements, making corruption widespread.[38] Merchants who participated in trade fairs in Kyiv, Uman, Pereiaslav, and Bila Tserkva frequently bribed border guards and brought textiles, furs, and clothes to left-bank Ukraine. This merchandise bypassed the quarantine facilities and disinfection procedures, and was promptly sold in the markets in Kyiv.

The first cases of the disease were recorded in the borderland near the main quarantine facility in Vasylkiv; however, the pestilence continued to advance, reaching Kyiv in August 1770. The city's experience of the plague began in the Podil district, which served as a business center and hosted merchant communities from the Commonwealth, the Russian Empire, Crimea, and the Ottoman Empire (see Map 6.1).[39] The occurrence of the plague in Kozelets, Boryspil, Brovary, Pereiaslav, and many villages surrounding Kyiv in the fall of 1770 exposed the ineffectiveness of the Vasylkiv quarantine and anti-plague precautions in Kyiv. The city's physicians lacked experience in treating the plague and confused the malady with influenza, thus allowing its further spread. The failure to isolate Kyiv in a timely manner led to the spread of the plague by Russian soldiers, students of the Kyiv Academy, and merchants—all of whom fled the city in a great panic. Imperial authorities became increasingly alarmed about the danger of the situation when the plague broke out in Chernihiv and Nizhyn, from whence it spread to the inner regions of the empire via major trade routes, appearing in Moscow in December 1770.[40]

Johann Lerche, a surgeon who arrived in Kyiv from St Petersburg upon the request of the Kyiv governor general, concluded as early as late October 1770 that it would be impossible to stop the spread of the pestilence in the city.[41] Lerche was charged with the organization of anti-plague policies in Kyiv; however, he faced opposition from the city council, the merchant community, his fellow medical practitioners, the military, and the governor general himself. Lerche documented numerous violations of anti-plague precautions in Kyiv, thus uncovering widespread corruption and insubordination on the part of outpost guards, soldiers, and militiamen.[42] Instead of isolating the houses of the sick and the deceased and burning their infected belongings, as Lerche had requested, local militia appropriated and sold infected items at the local markets in the borderland, thus spreading the plague. According to Lerche, even though the plague abated in the late winter of 1771, these violations led to a new outbreak in May 1771. This time the plague ravaged not only the soldiers of the Kyiv garrison and militia, but also spread through all borderland towns and villages where the infected merchandise was sold.[43]

In winter 1771 imperial officials began to intervene more actively in anti-plague measures in Kyiv and investigate the problems with security and disease control in the borderland more generally. Catherine II sent Mikhail Shipov to assist Dr Lerche. A Major in the Life Guards, Shipov came to left-bank Ukraine from St Petersburg charged with conducting a thorough investigation and improving anti-plague precautions. His main goal was to organize an effective defense of the Kyiv and Smolensk provinces against the plague. He was responsible for reforming the border infrastructure, quarantine system, and government of the borderland.[44] Shipov's first reports to the Medical and Military Colleges in St Petersburg described the ineffectiveness of border security and proposed the construction of new border quarantines and hospitals. Russian surgeons, officials, and engineers sent to the borderland characterized it as an unruly region that undermined the economy and endangered the population of the Kyiv province. Shipov and Lerche's conclusions were that only complete isolation of the imperial province during plague outbreaks would protect the region and the empire from the malaise. Their reports and suggestions formed the basis of major reforms in the borderland and the Kyiv province in the 1780s.

Border Protection and Public Health

Reforms initiated by Russian imperial authorities in the borderland aimed to increase the defensive capacity of the Kyiv outpost line and improve the control of the imperial officials in Kyiv over the border. These reforms coincided with the first Partition of Poland (1772); however, the Partition did not influence the Russo–Polish border in this region. The plague outbreaks of 1770–71 served as a catalyst for the establishment of a new system of border security. They motivated the imperial authorities to add new outposts to the zigzagging and porous borderline, eliminate breaches in the border, and streamline its design.[45] The Russian Senate required the Kyiv governor general to increase the numbers of border guards and militia in the outposts. After the plague outbreaks, pickets, moats, and barriers were constructed between border outposts to reduce illegal border crossings that bypassed the outposts. These measures allowed the empire to establish a borderline, which was officially demarcated and accepted by the two states in 1781–83.

The Russian Senate also approved additional funding for the construction of new quarantines and improvements in border security in the Kyiv province. In 1771–72 alone, the Kyiv chancellery allocated 8172 rubles for the construction and hired additional medical practitioners, guards, and quarantine officers.[46] Upon request from the imperial representatives in Kyiv, the Military and Medical Colleges turned over control of the major outpost and quarantine complex in Vasylkiv to the governor general of Kyiv. This reform allowed the empire to decrease corruption among quarantine officers and redirect revenues from taxes and fees collected in Vasylkiv to the Kyiv treasury. In 1771–74, the main quarantine facility in Vasylkiv was expanded to accommodate about five hundred persons at any one time.[47] In addition, three smaller quarantines were constructed near Vasylkiv in Kuzyn Khutir, Kozelets, and Borky. This increased the capacity of the major border crossing in the Kyiv province.

Russian imperial authorities planned to enhance the Kyiv borderline with several new permanent quarantine facilities. The number of quarantine houses allocated to the Kyiv line was higher than the two quarantines recommended by the Medical College for each Russian province in 1771. In addition to that at Vasylkiv, Major Shipov established quarantine facilities in Sorokoshychi on the border of the Kyiv province of the Russian Empire. Shipov also insisted on the maintenance of the quarantine facility in Sloboda Dobrianka. These quarantines improved protection from the plague in the upper Dnieper area. In the lower part of the Kyiv line, Shipov requested that a quarantine facility be constructed in the small town of Perevolochna—a crossing-point for important routes connecting left-bank Ukraine, the Zaporozhian Host and Crimea.[48] Finally, the imperial government renovated the existing quarantine house in Kremenchuk and opened a new quarantine facility in Pereiaslav. Hence, after the reform, the Kyiv line was reinforced with the highest number of permanent quarantine facilities in its history.

In order to decrease corruption among quarantine officers, the imperial government in St Petersburg proposed that these workers be provided with the status of state employees.[49] To improve control over the activity of Vasylkiv personnel, Major Shipov required that all documents and information about travelers be presented to him and the Kyiv governor general. All travelers were able to cross the border only after acquiring permissions from the two officials and obtaining travel documents with Shipov's or the governor general's original signatures. This reform decreased the amount of border crossings with falsified travel documents and curtailed corruption among outpost officers.[50]

The Russian Senate decrees of 1771–72 established stricter requirements for the personnel of border outposts and quarantines. In particular, new heads of quarantine facilities were required to possess a good knowledge of at least one foreign language (German or French).[51] Each quarantine facility received a medical team to examine all newcomers. This team consisted of a main physician, his apprentices, nurses, and an apothecary. Shipov requested that the Medical College dispatch six additional physicians from St Petersburg to Kyiv, thus doubling the number of highly qualified medical professionals in the borderland. These surgeons were permanently assigned to the quarantine facilities and were absolved from the obligation to serve the Russian army or the population of neighboring towns. The Medical College established a standard salary for the quarantine physicians which was partially funded by quarantine fees and supplemented by the government. In particular, quarantine physicians were paid 180 rubles per year and were supplied with food, forage, office paper, and other essentials.[52]

In the 1780s, the Kyiv line included several large outpost and quarantine complexes as well as numerous small outposts located between them. Each of the complexes had customs, hospitals, shops, stables, storehouses, police units, as well as postal and currency exchange offices. Large teams of medical and custom personnel, clerks, and soldiers were permanently employed by the complexes and lived in special houses attached to them. The active support of the provincial government, a high concentration of travelers, and numerous employment opportunities attracted the local population and turned the areas that hosted the quarantine complexes into rapidly developing borderland towns. Travelers who were required to stay in the quarantine complexes for three to six weeks paid high fees for medical examinations, the disinfection of baggage, the provision of food and housing of various levels of comfort, as well as sundry other services.

Fully controlled by the provincial government, the new border infrastructure not only stopped illegal migration and protected the Kyiv province from the plague but also increased imperial revenues from the quarantine facilities and customs. Even after imperial boundaries were shifted westward as a result of the Partitions of Poland in 1793 and 1795, the Commerce and Medical Colleges of the Russian Empire decided to maintain the quarantine and custom complexes as they proved to be very effective in guarding the Kyiv province.[53] Furthermore, after officially losing their status as border institutions in 1797, the quarantine facilities and hospitals of the former Kyiv line successfully contained two new outbreaks of the plague in the Kyiv and Podolia provinces in the late 1790s.[54] These outbreaks claimed fewer lives and were stopped more quickly than the plague of 1770–71 because of the anti-plague precautions jointly organized by the governor general of the province, medical practitioners, the military, and the police. The outbreak of 1770–71 became the last documented major epidemic of the plague—and indeed the last documented medical catastrophe of any kind—in the Kyiv province in the eighteenth century.

Administrative Reforms and Population Migration

The sweeping reforms in border security and the introduction of permanent quarantine complexes were followed by the publication of new Senate laws, which regulated the government of the Kyiv province and the lives of its inhabitants. The new legislation emphasized the need for cooperation with imperial authorities, represented by the governor general and the nobility, to counteract outbreaks of the plague, and improve living standards. The Russian Senate required the governors of all provincial towns and districts to establish and maintain permanent police, medical, and postal services, all of which were gradually adjusted to conform to general imperial standards after 1775. Town and provincial governments received detailed instructions from the Senate regarding the control and improvement of sanitary conditions in town streets, markets, and other public areas. These instructions even ordered the draining of swamps and bogs on the outskirts of towns and villages in the belief that they increased humidity and precipitated the spread of the plague. The Senate also obliged local governors to circulate recommendations urging local residents to avoid living in damp houses and to disinfect all groceries and textiles bought at street markets.[55]

The Senate decree of September 25, 1771 required governor generals and the nobility to remain in office or on their estates and to lead the anti-plague precautions in their jurisdictions. The nobility was made responsible for control of population movement and the protection of their peasants and servants from the plague. Special instructions of the Senate obliged the nobility to supervise the disinfection of their peasants' homes during plague outbreaks. Later, further legislation charged the nobles with monitoring the everyday life of the local population.[56] These new regulations bound and subjugated the peasantry to their lords, limited their mobility, and created conditions for their enserfment. In order to curtail illegal migration to Poland-Lithuania, Russian authorities introduced monetary rewards for border guards who apprehended and returned runaway peasants to their landlords.[57] Another Senate decree encouraged the practice of reporting sick and suspicious individuals to the authorities. During plague outbreaks, the Senate authorized anti-plague commissions to solicit reports from the population and award ten rubles for information about any infected persons evading hospitals and quarantine facilities, and twenty rubles for apprehending individuals selling infected merchandise or household items.[58]

Beginning in the early 1780s, border outposts implemented a passport control system. Border guards were made responsible for checking passports of all travelers and stopping any of them from crossing the border without these documents. Anyone traveling abroad was required to contact the appropriate noble or district representative, governor general, or church authority to ask for permission to leave their place of residence; they were also required to obtain a special document confirming their identity and health status. Passports allowed for one or more journeys abroad. They served as a visa, not a permanent identification document. Usually, the chancellery issued not individual but group passports. (For example, a merchant and his retinue received a common passport.) Clergymen, diplomats, or foreign workers hired by the empire were granted travel passports upon presenting documents confirming their status. Merchants and peasants applied directly to the chancellery and explained the reasons for their trips abroad. Constantly present on the border and in provincial towns, the police inspected the documents of all newcomers or suspicious individuals.[59]

Improved control by the imperial authorities over the Kyiv province and the borderland allowed Catherine II to introduce a new vicegerency system (*namestnichestvo*) in the region. This new administrative order increased the power of imperial representatives, viceregents (*namestniki*) appointed by St Petersburg to all provinces of the empire, and decreased the power of local elites.[60] The earlier advancements in border control, administration, and the management of population mobility prepared the province for this reform. The introduction of *namestnichestvo* concluded the absorption of left-bank Ukraine into the empire and the transformation of this semi-autonomous region into a standard imperial administrative unit.[61] The border with its infrastructure and quarantine complexes became fully subordinated to the new Kyiv provincial governor.

The system of vicegerencies established a new hierarchy of administrators recruited from a circle of reliable imperial governors or the local nobility. These administrators were required to maintain records and report population counts for each district of the province to higher officials. The initial decree that created the Kyiv *namestnichestvo* had not precisely defined the borders of the new administrative unit. The new imperial administration in Kyiv was therefore required to determine these borders and send this information to the Russian Senate for final approval.

Following the curtailment of the peasant movement and the clear separation of society into groups and classes, it became increasingly difficult for anyone to remain in the province unnoticed for long. Beginning in 1785, the Kyiv governor's chancellery was required to track all foreigners, legal or illegal newcomers, and homeless persons in the province. Special police units kept vigil over the homeless and the poor in each district of Kyiv and in provincial towns. All newcomers without passports or other travel documents who refused to leave the province were put to forced labor, repairing bridges and roads. The local population was obliged to supply these workers with food, which discouraged local residents from welcoming and supporting illegal newcomers. The new local police system also acted against the integration into local communities of runaways and other illegal immigrants. Outsiders were not able to acquire a good job or receive a loan. Property owners harboring runaways had to report them to the authorities on pain of repaying all the debts of their guests.[62]

Despite the closure of the Russo–Polish border, the Russian provincial administration continued to interact with the Polish nobility and the population of right-bank Ukraine. During plague epidemics and crop failures the Kyiv governor generals purchased grain, flour, and food supplies from Polish landholders and settlers in right-bank Ukraine. Even after the Partitions of Poland, the right-bank territories were still referred to as 'Polish Ukraine.' Some representatives of Polish magnate families (that is, the Potockis, Branickis, and Lubomirskis) developed dual Polish and Russian political loyalties. These loyalties were enhanced through marriages to members of Russian aristocratic families, conducting grain trade in the new Russian ports of the Black Sea region, and the large-scale supply of produce to the imperial army participating in the Russo–Ottoman wars.[63] Finally, the nobility of right-bank Ukraine confirmed their loyalties to Catherine II during the Confederation of Targowica, a movement which opposed the new Polish Constitution of May 3, 1791 and precipitated the second Partition of Poland in 1793.[64]

In the 1780s, Catherine's policies of repatriating Russian runaway peasants and settlers from right-bank Ukraine to the empire finally came to fruition. Russian manifestos and leaflets promising returning peasants tax incentives, individual freedom, and land in New Russia or other low-population-density provinces attracted Russian runaways in the Polish-Lithuanian Commonwealth.[65] Kyiv became the major admission location for many repatriates, the place where they received their new Russian passports and negotiated with local imperial authorities about their resettlement in the empire. The improvements in the security of the border with the Commonwealth prevented these newly admitted Russian subjects from future attempts at escape to Poland-Lithuania.

Actions by the provincial imperial governments increased public awareness of the empire's civilizing role in public health, economic stability, and the welfare of the local population generally. The image of the empire presented by the new cadre of physicians, clerks, and administrators influenced the political views and loyalties of the borderland population. For example, one resident of Kremenchuk, a town that belonged to the Kyiv outpost line, described how effectively the local government, medical teams, and militia dealt with a severe outbreak of the plague, which occurred in the town in February 1785. The plague in Kremenchuk claimed only several lives because the local government organized all anti-plague precautions in a timely manner. Full of optimism, the Kremenchuk inhabitant advised his friend not to sprinkle his letter with vinegar, and revealed new patriotic sentiments which, according to him, were shared by the local population: 'We are fortunate to live in an age when no trouble can overcome us. God gave us the empress who always rescues her people from misfortunes. Her thoughtful laws and powerful institutions are a blessing for us, and the anti-plague measures saved our lives.'[66] Aside from this praise of the empress, the letter also saluted Prince Grigory Potemkin, whose soldiers had surrounded the town and helped stop the plague.

Conclusion

At the beginning of Catherine II's reign in 1762, the Polish-Lithuanian Commonwealth and the Russian Empire were officially separated by a border documented in the Eternal Peace Treaty of 1686. In practice, the Russian cartographers could not draw a clear-cut line on the maps and had limited knowledge about the borderland's physical and human geography. The only obvious sign of the existing border, the Russian line of outposts, did not function effectively and was often broken or ignored by new settlements and colonization by Russian and Polish subjects. Lack of clearly defined and policed barriers between the Russian Kyiv province in left-bank Ukraine and the Polish palatinates in right-bank Ukraine led to massive peasant flight which undermined the economy of the empire, lowered tax revenues, and decreased military conscription.

The bubonic plague, which ravaged the borderland in 1770–71, alarmed Russian imperial authorities about the problem of border security, migration control, and public health in the Kyiv province. Profound reforms of the quarantine facilities and outpost complexes that followed the outbreak of the plague in Kyiv changed the built landscape of the borderland and established a constant presence of imperial officials in the region. These advancements not only facilitated the demarcation of the border with the Commonwealth, but also stimulated a reform of the provincial government in left-bank Ukraine in 1782. As a result of this reform, a new administrative unit, the Kyiv *namestnich-estvo*, was created in the Russian Kyiv province. This province was integrated into the core of the Russian state and developed a new, standardized and centralized provincial bureaucracy, and introduced peasant indentureship to land, a passport-control system, and a well-organized hierarchy of local administrators. The new border infrastructure made peasant flight to the Polish palatinates more difficult and protected Russian subjects from the plague. These reforms improved the living standards and economy of the borderland and increased the support of the Russian government by local elites. Later in the century, in 1796, the empire's civilizing expansion in its western borderlands was concluded with the creation of a new province in right-bank Ukraine, which encompassed the territory of the former Polish palatinates acquired by the empire after the second Partition of Poland.

This essay contributes to the literature on the Partitions of Poland (1772–95) and the history of East European borderlands in the eighteenth century. It demonstrates that advancements in public health and migration control were as important in establishing control over these borderlands as were ideological and political confrontations and military interventions. It also demonstrates that intense competition among states and empires for land and population could be resolved—and resolved effectively—via noncoercive methods which themselves could change the loyalties of borderland populations and facilitate territorial acquisitions. The populations of the Polish Kyiv palatinate indirectly exposed to Russian modernizing reforms in the 1780s became subjects of the empire after the Second Partition of Poland in 1793, and did so without noticeable protests or confrontations with the Russian military. These newly minted Russian subjects often assisted imperial authorities in the creation of new provinces in the acquired Polish territories.

Examination of Russian provincial development in right-bank Ukraine sheds new light on the reasons for the Partitions of Poland. It demonstrates that the Partitions were influenced by Russian domestic policies in its western borderlands as well as diplomatic intrigues and negotiations between Prussia, the Habsburg Monarchy, and the Russian Empire. Thus, the eighteenth-century borderlands were not only subjects of political and diplomatic maneuvers, but also physical entities prone to medical and natural cataclysms requiring government attention and large investments of infrastructure and personnel. Their populations were frequently inclined to switch their political loyalties in favor of a competitor state which seemed to offer them better living standards, border security, and health care.

Notes

1. See D. Power and N. Standen (eds), *Frontiers in Question: Eurasian Borderlands, 700–1700* (New York, 1999), pp. 5–9.
2. According to Robert Jones, problems related to the borderland became an important reason for the Partitions of Poland in 1772–95. Jones argues that peasant flight from the Russian Empire to Poland-Lithuania motivated the empire to participate in the Partitions and incorporate Polish territories harboring Russian fugitives. See R.E. Jones, 'Runaway Peasants and Russian Motives for the Partitions of Poland,' in H. Ragsdale (ed.), *Imperial Russian Foreign Policy* (Cambridge, 1993), pp. 103–16.
3. Rossiiski Gosudarstvennyi Arkhiv Drevnikh Aktov (RGADA), F. 248, op. 113, delo 540, pp. 1–3 (hereafter, in form, RGADA, 248-113-540: 1–3).
4. RGADA, 248-113-540: 1–3.
5. Z. Zielinska, *Polska w okowach 'systemu Polnocnego': 1763–1766* (Cracow, 2012); P. Stegnii, *Razdely Polshi i diplomatiia Ekateriny II* (Moscow, 2002); I.D. Armour, *A History of Eastern Europe 1740–1918: Empires, Nations, and Modernisation* (London, 2012); J. Burbank and F. Cooper (eds), *Empires in World History: Power and the Politics of Difference* (Princeton, 2010).
6. I. de Madariaga, *Russia in the Age of Catherine the Great* (London, 2002); O. Eliseeva, *Ekaterina Velikaia* (Moscow, 2010).
7. For the participation of the nobility of left-bank Ukraine in the Legislative Commission, see Z.E. Kohut, *Russian Centralism and Ukrainian Autonomy: Imperial Absorption of the Hetmanate, 1760s–1830s* (Cambridge, MA, 1988), pp. 125–90.
8. The Geographic Department of the Russian government collected information to address M. Lomonosov's criteria regarding the drafting of maps and atlases, which included topography, economics, and natural resources. See S.A. Riabov, *Zdes gosudarevym 'ukrainam' bylo berezhenie: Rossiiskoe pogranich'e—osobyi ob'ekt kulturnogo naslediia* (Moscow, 2007), p. 67.
9. A.A. Bibikov, *Zapiski o zhizni i sluzhbe Aleksandra Il'icha Bibikova* (Moscow, 1865), pp. 32–4.
10. B. Nosov, *Ustanovlenie Rossiiskogo gospodstva v Rechi Pospolitoi, 1756–1768* (Moscow, 2004), pp. 223–6.
11. In Polish historiography, the treaty is called *Pokój Grzymułtowskiego* or the *Grzymułtowski Peace Treaty*—named after Krzysztof Grzymułtowski, the Polish diplomat and envoy to Russia who signed the document. For the Polish and Russian versions of the treaty, see *Traktaty pokojowe pomiędzy Rzecząpospolitą a Rosją w XVII wieku* (Krakow, 2002), pp. 48–65.
12. Before the Partition of Ukraine secured by the Eternal Peace Treaty of 1686, the Russian and Polish sections of the borderland belonged to the Ukrainian Cossack Hetmanate, which united them into a political, cultural, and administrative polity. Located between the Commonwealth and the Muscovite state, the Hetmanate was subject to competition between its powerful neighbors, and was the key factor in seventeenth-century Russo–Polish relations. The Pereiaslav Treaty of 1654 between the Hetmanate and Muscovy granted autonomy to the left-bank Hetmanate under the Muscovite protectorate. Later, leaders of the Hetmanate attempted to reverse the treaty and to bring the Hetmanate under the Polish protectorate. The civil wars that ensued led to the devastation of the state and its political decline. The Eternal Peace Treaty of 1686 sealed the division of the Hetmanate between the Russian Empire and the Polish-Lithuanian Commonwealth.
13. In theory, the border was regulated by border courts comprised of Polish and Russian representatives. These courts were created to resolve land conflicts and punish cross-border crimes. In reality, the courts did not function properly and for many months

were not even able to convene. The work of the courts was inadequately supervised and not satisfactorily funded. As a result, the majority of border conflicts remained unresolved, and the aggrieved parties appealed to the governors of the border provinces.

14. J. Michalski, *Stanislaw August Poniatowski* (Warsaw, 2009); K. Zienkowska, *Stanislaw August Poniatowski* (Wrocław, 1998); N.R. Bain, *The Last King of Poland and his Contemporaries* (London, 1909).

15. Boris Nosov argues that the treaty played a defining role in Russo–Polish relations, helping the Russian Empire to promote its demands without signing a new agreement between the two states. See B. Nosov, *Ustanovlenie Rossiiskogo gospodstva v Rechi Pospolitoi, 1756–1768* (Moscow, 2004), pp. 244, 252.

16. Many epidemics of plague were brought to Ukraine from Crimea and the Ottoman Empire, thus generating a negative stereotype about these territories among the local population.

17. M.K. Borodii, 'Do istorii borotby z chumoiu na Ukraini v XVIII st,' *Ukrainskyi istorychnyi zhurnal*, 5 (1984), 82–90.

18. E. Romer, 'O wschodniej granicy Polski z pred r. 1772,' *Ksiega Pamiatkowa ku czci Oswalda Balzera* (Lwow, 1925), II, pp. 6–7 [360–1].

19. RGADA, 248-113-1499: 32.

20. The word Haidamak can be traced to the Turkic words *hada* and *hajdemak,* which mean 'to harass' and 'to pursue.' See J. Pelenski, 'Haidamak Insurrections and the Old Regimes in Eastern Europe,' in Pelenski (ed.), *The American and European Revolutions, 1776–1848: Sociopolitical and Ideological Aspects* (Iowa City, 1980).

21. RGADA, 13-1-11.

22. M. Tyshchenko, 'Forposty, mytnytsi ta karantyny na zakhidnomu pohranychchi u zv'iazku z zovnishnioiu torhivleiu Ukrainy v XVIII st,' in Tyshchenko, *Narysy z istorii zovnishnioi torhivli Ukrainy v XVIII st* (Bila Tserkva, 2010), p. 11.

23. S. Shamrai, 'Misto Vasylkiv,' *Istoryko-Heohrafichnyi Zbirnyk*, 3 (1929), 40.

24. I. Riabinin, 'K voprosu o pobegakh russkikh krestian v predely Rechi Pospolitoi,' *Chteniia v Imperatorskom Obshchestve Istorii i Drevnostei Rossiiskikh pri Moskovskom Universitete*, vol. III (1911), pp. 12–20; V.A. Markina, *Krestiane Pravoberezhnoi Ukrainy* (Kiev, 1971), pp. 15–29.

25. A. Kaminski, 'Neo-Serfdom in Poland-Lithuania,' *Slavic Review*, 34 (1975), 257.

26. RGADA, 13-1-73.

27. RGADA, 13-1-73; G.A. Maksimovich, *Deiatelnost Rumiantseva po upravleniiu Malorossiiei* (Nezhin, 1913), I, pp. 44–8.

28. RGADA, 248-113-1488.

29. RGADA, 248-113-1488: 124–7.

30. RGADA, 248-113-1488; see also RGADA, 248-113-1491: 254 and 248-113-1486: 1–20.

31. Riabinin, 'K voprosu o pobegakh russkikh krestian v predely Rechi Pospolitoi,' pp. 12–20.

32. RGADA, 12-1-141.

33. RGADA, 12-1-141.

34. W. Konopczynski, *Konfederacja barska* (Warsaw, 1991); B. Skinner, *The Western Front of the Eastern Church: Uniate and Orthodox Conflict in 18th-Century Poland, Ukraine, Belarus, and Russia* (DeKalb, 2009), pp. 129–34; Z.E. Kohut, 'Myths Old and New: The Haidamak Movement and the Koliivshchyna (1768) in Recent Historiography,' *Harvard Ukrainian Studies*, 1 (1977), 359–78; O. Lola, *Haidamatskyi rukh na Ukraini* (Kyiv, 1965).

35. RGADA, 248-113-505; 'Litopys monastyria Vasylian: rok 1770,' *Dilo*, 43 (1890), 2.

36. For more details on the precautions against the plague in the Russian Empire, see J.T. Alexander, *Bubonic Plague in Early Modern Russia: Public Health and Urban Disaster* (Baltimore and London, 1980), pp. 29–35.

37. N. Zakrevskii, *Letopis i opisanie goroda Kieva* (Moscow, 1858), I, p. 86; F. Doerbeck, *Istoriia chumnykh epidemii v Rossii s osnovaniia gosudarstva do nastoiashchego vremeni* (St Petersburg, 1905), p. 115.
38. RGADA, 248-113-430: 39.
39. This outbreak of the plague in Kyiv is described in Alexander, *Bubonic Plague*, pp. 110–15.
40. The outbreak of the plague in Moscow provoked a mass riot of the city's population against imperial authorities. See Alexander, *Bubonic Plague*, pp. 177–201.
41. Zakrevskii, *Letopis i opisanie goroda Kieva*, I, pp. 86–7.
42. Zakrevskii, *Letopis i opisanie goroda Kieva*, I, p. 91.
43. Doerbeck, *Istoriia chumnykh epidemii*, p. 116.
44. Zakrevskii, *Letopis i opisanie goroda Kieva*, pp. 88–90.
45. RGADA, 248-113-1499. According to a new regulation, the outposts were placed closer to each other to allow guards to see their colleagues and communicate using fire signals.
46. RGADA, 248-113-1605: 179.
47. RGADA, 248-113-1499.
48. Perevolochna was a town which enjoyed a special status as the major Dnieper crossing populated mostly by the Cossacks. In the mid-eighteenth century, the town consisted of a fort and additional barriers. It was incorporated into the New Russia province of the Russian Empire in 1764.
49. In the 1750s and the early 1760s, the empire leased border quarantine and custom complexes to large merchants or local landowners. Even though the complexes accommodated many travelers and merchants who paid large fees for their services, the leaseholders forwarded only a small fraction of their revenues to the state treasury. This led to the empire abolishing the practice of leasing customs and quarantine houses and the conversion of the complexes into a state property.
50. Zakrevskii, *Letopis i opisanie goroda Kieva*, I, pp. 89–90.
51. RGADA, 248-113-1499: 31.
52. RGADA, 248-113-430: 45.
53. Rossiiskii Gosudarstvennyi Istoricheskii Arkhiv (RGIA), 1341-1-1: 2–21.
54. RGIA, 1374-1-89: 2–31.
55. RGADA, 248-113-1605: 15.
56. Polnoe Sobranie Zakonov Rossiiskoi Imperii (PSZ), 13,662 (12/25/1771), 19: 315–16.
57. PSZ, 13.722 (12/24/1771), 19: 408; RGADA, 248-113-1518.
58. PSZ, 13.722 (12/24/1771), 19: 408.
59. PSZ, 13.674 (10/10/1771), 19: 327–28.
60. PSZ, 14.392 (11/07/1775), 20.
61. Kohut, *Russian Centralism*, pp. 209–18.
62. RGADA, 248-113-1518.
63. J. LeDonne, 'Geopolitics, Logistics, and Grain: Russia's Ambitions in the Black Sea Basin, 1737–1834,' *International History Review*, 28 (2006), 1–41.
64. J. Wasicki, *Konfederacja Targowicka i ostatni sejm Rzeczypospolitej z 1793 roku* (Poznań, 1952).
65. Tsentralnyi Derzhavnyi Istorychnyi Arkhiv u m. Kyevi, 59-1-9127 and 9131; E. Zagorovskii, *Organizatsiia upravleniia Novorossiei pri Potemkine: 1774–1791* (Odessa, 1913), pp. 1–33.
66. National Library of Russia in St Petersburg, Manuscript Division: 609-1-495.

7

Borders, War, and Nation-Building in Napoleon's Europe

Michael Rowe

The French Revolution, in its early stages, promised the creation of a Europe without borders. Yet borders within Europe were destined to endure. Indeed, they became more tangible in the nineteenth century in terms of their delineation, demarcation, and policing. This chapter analyzes the postponement of borderless Europe. Initially, the focus is on the decade after the outbreak of the Revolution in 1789, a period characterized in France by dramatic oscillations in many policy areas, including with respect to borders. The essay then examines the Napoleonic period proper (1799–1815), identifying in particular the policy contradictions inherited from the Revolution.

Conceiving of borderlands is problematic for the Napoleonic era. The borders of the French Empire, which reached its maximum size in 1812, were extensive and diverse. In the east, they stretched along northern Germany from Lübeck on the Baltic to the confluence of the rivers Rhine and Lippe, before then following along the Rhine to Switzerland. To the southeast, the French Empire included northwest Italy, bordering the Kingdom of Naples to the south, with the Apennines marking the border with the satellite Kingdom of Italy to the east. The Mediterranean coast and Pyrenees marked the southern border, and the Atlantic, Channel, and North Sea the border of the French Empire to the west and north. Some of these borders, notably the Pyrenees and the southern stretch of the Rhine separating France from the German southwest, were fairly well established. Others, such as the northeastern border with the satellite Kingdom of Westphalia, were only recent. The French state and its servants recognized the qualitative difference between borderlands such as the Pyrenees and Apennines, which were considered especially troublesome on account of their inaccessibility and perceived lack of experience of good governance, and those like the Rhineland, which straddled the border between the French Empire and states that increasingly resembled France itself.

144

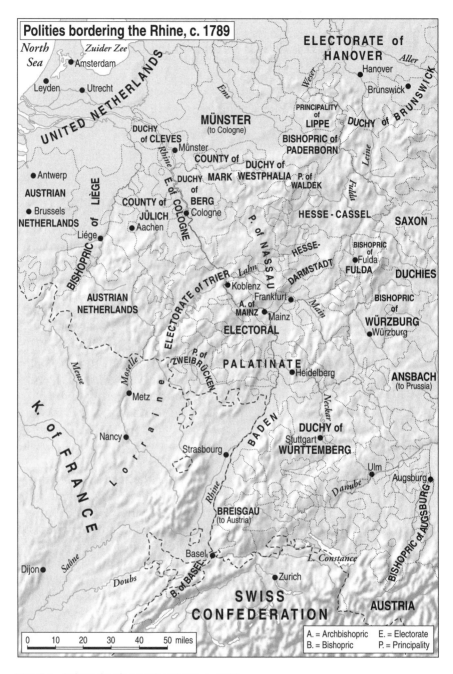

Polities bordering the Rhine, c. 1789

North Sea

Zuider Zee

Amsterdam

Leyden ● Utrecht

UNITED NETHERLANDS

Ems

ELECTORATE of
HANOVER *Aller*

Hanover ●

Brunswick ●

Weser

DUCHY of BRUNSWICK

MÜNSTER
(to Cologne)

PRINCIPALITY
of
LIPPE

BISHOPRIC of
PADERBORN

Leine

DUCHY
of CLEVES

Münster ●

COUNTY of
MARK

DUCHY
of WESTPHALIA

P. of
WALDEK

Fulda

● Antwerp

AUSTRIAN

● Brussels

NETHERLANDS

BISHOPRIC of LIÉGE

Liége ●

COUNTY of
JÜLICH

Aachen ●

DUCHY
of
BERG

Cologne ●

E. of COLOGNE

Rhine

HESSE-CASSEL

SAXON

BISHOPRIC
of
Fulda ●

FULDA

DUCHIES

AUSTRIAN
NETHERLANDS

Meuse

Moselle

ELECTORATE of TRIER

HESSE-

P. of NASSAU

Lahn

Koblenz ●

DARMSTADT

BISHOPRIC
of
WÜRZBURG

Würzburg ●

Frankfurt ●

A. of
MAINZ
Mainz ●

Main

ELECTORAL

P. of
ZWEIBRÜCKEN

PALATINATE

Metz ●

Heidelberg ●

ANSBACH
(to Prussia)

Nancy ●

Strasbourg ●

Neckar

BADEN

DUCHY of
Stuttgart ●
WÜRTTEMBERG

Ulm ●

Augsburg ●

K. of FRANCE

Lorraine

Rhine

BREISGAU
(to Austria)

Danube

BISHOPRIC of AUGSBURG

Dijon ●

Saône

Doubs

Basel ●

B. of BASEL

L. Constance

Zurich ●

SWISS
CONFEDERATION

AUSTRIA

0 10 20 30 40 50 miles

A. = Archbishopric   E. = Electorate
B. = Bishopric   P. = Principality

Map 7.1   Polities bordering the Rhine, c. 1789

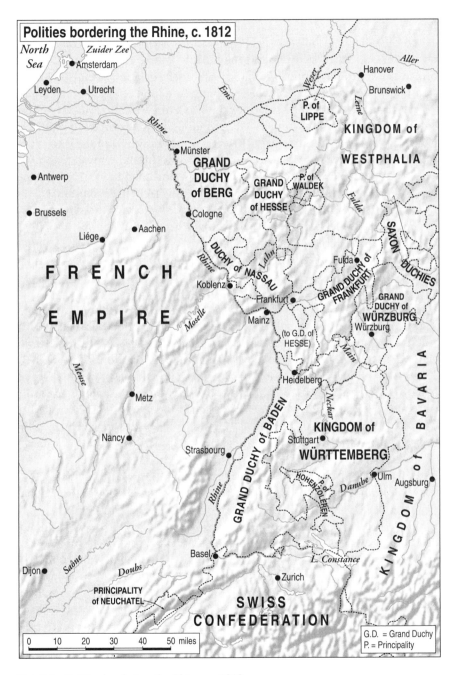

Map 7.2   Polities bordering the Rhine, c. 1812

1789: Toward a Borderless Europe?

The Napoleonic period followed on from the French Revolution of 1789, and this event profoundly changed the nature of Europe's borderlands. The Revolution and its progeny, the Revolutionary Wars starting in 1792, transformed the territorial landscape. The great variety of polities that had coexisted under the Old Regime, including composite monarchies, church states, and city states, experienced something akin to a 'mass extinction event' from which only the larger, territorially conterminous, states emerged. It would be misleading to assert that the French revolutionaries were uniquely to credit for this development, as the trend toward such territorial consolidation was evident throughout the eighteenth century. This included the various territorial exchange schemes hatched on the eve of the French Revolution by the Habsburg emperor, Joseph II. To self-consciously enlightened rulers like Joseph, the kind of territorial patchwork that distinguished Central Europe in particular, with its overlapping jurisdictions, enclaves, exclaves, and so forth, simply defied justification on grounds of public utility. Nonetheless, Old Regime balance-of-power politics blocked any substantial change of borders, especially within the Holy Roman Empire.[1]

In neighboring France, the century leading up to 1789 witnessed a degree of simplification of the border, notably with the elimination of both enclaves and 'foreign' jurisdictions in the eastern province of Alsace acquired during the reign of Louis XIV. The Bourbon government made efforts to demarcate more clearly the boundary line separating France from the Holy Roman Empire, thereby giving reality to the idea that the kingdom should be contained within 'natural' borders reinforced by Vauban-designed fortresses. Physical reality on the ground was bolstered theoretically by the concept of the sovereign state. It was reinforced further by Montesquieu's geographical determinism from the mid-eighteenth century, and then Rousseau's emphatic argument that Nature set limits to political nations.[2] These ideas reached Germany, but never took hold to the same extent as in France, not least because they were so at odds with the existing order.[3]

Initially, the French Revolution's impact was largely contained within France. In territorial terms, the major developments included the replacement of the old provinces with departments in 1790, and the realignment of the borders of other jurisdictions such as dioceses, court circuits, and military divisions so as to conform to this new administrative unit. Of greater significance for France's relations with the outside world was the elimination of the remaining 'foreign' enclaves located within 'French' territory, including Papal Avignon and Comtat Venaissin. This reordering, though at first sight comparable to Old Regime exchange schemes, was justified not primarily on utilitarian grounds, but on the revolutionary concept of the nation 'one and indivisible,' to use the phrase henceforth employed in French constitutional law. This concept of absolute and exclusive sovereignty, resting on the basis of national self-determination, was incompatible with the kind of multiple, overlapping and shared jurisdictions as had existed within the historic provinces of France, especially in border regions like Alsace where the existing order was additionally buttressed by international law.[4]

Before the Revolution, Alsace constituted but one province governed by the French Crown, but even more than the others, it was distinguished by difference in terms of its legal system, confessional politics, and language. For example, Protestants enjoyed a measure of toleration unique to France, as Alsace was exempted from Louis XIV's Revocation of the Edict of Nantes. And while urban elites spoke French, the mass of rural people spoke a German dialect that would have been more intelligible to the inhabitants of the wider Rhineland region that formed the western extremity of the neighboring Holy Roman Empire than to 'Frenchmen' west of the Vosges.[5] None of this had been especially problematic before 1789, but thereafter this difference became incompatible with at least one vision of how France should look. This was the vision of cultural and especially linguistic uniformity, as expressed most famously by the Abbé Grégoire.[6] As with territorial consolidation, so with linguistic uniformity the revolutionaries were not entirely original: rulers such as Joseph II had previously attempted to move in this direction on grounds of administrative efficiency. Where the revolutionaries differed was in their justification for uniformity. In Grégoire's formulation, linguistic diversity was incompatible with participatory democracy, which was predicated on the state having an unmediated connection with its citizens. Such a connection demanded a common language. Where this was absent, Grégoire warned, bicultural intermediaries, especially nobles and clergy hostile to the new order, interposed themselves between the state and its citizens, and thereby corrupted the body politic.[7] This argument did not apply solely to the borderlands, but to France as a whole, an entity where French was far from constituting the mother tongue of all. However, it was especially applicable to lands like Alsace, Flanders, and the Pyrenees where linguistic minorities straddled the border.

Grégoire's was not the only voice raised on the issue, which like so many others divided rather than united opinion. The early Revolution encompassed powerful forces that opposed the construction of a culturally uniform, politically centralized nation-state, hermetically sealed by well-policed borders. Not surprisingly, these voices were audible especially within the multilingual borderlands. In Alsace, for example, the provincial intelligentsia concentrated in Strasbourg had little difficulty in combining political allegiance to revolutionary France with a celebration of linguistic diversity. Indeed, diversity that allowed for the preservation of minority languages spanning the borders of France helped promote international cooperation. At least this was the argument posited in one of the numerous pamphlets produced on the subject in the first years of the Revolution.[8] These arguments met with much sympathy in Paris, at least up to the outbreak of the Revolutionary Wars in April 1792, when the new regime was characterized by cosmopolitanism and tolerance of cultural difference. This was a period of great optimism when it came to the belief that the Revolution's principles were universally applicable and would be welcomed abroad. This universalism combined with a strong element of localism, which included the idea that France should constitute a gigantic federation of self-governing communities in which central government's role would be minimal. Logically, there was no reason why this vision might not encompass a borderless Europe, a federation of 'municipalities' stretching from Paris to St Petersburg, to paraphrase the President of the Paris Commune, Pierre Gaspard Chaumette.[9]

Borderlands, as zones with distinct policing and military functions, would have no place in this new order, which could not accept any difference between the rights and duties of citizens irrespective of where they lived. This promised welcome relief to inhabitants of such zones, notably France's maritime 'borderland' facing the Mediterranean, Atlantic, and English Channel.[10] Under the Old Regime, communities living within a specified distance of the French coast were required to provide quotas of men to serve in the *milices garde-côtes*, or paramilitary coastguard formations. The role of these units was to guard the coast against hostile foreign incursions, a function they had performed during the numerous Anglo–French wars since the time of Louis XIV. One of the arguments employed to justify entrusting natives of the borderlands with this specific military role was that they, like any private property owner guarding his estate, had a special vested interest in ensuring the job was well done. This was the view from the center. Natives on the maritime borderlands, in contrast, detested the commitment. This is not surprising given the trend, in the final decades of the Old Regime, to strengthen the coastguard system by imposing an ever-greater degree of militarization on coastal communities, including especially severe restrictions on movement: men serving in the coastguards were not allowed to leave their communities for longer periods without authorization. The degree of local dislike of the institution is apparent from the presentation of grievances (or *cahiers de doléances*) on the eve of the French Revolution. The new revolutionary center responded sympathetically, as the notion that natives of borderlands alone should shoulder the burden of defending the territorial integrity of France flew against the new idea that all French citizens share equally in the task. In institutional terms the new thinking resulted in the abolition (through the decree of September 9, 1792) of the coastguard, whose functions were now inherited by the National Guard. This was, as its name suggests, recruited nationwide. And when war between France and Britain erupted the following year, great care was taken by the French authorities to ensure that the burden of defense was shared equitably, with military units from local districts being rotated in such a fashion as to ensure that all local communities had sufficient manpower available to keep the civil economy functioning. The principle of an equitable distribution of the military burden was subsequently refined in the new system of military conscription introduced in 1798, and perfected under Napoleon.

The outbreak of war in 1792 prevented the realization of Chaumette's vision of a borderless Europe. Conflict between revolutionary France and the rest of Europe reinforced 'chauvinism' rather than fraternity. The problem was that the peoples living beyond the border failed to welcome the French revolution-ary troops as liberators—this despite the offer made by the French Convention of November 19, 1792 promising that France would assist all people struggling to obtain their liberty.[11] Rejection was not immediately apparent, however. Indeed, as French forces crossed the border into the Austrian Netherlands and German-speaking Rhineland there were signs that local peasants and burghers might welcome the French who promised 'peace on the cottages, war on the *chateaux.*' It was on the basis of these hopeful early signs that Georges-Jacques Danton, who served briefly in occupied Belgium as a commissioner appointed by the Convention, made his famous parliamentary address of January 31, 1793 calling for the expansion of France to its 'natural borders.' This was to occur through the annexation of the occupied lands on the grounds that this was what their inhabitants wished: 'The borders of France are marked out by Nature. We will obtain them in all quarters, namely the ocean, the Rhine, the Alps, and the Pyrenees.'[12]

As Danton spoke, however, these early hopes were in the process of being dispelled by French military reversals. The French lost, in rapid succession, Frankfurt am Main (December 2, 1792), Aachen (March 2, 1793), and Mainz (July 23, 1793), which they had occupied the previous autumn.[13] What made these setbacks so traumatic in ideological terms was the irrefutable evidence that the inhabitants of these towns and surrounding regions offered active support to the anti-French coalition forces. The wider conclusion to be drawn from this was that the border between France and the outside world separated areas of profound ideological difference: according to the Revolutionary ver-sion, on the one side a France that had liberated itself, and on the other, the rest of Europe that had failed to free itself when given the opportunity to do so.

The French government, which at this point and up to July 27, 1794 was dominated by the extreme Jacobin faction, adopted in response a policy that might be labeled 'revolution in one country.' Robespierre had already fore-shadowed this approach with his observation, made in early 1792, that no one likes armed missionaries. The new policy aimed solely at the preservation of the Revolution within France, and rejected any notion of spreading the ideals of 1789 beyond French borders, which assumed the status of a near-permanent ideological dividing line. Among the most prominent victims of this shift in policy were foreign revolutionaries resident within France—many of whom, including their self-proclaimed leader Anacharsis Cloots, were guillotined during the Terror. These foreign radicals, who ironically had been more vocal than French voices in calling for France to obtain her 'natural borders,' were not the only ones to experience the altered political climate. So were the inhabitants of the new Republic's border regions, which were subject to specific legislation that reversed the earlier move to treat borderlands like the French interior. For example, the famous Law of the Maximum (September 11, 1793) forbad the establishment of granaries within 6 km of the border.[14]

Even more dramatic were the moves to break cultural ties that spanned the border. In Alsace, for example, the Jacobin regime embarked upon a policy of propagating French and eliminating German, a language now associated with royalism and foreign enemies. In practical terms, the new policy included a campaign against the wearing of regional dress and the use of Gothic typefaces and German on public signs, and the purging of German speakers from local Jacobin Clubs. A few from within the regime called for even more drastic measures, including the deportation of the German-speaking population into the French interior where it could do less harm, and its replacement with French-speaking colonists.[15] Jacobin suspicions over the loyalty of the inhabitants of France's borderlands were not unreasonable: coalition forces never managed to invade Alsace (at least not until 1814), so this region was never tested; however, many inhabitants of the French Pyrenees welcomed the Spanish army when it invaded in 1793, despite their earlier protestations about being loyal to France. This had less to do about cultural or linguistic affinity with those across the border, and more about dislike of compulsory military service recently introduced in France but not operative in Spain.[16] However, the Jacobin regime in Paris interpreted the causes differently, and concluded that the propagation of French at the expense of 'foreign' languages and dialects was essential for reasons of national defense.

The Jacobin regime fell before it could secure, to its satisfaction, France's borderlands through population transfers and elimination of linguistic minorities. Its schemes, though aborted, foreshadow the solutions characteristic of twentieth-century Europe. The French regime that followed, the Directory (1795–99), took a more pragmatic approach to the borderlands, just as it did with other issues. It drew up fairly impressive schemes for public education, and finally implemented the republican festive cycle and decimalized calendar throughout France. These were all measures designed to forge a new type of French citizen, including in the departments on the Republic's borders.[17] However, where precisely the borders should be located remained contentious. Republicans of the 'revolution in one country' stamp, like Lazare Carnot, opposed annexations, in part because people in these regions had proved themselves unworthy of being

French, and were unlikely to embrace the Revolution anytime soon. Others favored expansion either through direct annexation or else the establishment of 'sister republics.' Whatever their motivation (and for some, issues of patronage and personal gain were undoubtedly to the fore), their public rhetoric drew on the idea of 'natural borders,' albeit now shorn up by the need to ensure strategic defense rather than arguments about national self-determination favored in 1789. In essence, the policy was no different from that enunciated by the French foreign minister, Charles-François Dumouriez, on April 18, 1792, two days before the outbreak of war, namely that France could have no lasting security without the Rhine.[18] Dumouriez, it might be noted, was a native of Cambrai, the archdiocese of which included the Flemish-speaking borderland spanning France and the Austrian Netherlands. Later, under the Directory, bilingual (in French and German) natives of the Alsatian borderland, notably Jean-François Reubell, were among the prominent proponents of this expansionist vision. Reubell, who became the most influential individual in the formulation of French foreign policy following the coup of September 4, 1797, ensured that the entire left bank of the Rhine was effectively incorporated into France in 1798, rather than turned into a satellite state as others had planned. Reubell appointed another native of Alsace, François Joseph Rudler, as the senior French official in the new Rhenish territories with the responsibility of integrating them into the Republic.[19] At least in these terms the dream of Alsace acting as a bridge between France and Germany was coming true. More generally, natives of France's borderlands were not simply the passive victims of policies determined in Paris, but rather they played a disproportionately large role in shaping these policies.

Reubell's victory came at the expense of Lazare Hoche, one of the most promising of a new generation of generals whose meteoric rise in the French army was matched only by Napoleon Bonaparte. Hoche had dreamt of forging the occupied Rhenish territories into a 'Cisrhenan Republic,' rather along the lines of the Italian 'Cisalpine Republic' established by Bonaparte. The state-building activities along France's borders of these young military leaders (Hoche was born in 1768, Bonaparte in 1769) was the most obvious manifestation of the evolution of the French army from a force that was ostensibly an extension of the citizenry into a force that was essentially 'praetorianized.' By the late 1790s it owed its allegiance to its commanders rather than to the civilian politicians back in Paris.[20] This is relevant to the development of France's borderlands. Most obviously, the French army in the 1790s was concentrated in the borderlands, especially along the Rhine, where strong Habsburg resistance meant that the French did not make much progress in advancing east into Germany. The process of 'praetorianization' involved the French army perceiving itself as somehow uniquely embodying national virtues, a sentiment that bred a degree of contempt in civilians, whether they be French speakers in the core departments, the speakers of patois in border areas, or foreigners abroad. This indirectly undermined the Jacobin vision of a unified national community under siege, and with it the related campaign to impose linguistic uniformity. In any case, French commanders with their base of operations in borderlands were above all concerned about feeding their troops. This demanded pragmatic engagement with local elites, and the avoidance of ideologically driven 'Frenchification' that would have caused unnecessary offence. The sharp distinction drawn by the Jacobins, of a French Republic containing a free people separated from foreign lands whose inhabitants preferred slavery, broke down. If the world was to be divided between 'them' and 'us,' then the separation was between civilians and soldiers—many of the latter being non-native French speakers recruited from the borderlands where the army was concentrated.[21]

It was in this context that one must read the rise of Napoleon Bonaparte, who was to seize power in France in November 1799. He was, of course, a man of the borderlands through and through: to be born in Corsica in 1769 was to be born in an area that only a year previously had belonged to a different state (the Republic of Genoa), and whose official incorporation into the Bourbon monarchy marked the starting point of a long and bitter counter-insurgency campaign waged by the royal French army to master its new possession. As such one can realistically assume that Napoleon, from a tender age, knew more than most about the peculiarities of borderlands. He would add to this stock of knowledge as commander of French forces operating in north Italy in 1796–97, and in the Middle East in 1798–99.[22]

Upland Borders and the Extension of Napoleonic Civilization

Recent work in the field of 'Napoleon studies,' notably by Michael Broers, has drawn inspiration from Frederick Jackson Turner's 'frontier thesis' to emphasize the importance of the 'frontier' in shaping the Napoleonic Empire. Frontier, in this particular context, refers not to the border of the Napoleonic state in general, and especially not to those substantial segments that separated the French Empire from its smaller clone-like satellites across the Rhine in Germany. Rather, it refers to those stretches, including the Pyrenees and Apennines (respectively marking the borders with the kingdoms of Spain and Italy), that were perceived to separate 'civilization' from non-civilization. Service on this border, so goes the argument, shaped the collective mentality of those deployed to it in either a military or civilian capacity, and reinvigorated them with a missionary 'civilizing' zeal that had been lost under the Directory. This zeal, acquired on the periphery, was then to an extent reimported back to the center where it shaped the wider political culture of the French Empire, and indeed survived Napoleon's overthrow in 1815 (and possibly found an outlet in France's overseas empire after 1830).[23] The common civilizing mission furthermore helped 'amalgamate' the French elite, uniting the Old Regime nobility and the new elite of 1789 in a common purpose. The wilder peripheries of the Empire thereby stabilized France domestically.

Rousseau had argued before the Revolution that states should not attempt to go beyond the permanent borders, like the Pyrenees, that had been set by 'Nature.' However, such defeatism was not detectable among those determining the policy of Napoleonic France. Rather, the border dividing the civilized 'inner empire,' as Broers refers to it, from what lay beyond, was seen as a temporary condition to be overcome by effective institutions and willpower. The 'inner empire' in this context was in many respects synonymous with lowland Europe, and was characterized by higher population density and urbanization, greater affluence, sedentary populations with relatively high literacy rates, usually topped off with a Frenchified elite, and generally with a history of effective administration. What lay beyond was highland Europe, consisting of sparser, poorer but in sociological terms more homogenous populations, long-used to managing their own affairs without outside interference, whether it be by the early modern absolutist state or the post-Tridentine church.[24] A classic example of 'highland Europe' was the Tyrol, which distinguished itself by rising against Napoleon and his 'lowland' Bavarian clients in 1809.[25] Others included the least accessible mountainous areas of the Italian and Iberian peninsulas.[26] Napoleonic rule had less to offer highland Europe, but unlike its predecessors refused absolutely to adopt a policy of benign neglect, but instead acted with unprecedented vigor in asserting the state's monopoly of violence.[27]

The distinction between highlands and lowlands was reflected in the different Napoleonic institutions and personnel imposed on each: the former were subject to the mailed fist, including military government, states of siege, suspension of constitutional law, military tribunals, mobile columns, and firing squads. Typical of the personnel who ran this part of the empire was Jacques Norvins, Napoleon's police chief in the troublesome department of Rome which bordered the Kingdom of Naples, and whose formidable, uncompromising character is well projected in Ingres's portrait now hanging in London's National Gallery. The inner empire, in contrast, was treated to the benign methods and institutions that helped turn Napoleon into something of an icon for liberals and progressives in much of Restoration Europe. These methods included a civilian administration operating according to regular procedures, public courts with juries and with the right of appeal, equality before the law, and a network of consultative councils and chambers of commerce.[28]

Broers's reconception of the political geography of Napoleon's *Grande Empire* challenges older historiography focused on the high political nature of borders separating the various territorial entities that dominated Europe at this stage, namely the French Empire proper, the satellite states ruled by members of the Bonaparte clan, satellites under native dynasties, allied states, and enemies like Russia and Britain. Not only does this reconception make a great deal of sense, but it has also stimulated fruitful new research into the Napoleonic Empire. In thematic terms, it has refocused attention away from foreign relations, interstate war and high politics, and toward the sociology of the Empire. Also, it has shifted the emphasis from Napoleon himself, and his inner circle, toward the next tier down: the tens of thousands of 'notables' who ultimately made the Empire work on the ground. And while it might be fair to conclude that Napoleon himself lacked an overarching coherent idea that might have sustained the Empire ideologically, a fair proportion of the 'notables' who served in his administration were motivated in part by the belief that they were improving rather than just exploiting humanity. Ideology was not absent from the Napoleonic Empire, only from its apex, and Broers's work does this justice.

Nonetheless, there are elements of Broers's thesis that can be challenged. The most obvious is in his delineation of the 'inner' and 'outer' zones of the Empire. Broers conceives of these in very wide terms territorially, each covering entire regions of multiple departments. An alternative, more convincing model instead sees the French Empire divided into multiple inner and outer zones, often bisecting individual departments. Numerous borders existed throughout Napoleonic Europe, and not only wherever highlands met lowlands, but wherever factors—sociological, cultural, historical, as well as geographical—inhibited the ability of the Napoleonic state to operate effectively. A related criticism is that Broers's model privileges the most dramatic overt forms of resistance to the Napoleonic Empire, which were most characteristic of wild uplands, and neglects subtler varieties of nonviolent resistance, such as fraud and bureaucratic obstructionism from within the administration, that one might find elsewhere. An armed bandit, surviving in the hills and taking pot-shots at Napoleonic gendarmes might have captured the imagination of later romantic artists, but the corrupt mayor, medical officer, or customs official ultimately enjoyed a greater capacity to wreck the Empire from within.

However, it cannot be denied that Napoleonic officials who ran the Empire themselves obsessed about the division between highlands and lowlands. This was expressed most clearly in the statistical-topographical surveys that were commissioned by Jean-Antoine Chaptal who served as French Minister of the Interior for most of the Consulate (1799–1804).[29] The surveys, completed by the prefects in collaboration with local notables, covered the entire formal French Empire, which at this juncture included over one hundred departments. The findings were subsequently published as part of a series of volumes under the First Empire (1804–14/15). A brief reading of surveys conducted in departments located on France's formal international border reveals that prefects recognized not so much the existence of a specific 'borderland' type, but rather of distinct 'highlanders' and 'lowlanders.' The former, the surveys tended to agree, were distinguished by their superior physique fostered by the difficult terrain and, less tangibly, a more positive independent-minded outlook in part resulting from a lack of social stratification, something that also resulted in the absence of such vices as envy and love of luxury. However, they also possessed a tendency toward religious devotion and superstition. Lowlanders, in contrast, were portrayed as physically weaker, of smaller stature, with their characters (and finances) undermined by a variety of urban vices. At the same time, the surveys leave the impression that the distinction between these two worlds was weakening, not least through the spread of commerce whose flourishing state was lauded as a testament to the return of good governance under Napoleon.[30]

In this sense, the topographical surveys reflect the optimism that under-lay the Napoleonic civilizing mission: the belief that France, which under Napoleon combined the best of the Old Regime and the Revolution, might regenerate all of Europe, including such unpromising parts as the Illyrian Intendancies, Calabria, and Spain. Typically, 'regeneration' included the development of sedentary agriculture in place of transhumance, not least because the former was easier for the state to control and exploit. Acculturated notables from the Napoleonic borderlands, including Alsatians, Belgians, Piedmontese, and Rhinelanders, played a prominent role in this civilizing project, alongside 'Old Frenchmen' from the core departments. In this sense the border helped not only overcome the ideological divisions of the Revolution, but also facilitated the integration of the elites from the newly annexed departments.[31] The Napoleonic project was a European rather than a narrowly French phenomenon, involving Savoyards in central and southern Italy, Bavarians in the Tyrol, and Alsatians in western Germany. Collectively, they were engaged in Napoleon's 'Other War.'[32]

Napoleon's 'Fortress Europe'

This 'Other War' consisted of civilizing previously under-governed Europe, especially the borderlands of the south. The 'real war,' the run of conflicts against a series of coalitions of Europe's Great Powers, involved not ideology by the time Napoleon seized control of France in 1799, but rather represented a continuation of the great power rivalry that predated the French Revolution. It was concerned with the balance of power and the location of borders between relatively well-governed European states. In a few places, such as along the Pyrenees, formal interstate conflict on occasion intersected with the struggle to establish control of under-governed uplands described in the preceding section: the two forms of conflict were not always discrete, but did have lives of their own. This, for example, was the conclusion of a fact-finding mission, counting Lazare Carnot as one its members, sent by the newly instituted French Republic to the Pyrenees in early 1793 in anticipation of war with Spain. The mission reported back that conflict between mountain communities of 'primitive peoples' located on either side of the border over access to common pastures was endemic, that it had occurred since time immemorial, and irrespective of whether Madrid and Paris were allies or enemies.[33]

'Primitive' is of course a loaded term that jars against modern sensibilities. Peter Sahlins's research on the Franco–Spanish borderland, and especially the valley of Cerdanya at its eastern end, demonstrates that the inhabitants of the region of this relatively well-established border were far from primitive in exploiting their status as 'borderlanders' to derive maximum advantage in their dealings with Paris and Madrid. In so doing, they often employed the language of Spanish or French nationhood, as circumstances required, in order to convince the center that their particular interests were those of the state as a whole. Over time, this language took on a life of its own, with natives of the borderland internalizing a concept of the national interest that was also local, and of the local interest that was also national.[34] In this respect, nothing much would change with regards to the Pyrenees in the Napoleonic period, given the brevity of the episode as set against the long-term evolution of this borderland.

Elsewhere in Napoleonic Europe 'borderlands' straddled not impenetrable mountain ranges, but rather lowlands, river valleys, and larger expanses of water, including the English Channel. By the late eighteenth century they were fully integrated into an increasingly global economy.[35] As noted above, the early Revolution, in its optimism, had foreseen the dismantling of such borders as peoples beyond France embraced the principles of 1789. In some ways this would have meant an extension abroad of revolutionary policy within France itself, which had involved the dismantling of domestic barriers, including the infamous *Mur des fermiers généraux* that surrounded Paris.[36]

The Rhineland historically constituted a 'borderland' that divided France from the Holy Roman Empire, and French and German high cultures. Before its territorial reordering, the wider Rhineland region was subdivided into a multiplicity of small states, many of them ecclesiastical. It contained numerous small and often-overlapping jurisdictions. It constituted what one Alsatian termed a 'tangled and knotted land' ('pays mêlé et entrelacé').[37] From 1802, the entire left (or west) bank of the Rhine was formally annexed to France. Shortly thereafter, much of the right (or east) bank and its hinterland was bundled into the newly constituted Napoleonic satellite states: the Grand Duchy of Berg, created in 1806 and granted to Joachim Murat (Napoleon's brother-in-law), and the Kingdom of Westphalia, established in 1807 and awarded to Napoleon's youngest brother, Jérôme. Both of these new states were closely bound to France and adopted its administrative and judicial structures. The Rhine no longer marked the bounds of French power, institutions, or indeed culture, though it did form the international border.

Significantly, within this simplified order, many of the boundaries traced between newly created municipalities, districts, and departments into which

Berg, Westphalia, and the French left bank were subdivided, followed along the lines of the ancient boundaries between what had been theoretically independent states subject only to the Holy Roman Empire. As such, these 'borders,' though now internal rather than 'international,' retained a fairly strong presence in the consciousness of local people, especially if they marked also a confessional boundary, something that was often the case. Rivalries between distinct territories lumped together within the same department or district became apparent in the arguments that erupted over the location of administrative bodies, law courts, and institutions of higher education, and were often played out within the representative councils that ran alongside the local executive structures. It is clear from the evidence that such internal subdivisions enjoyed a quality far different from, say, the 'border' between two London boroughs. Despite such bickering, the greater Rhineland region proved eminently suited to expensive, intrusive Napoleonic-style government. The region was wealthy enough to support the costs, and also stood to benefit from key Napoleonic reforms, including the abolition of seigneurialism, the auctioning off of church lands, the introduction of modern property law, a more equitable taxation system, and a more rational administrative and legal structure.[38]

Given that Napoleon's power bounded the Rhine, it might have been supposed that the river would cease to function as a border, with the wider Rhineland shedding its borderland status to become instead the core of a new Carolingian Europe.[39] Such was not the case. The main reason was the ongoing global Anglo–French war, and more specifically the strategy Napoleon adopted to fight it. After the Battle of Trafalgar in 1805, this conflict took on an increasingly economic dimension, as both belligerents strove to wreck each other's commerce through blockade and counter-blockade. Key to the French strategy was the restriction of the importation of British merchandise into France. This was to be achieved through thorough policing of the formal international border of France, along which now ran a tariff barrier. France's eastern tariff border was briefly established on the River Maas in 1795, following the annexation of Belgium, and then, three years later, shifted to the Rhine. From a geographical perspective the Rhine was better than the Maas: its breadth made it easier to police and to restrict movement, including of English merchandise. 'Natural borders' thinking was now justified by pragmatic considerations about policing, and specifically the interception of contraband.[40]

The Rhine tariff provided one of the most visible manifestations of Napoleonic rule in the eastern border region, running from the North Sea to the Alps. The tariff barrier was administered by the customs agency, or Direction des Douanes. This agency, subordinate to the Ministry of Finance, numbered close to 30,000 personnel by the end of Napoleon's rule, and the majority of these were field agents, many mounted on horseback, responsible for patrolling France's international borders. The border region was marked by a double line of customs posts, located at points where people were authorized to cross subject to inspection. The first line was located on the international border, or rather, on the left bank of the Rhine, as technically the border between France and its German satellites ran along the middle of the river. The second line ran parallel to the first but several kilometers within France, with the area in-between constituting a strip of territory subject to more or less rigorous patrolling by the Direction des douanes who could, in addition, call upon the newly created Gendarmerie for support. Only a tiny proportion of the *douaniers* were recruited from the borderlands, while the vast majority of those whom they inconvenienced through their activities were locals. Not surprisingly, therefore, customs officials became an unpopular symbol of outside domination.[41]

Confronting this apparatus was a whole smuggling infrastructure that quickly sprung up, and that involved inhabitants from both sides of the border drawn from across the social spectrum: carriers, carters, and boaters, often including women and children as well as men, who transported the contraband across the border; tavern owners, whose premises served as storage facilities and operational command centers; local government officials who provided bogus documentation and sometimes tip-offs of raids by *douaniers*; and wealthy merchants and bankers who ultimately financed and organized everything from behind the scenes.[42] Given that, as noted above, Old Regime polities had often spanned the Rhine, smugglers could comfortably draw on connections and networks straddling the tariff barrier. However, it would be inaccurate to conclude that opinion within the Rhineland was uniformly hostile to the tariff. Chaptal was essentially correct in his assessment that any government tariff policy, be it complete laissez-faire or aggressive protectionism, would please some and offend others given the multiplicity of interests at stake.[43] Supporters of free trade were prominent in Paris. Voices within the Rhenish borderlands included those who supported a protective tariff. When important business interests located in Berg petitioned Napoleon in 1811 for the Grand Duchy's annexation to France so as to be included within the imperial tariff, they faced opposition not only from manufacturers within the French interior, but also from their competitors only a stone's throw across the Rhine, notably Cologne's Chamber of Commerce. These competitors too knew how to employ the language of French national interest when lobbying the center for their own advantage, marking perhaps the beginning of that process of the nationalization of the local that Sahlins describes in the Pyrenees.[44] This example makes it misleading to speak of regional solidarity, or of a straightforward conflict between center and periphery, and still less of a national struggle between French and Germans. Nor should one conclude that 'borderlanders' were victims whose interests were consistently ignored by central government. The reality was more complicated.

Napoleon's Continental Blockade, which was aimed at destroying Britain's economy and hence capacity to wage war, was a subcomponent of the larger 'Continental System.' Its objective was to establish French economic hegemony on the continent through the creation of what Geoffrey Ellis labels an 'uncommon market' that favored manufacturers based in France over those established abroad.[45] When it came to economics, Napoleon, wedded to crude mercantilism, thought in French rather than European terms.[46] The satellite states, whether German or Italian, according to his vision, would be reduced to economic colonies supplying raw materials and markets for French manufactures. The border along the Rhine was central to this purpose. Informed commentators argued that success required not merely the creation of well-policed barriers that restricted trade flows, but also that commercial relations within the French borderlands be reorientated toward the center of the Empire.[47] Here, nature often got in the way. The new French departments on the left bank of the Rhine constituted a viable whole, but one that was separated from the French interior by natural barriers such as the Ardennes and Vosges. Improved transport infrastructure offered the answer in the form of roads and canals linking the eastern borderlands with the rest of France. Some of these schemes were laid out in the law of December 23, 1809, which envisaged the construction of the Napoléon canal linking the Doubs, and by extension the Rhône, to the Rhine.[48] However, these schemes represented just one more instance of Napoleonic ambition outrunning the capacity of France. Few of them were completed before the collapse of the Empire, and in any case, the real game-changer in transport technology, namely railways, lay several decades in the future.

In respect of international trade, Napoleonic mercantilism demanded that France run a permanent surplus, thereby increasing its stock of capital. This meant discouraging the import of a whole range of manufactures and commodities, and also limiting some exports. This impacted upon Napoleonic France as a whole, but especially regional economies dependent upon cross-border trade. Similarly, the Napoleonic state sought to manipulate, for its own advantage, the flow of people across the border. Napoleon's Empire, like its rivals, regarded population as a measure of power, and hence sought to prevent emigration but encourage immigration. Foreign manufacturers and skilled craftsmen, preferably in possession of trade secrets, were in particular demand. Again, this policy had an impact on the Empire as a whole, but was of particular significance to borderlands where existing ties facilitated the resettlement of manufacturers across the border. Cologne's Chamber of Commerce, in its memorandum on the disadvantages to France of the incorporation of Berg into the Empire, struck the right chord by reminding Paris that it would be far more advantageous to attract businessmen based in Berg across the border to the newly annexed Rhenish departments. Indeed, it provided a degree of historical justification for such a move, informing the French government that this would simply be reclaiming talent 'lost' in the previous century, when manufacturers had moved to Berg from Cologne, a city then infamous for its intolerance of both religious minorities and innovative business practices.[49]

The obverse of encouraging immigration was discouraging emigration. Early French legislation on the subject bore the imprint of the emigration of the nobility in the 1790s: emigration at this stage was interpreted as counter-revolution. Under Napoleon, this ideological dimension evaporated, to be replaced by mercantilist fears that emigration represented a loss of France's stock of human capital. Again, the resulting legislation aimed at stemming the outflow was applicable to the French Empire as a whole, but its impact was greatest in the regions along the newly established borders, where the restrictions impeded well-established migratory flows. One might view this centrally imposed restriction on movement in terms of the modern state's preference for sedentary populations that are easier to police and exploit.[50] This was certainly a factor, with the Napoleonic state viewing the emigration of young men of military age—often apprentices learning their craft—as a particularly serious threat that needed to be countered. The authorities therefore required men of military age to carry documents, provided by local authorities, stating they had fulfilled their military obligations. Young men needed to produce these if they wished to travel beyond departmental borders. Travel across the international border required a similar document, but one approved by the prefect himself, thereby making tangible the new hierarchy of borders. Movement by larger groups, notably pilgrims seeking to visit a religious shrine across the border, was generally banned by the authorities. During times of war, as in 1813, the Napoleonic authorities effectively sealed the border, to prevent the infiltration of the enemy, stop people crossing to read proscribed anti-Napoleonic pamphlets and foreign newspapers, and above all to prevent draft dodging.[51] Through such restrictions, including the need for documentation and checks, the border between France and the world beyond became much more tangible for those living in its vicinity.

At the same time, natives of the borderlands were encouraged to look toward the interior. Napoleonic France worked hard to entice men from the peripheries of empire, including the borderlands, into state service. This was an essentially elitist policy, aimed at the so-called 'notables,' those in possession of a substantial independent income that allowed them to live comfortably and wield influence. This Napoleonic policy often encountered considerable opposition, especially in the newly annexed Rhenish departments, where elites preferred the German universities across the border. Here, the regime ultimately resorted to tactics indistinguishable from military conscription in forcing notables to send their sons to schools in the French interior rather than across the border. At the same time, towns and cities located within the borderlands on the French side used this as an argument in attempting to persuade the central government to spend more money on educational establishments in their regions: it would be in the national interest, as such investment would discourage 'French' youth from going abroad, and might in addition attract talent from across the border.[52]

The aim of the government's 'containment' policy was to forge a culturally homogeneous population within the French Empire. Revolutionaries like Grégoire favored this in the interests of participatory democracy. For Napoleon, by contrast, administrative efficiency weighed more heavily, though the wider 'civilizing' agenda cannot be dismissed entirely. The ultimate expression of the drive toward homogeneity came with what can best be termed Napoleonic population transfer schemes. In scope these were far removed from the extreme policy suggestions made by Jacobins in Alsace in 1793 and 1794, but the ultimate goal was the same. The Napoleonic scheme involved the establishment of military colonies in the German and Italian borderlands.[53] Inspired by ancient Greek and Roman examples, the colonies were to be populated by veterans of the French army originating from the interior. The government expected them to marry local women, something it encouraged through the provision of endowments. This, the government hoped, would bring closer the day when there would be, 'stretching from the Rhine to the Pyrenees, only one spirit, one language, and one sentiment,' to quote one official.[54] Just five relatively modest colonies were established in the end, however, and their existence, like that of the Napoleonic Empire as a whole, was of insufficient duration to reshape borderland identities. They were, nonetheless, a harbinger of things to come.

Napoleon's Legacy: A Europe of Borders

The land borders of Napoleon's Empire were redrawn following his defeat in 1815. In territorial terms France went back to its position in 1790, losing its gains in the Low Countries, Rhineland, and northeastern Italy. Borderlands, as areas with distinct political cultures, require time to develop and this Napoleon's Empire failed to afford them. What this period did witness, however, was a step-change in the development of institutions to police borders. Most obviously, this included a vastly expanded customs agency to control the flow of goods across the border. It also encompassed a system of passports, policing, and nationality law as mechanisms to regulate the movement of people. Ironically, this impressive set of Napoleonic institutions was applied to borders that stabilized and endured only as a consequence of the destruction of the Napoleonic Empire.

It was in the century after 1814 that the combination of strengthened institutions and territorial stability allowed for the emergence of especially distinct borderlands around France's periphery. On the other side, the newly forged

Italian and German states that emerged from the wreckage of the Empire cooperated with post-Napoleonic France in tracing their mutual borders with greater precision. This exercise, and especially the erection of markers indicating the boundaries of the state, reflected the determination of these polities to assert and display their sovereignty. Within their delineated territories Europe's states proceeded to 'nation-build' in a project that distinguished the remainder of the nineteenth century.

Notes

1. For Joseph II's attempts to realign diocesan boundaries so as to conform to the borders of the Habsburg Monarchy, see D. Beales, *Joseph II. Vol. 2: Against the World, 1780–1790* (Cambridge, 2009), pp. 75–76, 277–280, 288–289.
2. P. Sahlins, 'Natural Frontiers Revisited; France's Boundaries since the Seventeenth Century,' *American Historical Review*, 95 (1990), 1423–51, at 1434–6.
3. H. Boldt, '"Souveränität": 19. und 20. Jahrhundert,' in O. Brunner, W. Conze, and R. Koselleck (eds), *Geschichtliche Grundbegriffe. Historisches Lexikon zur politisch-sozialen Sprache in Deutschland* (8 vols, Stuttgart, 1972–97), I, pp. 129–30.
4. T.C.W. Blanning, *The Origins of the French Revolutionary Wars* (London, 1986), p. 75; for an older, but fuller account of the Alsatian issue see T. Ludwig, *Die deutschen Reichsstände in Elsaß und der Ausbruch der Revolutionskriege* (Strasbourg, 1898).
5. D.A. Bell, 'Nation-Building and Cultural Particularism in Eighteenth-Century France: The Case of Alsace,' *Eighteenth-Century Studies*, 21 (1988), 472–90.
6. A.G. Sepinwall, *The Abbé Grégoire and the French Revolution: The Making of Modern Universalism* (Berkeley, 2005), pp. 90–7.
7. Grégoire subsequently presented his thinking on the subject to the National Convention on June 4, 1794. See *Archives parlementaires de 1787 à 1860: recueil complet des débats législatifs & politiques des chambres françaises imprimé par ordre du sénat et de la chambre des députés sous la direction de J. Mavidal … et de E. Laurent* (Paris, 1862–), 1st series (1787–99), vol. 91, pp. 318–27.
8. This pamphlet was written in March 1790 by Wilhelm Christian Koch, a prominent member of the Société des amis de la constitution de strasbourg. It was subsequently presented as an official address to the French National Assembly: Bell, 'Nation-Building,' 483–5.
9. C.I. Keitner, *The Paradoxes of Nationalism: The French Revolution and Its Meaning for Contemporary Nation Building* (Albany, NY, 2007), pp. 103–4.
10. The following paragraph is based upon R. Morieux, *Une mer pour deux royaumes: la manche, frontier franco–anglaise (xviie–xviiie siècles)* (Rennes, 2008), pp. 133–48.
11. For the original text of the declaration of Nov. 19, 1792, see *Archives parlementaires*, 1st series, 53, pp. 472–4.
12. 'Les limites de la France sont marquées par la nature. Nous les atteindrons dans leurs quatre points, à l'océan, au Rhin, aux Alpes, aux Pyrénées,' *Moniteur*, Feb. 1, 1793, XV, pp. 322ff.
13. J. Hansen, *Quellen zur Geschichte des Rheinlandes im Zeitalter der Französischen Revolution 1780–1801* (4 vols, Bonn, 1931–38), II, pp. 621 (Frankfurt), 773–7 (Aachen), and 884–8 (Mainz).
14. Morieux, *Une mer*, p. 177.
15. Bell, 'Nation-Building,' 486–7.

16. P. Sahlins, *Boundaries: The Making of France and Spain in the Pyrenees* (Berkeley, 1989), pp. 176–80.
17. Under the Old Regime, holidays varied from diocese to diocese, and sometimes within dioceses. See N. Shusterman, *Religion and the Politics of Time: Holidays in France from Louis XIV through Napoleon* (Washington DC, 2010).
18. Apart from Sahlins, 'Natural Frontiers,' see S.S. Biro, *The German Policy of Revolutionary France: A Study in French Diplomacy during the War of the First Coalition 1792–1797* (2 vols, Cambridge, MA, 1957), I, p. 25.
19. Reubell was a native of Colmar, and Rudler of Gebweiler: Hansen, *Quellen*, p. 3* n. 2.
20. J.A. Lynn, 'Toward an Army of Honor: The Moral Evolution of the French Army, 1789–1815,' *French Historical Studies*, 16 (1989), 152–73.
21. France's eastern borderlands provided a disproportionately high number of recruits to the French army even before the outbreak of the French Revolution: S.F. Scott, *The Response of the Royal Army to the French Revolution: The Role and Development of the Line Army, 1787–93* (Oxford, 1978), pp. 12–13.
22. P. Dwyer, *Napoleon: The Path to Power, 1769–1799* (London, 2007), esp. chs 1–4.
23. For Broers's most important contributions in this area see, in reverse chronological order, M. Broers, *Napoleon's Other War: Bandits, Rebels and Their Pursuers in the Age of Revolutions* (Oxford, 2010); *The Napoleonic Empire in Italy, 1796–1814: Cultural Imperialism in a European Context?* (Basingstoke, 2005); *Europe under Napoleon 1799–1815* (London and New York, 1996). For the Napoleonic civilizing mission more generally, see especially the work of Stuart Woolf, including 'French Civilization and Ethnicity in the Napoleonic Empire,' *Past & Present*, 124 (1989), 96–120, and also his *Napoleon's Integration of Europe* (London and New York, 1991).
24. M. Broers, 'Revolution as Vendetta: Patriotism in Piedmont, 1794–1821,' *Historical Journal*, 33 (1990), 573–96; Broers, 'The Police and the *Padroni*: Italian *Notabili*, French Gendarmes and the Origins of the Centralized State in Napoleonic Italy,' *European History Quarterly*, 26 (1996), 331–53. The upland/lowland divide in Napoleonic Europe in this context can usefully be compared to that existing in Asia, as analyzed by James C. Scott, *The Art of Not Being Governed: An Anarchist History of Upland Southeast Asia* (New Haven, 2009).
25. A flurry of scholarly and popular publications and exhibition catalogues has greeted the bicentenary of the Tyrolean Uprising, many focusing on that quintessential *homines alpine*, Andreas Hofer. Amongst the most informative is A. Oberhofer, *Der Andere Hofer. Der Mensch hinter dem Mythos* (Innsbruck, 2009). Older accounts include F.G. Eyck, *Loyal Rebels: Andreas Hofer and the Tyrolean Uprising of 1809* (Lanham and London, 1986); J. Sévillia, *Le chouan du Tyrol: Andreas Hofer contre Napoléon* (Paris, 1991); L. Cole, 'Nation, Anti-Enlightenment, and Religious Revival in Austria: Tyrol in the Late 1790s,' *Historical Journal*, 43 (2000), 475–98.
26. For Italy, apart from Broers, see J.A. Davis, *Naples and Napoleon: Southern Italy and the European Revolutions, 1780–1860* (Oxford, 2006); and on Spain, C.J. Esdaile, *Fighting Napoleon: Guerrillas, Bandits and Adventurers in Spain, 1808–1814* (London, 2004).
27. See esp. Broers, *Napoleon's Other War*. Also, on Napoleonic policing, see H.G. Brown, 'From Organic Society to Security State: The War on Brigandage in France, 1797–1802,' *Journal of Modern History*, 69 (1997), 661–95.
28. For an analysis of these institutions in action, see M. Rowe, *From Reich to State: The Rhineland in the Revolutionary Age, 1780–1830* (Cambridge, 2003), pp. 87–115.
29. *Recueil des lettres circulaires, instructions, arrêtés et discours publics, émanés des C.^ens Quinette, Laplace, Lucien Bonaparte et Chaptal, ministres de l'intérieur, depuis le 16 Messidor an 7 jusqu'au Ier Vendémiaire an 10*, Vol. 3 (Paris, 1802). Chaptal, in his

instruction to prefects who were responsible for completing the surveys in their departments, directed that the data be arranged under the following chapter headings: 1. Topography; 2. Population; 3. History and Administration [under which heading was included sociological and anthropological observations]; 4. Agriculture; 5. Industry.

30. *Statistique générale de la france, publiée par ordre de Sa Majesté L'Empereur et Roi, sur les mémoires addresses au ministre de l'intérieur, par MM. les préfets. Département de l'ain. m. bossi, préfet* (Paris, 1808), pp. 290–325; *Mémoire statistique du department du doubs, adressé au minister de l'intérieur, d'après ses instructions. Par M. Jean Debry, préfet de ce département. publié par ordre du gouvernement* (Paris, an xii), pp. 67–8; *Statistique générale de la france, publiée par ordre de Sa Majesté L'Empereur et Roi, sur les mémoires addresses au minister de l'intérieur, par MM. les préfets. Département du mont-blanc. M. de Verneilh, ex-préfet de la corrèze et du mont-blanc* (Paris, 1807), pp. 276–9, 285–6.
31. Woolf, *Napoleon's Integration of Europe*, pp. 76–81.
32. To cite the title used by Broers for his recent book on the subject: Broers, *Napoleon's Other War*.
33. Convention Nationale, *Rapport fait à la convention nationale par ses commissaires Carnot, Garrau et Lamarque. Envoyés par elle aux frontiers des pyrénées; présenté à la convention le 12 Janvier 1793, l'an second de la République* (Paris, 1793), pp. 2–3.
34. Sahlins, *Boundaries*, esp. pp. 7–9, 37, 110–13, 155–6, 164–5, 276.
35. For the English Channel as a border separating Britain from France, see Morieux, *Une mer*. See also P. Readman, '"The Cliffs Are not Cliffs": The Cliffs of Dover and National Identities in Britain, c.1750–c.1950,' *History* (forthcoming).
36. The *Mur des fermiers généraux* was a wall constructed around Paris in the 1780s for the sole purpose of levying duties on merchandise entering the city. The revolutionary rationale behind the elimination of domestic barriers to movement and trade can be seen in *Rapport fait a l'Assemblée Nationale, au nom du Comité du Commerce et d'Agriculture, sur la suppression des droits de traits percuss dans l'intérieur du Royaume, le reculement des douanes aux frontiers, & l'établissement d'un tarif uniforme* (Paris, 1790).
37. *Lettre a l'auteur des considérations sur les droits particuliers et le véritable intérêt de la province d'Alsace, dans la présente situation politique de la France. Par un citoyen d'Alsace* (Strasbourg, 1789), pp. 2–3.
38. Rowe, *Reich to State*.
39. For the deconstruction of the Franco–German border since World War II, see Michael Loriaux, *European Union and the Deconstruction of the Rhineland Frontier* (Cambridge, 2008), pp. 257–99.
40. Napoleon's Continental Blockade is neglected in the more recent historiography. The best account in English, despite its Alsatian focus, remains G. Ellis, *Napoleon's Continental Blockade: The Case of Alsace* (Oxford, 1981). For the English Channel, see Morieux, *Une mer*, pp. 241–73.
41. J. Bordas and F. Gambini, 'Douanes,' in J. Tulard (ed.), *Dictionnaire Napoléon* (2 vols, 2nd ed., Paris, 1999), I, pp. 659–65.
42. R. Dufraisse, 'La Contrebande dans les départements réunis de la rive gauche du rhin à l'Époque Napoléonienne,' *Francia*, 1 (1973), 508–36. See also M. Rowe, 'Economic Warfare, Organised Crime, and the Collapse of Napoleon's Empire,' in K. Aaslestad and J. Joor (eds), *The Napoleonic Continental System: Local, European, and Global Experiences and Consequences* (forthcoming).
43. J.-A. Chaptal, *De l'industrie française* (Paris, 1819), p. 412.
44. C. Schmidt, *Le Grand-Duché de Berg (1806–1813): Étude sur la domination française en Allemagne sous Napoléon Ier* (Paris, 1905), pp. 492–7.

45. Ellis, *Napoleon's Continental Blockade*, p. vii.
46. An alternative, more 'European' strategy was proposed by Coquebert de Montbret, whilst serving as director of the imperial statistical office: he proposed the creation of a single market including the French Empire proper and the satellite states. See Morieux, *Une mer*, pp. 178–9.
47. J.J. Eichhoff, *Mémoire sur les quatre départements réunis de la Rive Gauche du Rhin, sur le commerce et les douanes de ce fleuve* (Paris, Year X [1801–02]), pp. 35–9.
48. M. Merger, 'Canaux,' in Tulard, *Dictionnaire*, pp. 380–1.
49. Schmidt, *Grand-Duché de Berg*, pp. 492–7.
50. Again, one is struck by the parallels with the attitude of the state in Asia, including the British Empire, to uplands, as described in Scott, *The Art of Not Being Governed*, pp. 39, 59–61, 98–105.
51. *Recueil des actes de la prefecture du department de la roer* (11 vols, Aachen, 1802/03, 1813), XI, pp. 257–9.
52. For the Franco–German borderlands, see Rowe, *Reich to State*, pp. 132–7; for Italy, see Broers, *Napoleonic Empire in Italy*, pp. 265–72.
53. Napoleon I, *Correspondance de Napoléon Ier Publiée par ordre de l'Empereur Napoléon III* (32 vols, Paris, 1858–69), VIII, pp. 40–1.
54. Rowe, *Reich to State*, pp. 130–1.

Part IV
National Identities and
European Borderlands

8

Living a British Borderland: Northumberland and the Scottish Borders in the Long Nineteenth Century

Paul Readman

On August 26, 1850, Queen Victoria visited the town of Berwick to open Robert Stephenson's Royal Border Bridge, a grand 2160-foot viaduct spanning the river Tweed. The Queen received a loyal address from the Lord Mayor before an honor guard and audience of several thousand people. To add dignity to the event, a triumphal arch had been erected over the railway line near where the presentation of the address took place. On the arch, picked out in gold lettering, were the words 'The last act of the Union.'[1] In so referencing the 1707 Act of Union—which had joined England and Scotland together in one state—the new bridge was represented as reinforcing and embodying the ties between the two countries. Strictly speaking, of course, Berwick was not in Scotland: the Royal Border Bridge did not bridge the formal border between the two nations. Yet for all that, it was not misnamed. In crossing the Tweed, it crossed what was widely believed to be the natural boundary line on the eastern seaboard, before giving way to the Cheviot Hills further west. Traveling north as a young man towards the end of the nineteenth century, the English writer A.G. Bradley thought he 'was entering Scotland' as his train rumbled over the Border Bridge into Berwick, the long-standing impression that the town was only uncertainly English proving ineradicable.[2] This was true of other parts of the nineteenth-century Anglo–Scottish Borderland. Writing about the bleak moorland around Kielder in north Northumberland in 1835, William Chatto found that 'the precise boundaries of each kingdom [were] rather "*ill to red*",' a judgment with which Bradley would concur three-quarters of a century later when he visited the same district.[3]

   The surprise experienced by these observers illustrates that borderlands are as much sites of ambiguity and interblending as demarcation and separateness. The lines on the map often bear uncertain relation to the lived reality of the borderlands they define. This is true even where the cartographer's pencil follows supposedly natural lines of division, such as rivers and mountain ranges. It is perhaps especially pronounced where the borderland is not that of two

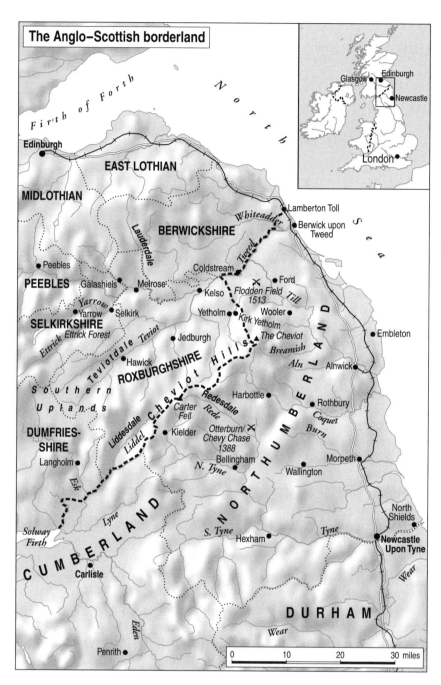

Map 8.1 The Anglo–Scottish borderland

sovereign nation-states, as in the Anglo-Scottish case after the 1707 Act of Union. In his *Visits to Remarkable Places* (1840), William Howitt recorded his impressions of Flodden Field near Branxton in Northumberland, site of a battle in 1513 which had seen the devastating defeat of the Scots and the death of their king. To Howitt's surprise, the once blood-soaked moorland now offered vistas of 'plentiful corn-fields and comfortable farms,' a transformation that he attributed to 'the signal effects of the Union.' So great, indeed, had been the transformation, Howitt concluded, that the two countries had 'blended' into each other in peace and prosperity.[4]

Yet, however felicitously interblended the Anglo–Scottish Borderland might have seemed to some contemporaries by the mid-nineteenth century, its land-scape remained a powerful memorial to centuries of conflict. As one writer put it in 1861, until the union of the two kingdoms, it 'had been the con-stant scene of invasion, reprisal, battle, fire and plunder ... Castles were fired, monasteries plundered, villages razed, and crops destroyed, with a vigor that generation transmitted to generation unimpaired.' This had resulted in a dis-tinctively martial style of architecture: 'The many-gabled, picturesque timber-houses of the south found no place here.'[5] There were castles of nobles such as the magnificent Alnwick, seat of the dukes of Northumberland; the pele towers of gentry and clergy; the bastel-houses of smaller proprietors, with their doors several feet off the ground to deter marauders. There were even fortified barns for cattle. The turbulent past was ubiquitous, written into the environment it had shaped.

Largely thanks to the writings of Sir Walter Scott, especially the *Border Ballads* and his epic poem *Marmion*, set at the time of Flodden, this legacy of violence was increasingly consumed as romantic heritage. As with Constable and East Anglia, or Shakespeare and Stratford-Upon-Avon, Scott altered the geographical imaginary.[6] He helped create a cultural 'Border country' straddling the national divide. More could be said about this Scott-inspired representation of the Borderland. Indeed, the recovery and analysis of literary, touristic, and heritage landscapes generally has preoccupied much excellent and often interdisciplinary scholarship in recent years.[7] Yet almost always missing from such research is consideration of the social experience of the inhabitants of the locales so valorized. In the case of borderlands we often know more about the meanings attached to them from external perspectives than we do about the realities of life in such environments. More particularly, we know relatively little about the ways in which social experience in border-land regions was shaped by the physical reality of their being borderlands. In part, this is a reflection of academic fashion, the focus on language, dis-course, and the recovery of meaning now being in the ascendancy, arguably to the detriment of historians' traditional preoccupation with causation.[8] But whatever the reasons may be, the post-1707 Anglo–Scottish Borderland is no exception to the rule. Indeed, its status as an intrastate borderland might support an assumption that little of historical significance can be said on the subject, certainly in comparison to what can be said about the social history of the Borders in the medieval and early modern periods, before the coming of peace and Union, on which considerable work has been done.[9] Any such assumption is mistaken. Focusing on north Northumberland and the adjoining Scottish border counties, this essay seeks to demonstrate that British borderlands continued to matter after 1707, and not simply in terms of heritage or representations. Intrastate the Anglo–Scottish border may have been, but it had a real presence for those who lived in its hinterlands, and one which acted to shape social, regional, and national identities.

Borderland Society and Identities

The violent past cast a long shadow over conditions of life in the Borderland. Into the nineteenth century, commentators asserted that the moorland wilderness and its remoteness from civilization and state power had combined with age-old national and clan animosities to encourage a pervasive lawlessness, traces of which were still found amongst the people of the region.[10] No doubt such claims were colored by prejudice, but they nonetheless reflected a generally accepted truth: border traits nurtured by centuries of conflict died hard. Martial prowess, for example, still appears to have been celebrated by Borderlanders. Mid-Victorian accounts testified to the prizing of 'Manly strength ... among the Northumbrian shepherd families at the present day.' As one related,

> In the district between the Cheviot Hills and the head of the [river] Coquet, a young man was, not a great many years since, courting a lass named Hedley, whom he wished to marry. 'Let him in among us,' said the mother when the proposal came to be deliberated; 'he's a grand fighter.'[11]

Rowdiness at fairs and football matches was common. At Harbottle Fair there would often be fighting between Redesdale and Coquetdale men. As 'a Redesdale man was heard disconsolately to remark' at one fair in the 1840s, 'Aw nivor seed sic a fair I' maa life, past 'leven o'clock o' the forenoon an' nivor a broken heid.'[12]

A pugilistic spirit was just one aspect of the legacy bequeathed to the Borderland by its history of conflict. For centuries the Anglo–Scottish border had defined a zone liminal in behavioral as well as geopolitical terms, where the rule of law was routinely flouted by a population for whom raiding and feuding had been a way of life. Vestiges of this transgressive lifestyle persisted into the nineteenth century, taking the form of robust disdain for authority. This was a source of regret for those who saw it as their business to promote habits of right living. In his submission to the *New Statistical Account of Scotland* in 1839, the Minister of Hawick lamented some aspects of 'the border spirit' still extant among inhabitants of his parish. While admitting that his flock benefited from much inherited energy, he found their independence shaded into social insubordination. 'Anything like a spirit of vassalage to any man, or to any class of men, how elevated soever in rank, is what they cannot brook … There are … few places where less attention is paid to the ordinary distinctions of rank.'[13]

If the prevalence of poaching was anything to go by, the minister had a point: many ordinary people—women as well as men—were enthusiastic filchers of the abundant salmon preserved by their social superiors, and indeed poaching of the fish was still widespread in some areas as late as the 1920s.[14] The persistence of the practice was routinely described as a residue of past lawlessness. It certainly had deep roots in many places, being carried out in an organized fashion with the connivance of the local community (even gamekeepers sometimes turning a blind eye), and in accordance with inherited traditions of honor.[15] Writing in 1898, Edmund Bogg reported in his travelogue account of 'wandering in the Border Country' that 'There seems to be a kind of freemasonry amongst the men who indulge in this illicit sport, and they will suffer imprisonment rather than inform of each other.'[16]

Another legacy of past years of strife was a sizeable gypsy settlement at Kirk Yetholm, an isolated village just across the border on the Scottish side. Numbering about one hundred people in the 1830s, this community had its own system of social organization (including a leader, sometimes described as a 'king' or 'queen'). Its existence was a consequence of the limited reach of settled authority on the Borders before the Union; its persistence into the twentieth century was a function of the region's continued liminality. Although increasingly integrated with the settled population, the gypsies' refusal to conform to orthodox modes of behavior prompted censure from predictable quarters (such as the manse of Yetholm parish),[17] as well as a probably unjustified reputation for roguery. Even in the 1860s, an otherwise comparatively positive account could breezily affirm that the gypsies 'cannot now be accused of more theft or bloodshed than the generality of illiterate British subjects,' which given the rapidly declining rates of illiteracy by this time was scarcely much of a compliment.[18]

Differences in administrative arrangements between England and Scotland also had an impact. Notwithstanding the Union, separate legal, educational, and religious systems remained in place in each of the two countries. This had particular significance in the Borderland. One notable administrative peculiarity was the variation in duty paid on alcoholic spirits north and south, a variation which persisted until 1856, when the duties were finally harmonized. In 1823 the difference was more than 9 shillings per gallon.[19] This gave a great incentive to the smuggling of whisky from Scotland into England, where the duty was higher. Women and men were involved, the concealment of specially fashioned body-fitting tin cases under women's clothes being among the stratagems employed to thwart excise-men. In Yarrow on the Scottish side, illicit stills were set up, apparently with the connivance of local farmers, and in Yetholm perhaps one-fifth of the parish's population was associated with smuggling.[20] James Burn, an itinerant peddler, recalled in his memoirs the ubiquity of smuggling—in which he himself partook—all along the early nineteenth-century border: 'I could fill almost a whole volume with the numerous smuggling incidents I have witnessed.' And as Burn attested, smuggling—like poaching—enjoyed the active support of local populations: 'nothing could please the people better than to see an excise-officer outwitted.'[21]

Another administrative difference with significant implications for Borderland life related to marriage. Notoriously, a couple could tie the knot much more easily in Scotland than in England, simply needing to make a declaration in the presence of witnesses. Facilitated by road improvement from the late eighteenth century onward, irregular marriage venues proliferated along the border, not just at Gretna Green near Carlisle in the west, where between 300 and 400 ceremonies were being held annually in the 1830s,[22] but also at places such as Coldstream Bridge and Lamberton Toll to the east. Over 2500 couples were married by one man alone—Henry Collins—at Lamberton between 1833 and 1849, one local jeweler recalling how on market days and holidays he would sell between 12 and 18 rings every morning for use at the toll house.[23] Of the couples married by Collins one-third were Scots (almost all from Berwickshire and East Lothian), over one-quarter from Berwick-Upon-Tweed, and the rest from Northumberland.[24] One reason for irregular marriage was simple convenience, but religious pluralism and relatively high levels of secularization also played a part.[25] The many Roman Catholics, Anglicans, and Presbyterians who lived in the Borderland found irregular marriage a solution to the problems otherwise presented by interdenominational unions. Moreover, such unions were socially more acceptable here than elsewhere owing to the limited extent of regular engagement with organized religion.[26] For a population that was relatively more pluralist, yet also more secular, than other parts of Britain, irregular marriage provided a cheap and uncomplicated means of forming conjugal partnerships.

But if the Borderland was a place where inherited traditions of transgression persisted, it was progressively less transgressive as time passed. Poaching slowly declined. Irregular marriage was first limited by legislation in 1856, and then gradually went out of fashion before its final legal abolition in 1940. Smuggling was subject to a massive repressive effort on the part of the authorities in the 1820s and 1830s. In 1834 the Commissioners of Inquiry into the Excise Establishment concluded that cross-border smuggling no longer prevailed 'to any considerable extent' owing to the 'chain of officers' now employed to police the border, a judgment with which other observers concurred.[27] Broken heads were less common at fairs, and the legacy of border marauding was transmogrified by borderers themselves from a source of concern to a felicitous inheritance. As early as 1848, one ex-farm worker reckoned many of the virtues of Berwickshire people derived from their border past. Unlike the men of East Lothian, who did not live on the border, 'the men of Berwickshire still partake of the habits and character of their free-booting forefathers, so far as vivacity and energy of action are concerned.'[28] The asperity of the physical environment of the Borderland was now seen to have produced a hardy class of people, whose attributes—which in the past might have tended to roguery—could now be directed towards wholesome forms of behavior. This wholesomeness was well caught by the Northumbrian Howard Pease, who in 1899 personified the borderer of his county as 'a bare-headed gipsy lass, freckled with sun and wind, who "fends" for her living with strategies of hand and head,' an individual who stood in marked contrast to 'the well-dowered matron' of southern England, secure in her more placid pastoral comforts. The stern virtues of the former, Pease explained, were a product of her historic environment. 'Still,' he went on,

> in the northern blood, the heritage of the 'raid' and the 'fray' abides, and still, as of old, are the children of the Borderland nursed by the keen wind of the moorland and the sea. 'Hard and heather-bred' ran the ancient North-Tyne slogan; 'hard and heather-bred—yet—yet—yet.'[29]

This reconceptualization of the character of the borderer by borderers themselves was not solely focused on the English side. As in the tourist imaginary, border identities were predicated on the idea of the borderland as a discrete entity that overlapped national divisions. This was a function of social commonalities and interconnections north and south of the boundary line, the borderland being a physical reality as well as a cultural construct. While the years of cross-border conflict had been bound up with national animosities, they had also helped to give a separate and ultimately transnational identity to the region as a zone apart from more tranquil parts of Britain. In rural areas, their immediate legacy was an underdeveloped agricultural economy based on rough pasture, the lack of security engendered by all the warring having been a major disincentive to investment and improvement.[30] Yet, peace and union had wrought rapid changes, in many areas effecting a transformation from what the rural writer Richard Heath called 'almost a state of nature' to a relatively advanced agricultural landscape.[31] Enclosure and the spread of large capitalist farms had resulted in increased productivity, if at the cost of rapid depopulation. Those that remained employed on the land, however, enjoyed relatively good wages and diets, the vigorous lifestyles of Cheviot shepherds attracting particular approbation in contemporary accounts.[32]

In one notable respect, however, living standards in the Borders lagged behind: housing. Cottages in early nineteenth-century Yarrow were remembered as 'little better than dark, smoky hovels' by one local.[33] And while there was some improvement over time, cramped one-room hut-like buildings crudely fashioned from mud and plaster remained commonplace. In a pamphlet of 1841, the vicar of Norham pointed out that only 27 of the 174 cottages in his parish had two rooms.[34] A few years later, James Caird's influential survey of English agriculture found that Northumbrian laborers shared their residences with their livestock, the cows and pigs separated from the family's living quarters by a flimsy partition.[35] One reason for the poor standards of cottage accommodation was reluctance on the part of landlords to invest in new housing stock, but this is far from the whole story: once again, the peculiarity of the Borderland experience is crucial. Before the coming of peace, there had been little point constructing good-quality dwellings for ordinary people only to have them destroyed in the next raid.[36] This bequeathed an inheritance of poor accommodation, as well as certain practices that by the nineteenth century were criticized as inimical to wholesome home life. The cohabitation of animals and people, for example, was as much a relic of border conflict—when the stealing of livestock was rife—as it was a function of a lack of alternative accommodation.

Such unorthodox arrangements—and the relatively poor standard of cottages more generally—were also a function of a high level of labor mobility. This was another special feature of the Border region.[37] One man who had worked as a hind (a Border farmworker) at Branxton in the 1820s recalled how in Berwickshire, hinds would typically leave—or 'flit'—after one or two years' service, and would rarely stay in the same place beyond three.[38] Another estimate, made by a resident of Otterburn in 1893, was that at least one-third of hinds moved each year.[39] These footloose tendencies seem to have encouraged a relatively relaxed attitude towards the standard of accommodation—though emphatically not towards the terms of its tenure, the offer of a cottage being bound up with the offer of employment for the hind and his family. Given that he would only stay in one place for a short time, 'If there are four walls and a roof' the average hind would 'make no objection to the cottage when once he has been satisfied with the wages offered,' was one judgment offered in 1892.[40]

Flitting was a long-established feature of border life, the rector of the village of Ford thinking it a vestigial 'old nomadic trait in the character of the people.'[41] There may have been some truth in this, insofar as the disturbed conditions of the region, for so many centuries, probably contributed to a greater willingness to shift from place to place than elsewhere, this willingness crystallizing into a traditional practice not given up with the coming of peace. Yet whatever the deeper causes may have been, the practice was symbiotically related to the hiring fairs held every May. The fairs encouraged migration by allowing hinds to reach a better bargain with alternative employers, and because farmers hired families as whole units, removal often occurred when work could not be found for all family members. As oral testimonies collected in the 1970s suggest, there seems also to have been a strong desire to move for its own sake, as a means of asserting personal independence and gaining new experiences. One wife of a ploughman recalled the case of a hind who told his master he would leave because 'the wife's a bit weary o' the place an' never getting' away onywhere, and wanted a bit o' a change, and a flittin' ye' ken aye lets ye see another bit o' the country,'[42] while another witness simply put it down to the fact that 'ma faither jist liked tae move when he wanted to: it let the farmer ken he didna own him.'[43] If these considerations applied in the early decades of the twentieth century, they certainly applied in previous years, given contemporary attestations as to the assertive independence of the average borderer.

If flitting was a distinctive feature of Borderland life, so too was the oddly named 'bondager' system whereby the hind undertook to supply a woman worker, often not a family member, as part of his terms of service. The practice was criticized as offering opportunities for extramarital sex, with illegitimate children as one consequence; it also had demeaning connotations with indentured labor. Howitt thought 'bondage' fundamentally un-British, being akin to 'serfdom' and sounding 'like a Siberian word.'[44] It was not popular with hinds, either, who mounted organized campaigns against it in the 1860s and 1870s, and this activism helped promote its decline. By 1892 it had largely died out, while leaving intact the practice—again rather peculiar to the Borderland—of hiring out whole families en bloc.[45]

The shared institutions of flitting, hiring fairs, and bondage reflected more general affinities north and south of the national line. The Borderland had its own agricultural system, one distinct from those found elsewhere in Britain. There was more of a difference between upland and lowland agriculture in Scotland than between agriculture on either side of the border.[46] The Borderland functioned in some ways as a single socioeconomic unit, based on sheep pasturage in marginal upland areas, crops in the fields, and small amounts of cattle throughout. High levels of labor mobility—all that flitting—also encompassed significant cross-border migration within the Borderland region, the phenomenon being apparent from the late eighteenth century.[47] And labor exchange went hand-in-hand with the sharing of farming techniques north and south, again a phenomenon that was observable from quite early on. In the 1790s the *Statistical Account of Scotland* reported the introduction to the Scottish border of Leicestershire sheep (which the innovative Culley brothers in Northumberland crossed with local sheep to create the lucrative and widely popular Border Leicester breed) as well as the use of lime fertilizer in Berwickshire, adopted in the light of the 'amazing effects' observed through its use on the other side of the national line.[48]

These socioeconomic commonalities formed the bedrock of a discrete Borderland identity, one feature of which—outside of larger settlements such as Berwick and Kelso—was a relatively undeveloped attachment to the local. Labor migration illustrates this. Unless moving out for good (in which case the destination was typically an urban or overseas one), hinds and their families flitted freely within the Borderland, but often did not remain long in any one village. This did not encourage a conventionally settled home life, or indeed an established connection to any one specific locale: evidence presented to the Select Committee on Settlement and Poor Removal (1847) suggested that the hinds of Northumberland 'have no local attachment whatever.'[49] This was a matter of regret to social conservatives desirous of promoting community cohesion, clerics being among the most vocal.[50] Vicar of Embleton between 1875 and 1884, the historian and future Bishop of London Mandell Creighton found flitting obstructed his pastoral work, reckoning it difficult to exert any influence over the moral and religious behavior of laborers who 'felt they belonged to no particular place.'[51]

Such concerns went hand-in-hand with a feeling that the relatively weak sense of village-level identity promoted secular tendencies. In running counter to sentiments of local attachment, high levels of population mobility discouraged religious observance, many people feeling no special connection to any one parish church. This caused frustration in predictable quarters. H.M. Neville, rector of Ford, damned the 'annual exodus' in 1909 as

> the negation of the idea of home. It keeps the people in a restless unsettled state of mind. The clergy find it a great obstacle in their work, for no sooner do they become acquainted with the people, and the people settle down as church-goers, than they have to say good-bye to them, and welcome and visit those who take their place. Congregations are unsettled long before the fatal day, and for some time after you do not get regular attendance from the farms.[52]

Modern studies have confirmed that rates of church attendance and denominational Sunday School attendance in the Anglo–Scottish Borderland were indeed significantly below those found elsewhere in Britain.[53] Remoteness and low population density had some effect, but—with the exception of the Welsh Borders—similarly isolated and sparsely populated regions in other parts of Britain did not feature such high rates of secularization.

Relatively weak local ties and levels of religious observance coexisted with relatively extensive levels of denominational diversity. By comparison with other areas of Britain, the Borderland was religiously intermixed to a high degree, harboring significant Anglican, Roman Catholic, and Presbyterian communities. Some families clung to Catholicism in isolated upland areas such as Coquetdale.[54] Elsewhere, Presbyterianism was prominent. It was the dominant faith in the Scottish Border counties and exerted a longstanding influence on Northumberland.[55] There were 68 Presbyterian churches in the county in 1851, with adherents having an especially strong presence in the north of the county—doubtless as a consequence of the proximity of the borderline.[56] In some places on the English border, Presbyterians outnumbered Anglicans.[57] The shepherding communities of the Cheviot Hills were almost all Presbyterian, for example, and from his rectory at Ford, Neville estimated in 1896 that overall three-quarters of the inhabitants of north Northumberland were of that denomination.[58]

Presbyterianism, pluralism, and relatively high levels of indifference to organized religion were significant factors in the construction of a discrete Borderland identity, transcending but not necessarily incompatible with national loyalties. Ballad culture was another important influence here. It is tempting to see nineteenth-century engagement with balladry as simply an element of elite culture, encouraged by Scott and other revivers of the genre. Yet while Scott's influence in particular was certainly profound, this was no tradition invented out of nothing, akin perhaps to the tartan craze.[59] It had a living presence in the cultural life of ordinary people. Scott had paid regular visits to Liddesdale from 1792 to collect Border ballads from crofters and shepherds, material he drew upon in preparing his *Minstrelsy of the Scottish Border* (1802–03),[60] and traditional song remained an important feature of quotidian rural life, passed on from old to young and also via the agency of wandering peddlers, tinkers, and musicians. In her memoirs, published in 1894, Janet Bathgate recalled that when, as a seven-year-old girl, she flitted with her family from their farm at Philiphaugh in Selkirkshire, they sang ballads commemorating the suffering of the Border covenanters, as well as 'Flowers of the Forest,' the great lament for the Scottish tragedy at Flodden.[61]

The songs sung by Bathgate's family as their cart moved off indicates the integration of popular balladry with the history, culture, and social reality of the Borderland. Closely linked to the border landscape, its castles, ruins, and moors, the typical ballad celebrated the tragedy and romance of that landscape's turbulent past, and 'the bloody and admired deeds' of its heroes.[62] As indicated by the fact that the same ballads—or variants of the same—were often sung on both sides of the border, this was a shared folk culture, one that reflected the sharing of musical culture more generally.[63] It thus provides more evidence of north/south affinities, being proof for some of the existence of something like a unitary people of the Borderland. G.M. Trevelyan, himself a borderer, was one of those convinced, suggesting late in life that the common ballad culture was a mark of separate identity: 'The moss-troopers of Liddesdale in Scotland and of North Tyne and Rede in England were much more like each other in life and thought than either were like the south English, and they had a ballad tradition in common equally distinct from that of the south.'[64] Moreover, as Trevelyan had suggested in 1905, the enmities of the past had been less about national divisions than might be thought, internecine clan violence and opportunist roguery having been as much a feature of border lawlessness as anything else.[65] Other accounts presented similar pictures, emphasizing how moss-trooping clans had often been as likely to raid on their own side of the border as across it.[66]

In some ways these were comfortingly Unionist stories to tell, hinting that Borderlanders had at best a dilute sense of national belonging before the caesura of 1707 transformed them into stout British patriots. Yet they were stories that expressed a specific border identity, one felt by local inhabitants, and one that testified to the continued reality of the Borderland as a physical presence in their lives. One notable manifestation of this border identity was the Berwickshire Naturalists' Club (BNC), established in 1831 'for the purposes of examining the Natural History and Antiquities of the county and its adjacent districts.' The reference to 'adjacent districts' was significant, implying a concern with Northumberland and Berwick-Upon-Tweed in addition to Berwickshire. The BNC's membership was drawn from both Scotland and England; its meetings were held alternately in English and Scottish venues; its presidents were elected annually, Scotsman being followed by Englishman and vice versa.[67] Well might the Rev. T. Knight note in his presidential address of September 1839 that the club—having visited Flodden Field ('now covered with peaceful flocks and golden corn-fields')—could reflect on how 'The ruder times of our forefathers, thank Heaven! have passed away; and now parties from the two countries, can meet for other purposes than that of bloodshed.'[68]

The BNC's journal promoted such sentiments of cross-border affinity, as well as providing a forum for articulating a shared sense of separate Borderland identity. Indeed, one study it published in 1870 found that the height and weight of Northumbrian men correlated with those of Scotsmen, who on average were taller and heavier than Englishmen. More specifically, Berwickshire men seemed to have the same physique as the men of north Northumberland, suggesting a conclusion that they were of the same stock, having 'had a common origin.'[69] This approached a racial analysis, which another BNC account published the same year put in explicit terms, arguing that Northumbrians between the Tyne and Tweed were 'a quite different race from those of the county of Durham and the whole of Yorkshire,' the latter being 'Dano-Saxon' in speech and physique, rather than 'Anglican' like the Scottish borderers. 'A man from Darlington or Morpeth at Alnwick' was 'as much a stranger in his tongue as an Irishman speaking English with a strong brogue', while 'Physically, the Northumbrian hind [was] a tall and handsome man' with a strong gait 'totally unlike the heavy waddling roll' of the peasantry of the rest of England. He was akin to the southern Scots in his possession of great martial qualities: tough, fit, and of 'pure Anglican race.'[70]

Claims for the racial identity of English and Scottish borderers had a long pedigree. Yet bolstered by the Victorian science of racial difference, confident assertions on these lines were accepted as received wisdom by the later nineteenth century, southern Scots being presented as racial Anglo-Saxons of a markedly pure type, far more like the border English than 'Celtic' Highlanders.[71] Such claims fed into arguments about speech and dialect. A proud native of the border valley of upper Teviotdale, the philologist and future founder of the *Oxford English Dictionary* J.A.H. Murray declared in the 1870s that 'ethnologically speaking,' the dialects of the southern Scots were 'not Scottish at all,' but in fact 'forms of the Angle, or English, as spoken by those northern members of the Angle or English race who became subjects of the King of Scots, and who became the leading race, and their tongue the leading language of the country.'[72] Rather glossing over the issue of the Northumbrian burr, a feature of speech not found north of the borderline,[73] Murray went on to suggest that

> The living tongue of Teviotdale, and the living tongue of Northumberland, would, in accordance with present political geography, be classed, the one as a Scottish, the other as an English dialect: in actual fact, they are the same dialect, spoken, the one on Scottish the other on English territory, but which, before Scottish and English had their political application, were all alike the Anglican territory of Northan-hymbra-land.[74]

Analysis of speech patterns suggested that Borderlanders were one historical people.

Such linguistic-racial arguments had a Unionist message to impart, or at any rate could be interpreted in Unionist ways. They were certainly compatible with the contemporaneous celebration of the Anglo-Saxon 'Teutonic' purity of the southern Scots, and their racial similarities with the English. And as Colin Kidd has suggested, this Teutonic racialism 'acted not only as a powerful and prestigious "unionist" counter-identity, but also as a restraint upon the Scottish nationalist imagination.'[75] But we should not push this point too far. The discrete identity of the Borderland made it sui generis, quite unlike the rest of Britain. The border Scots of the southern uplands of Berwickshire, Roxburghshire, and Selkirkshire were socially and culturally different from the lowland Scots of Lanarkshire and the Lothians, to say nothing of the so-called 'Celts' of the Highlands; a similar point could be made for the north Northumbrians in respect of inhabitants of Yorkshire or even Durham. Moreover, as will be shown below, the peculiarity of border society and the existence of a strong, autochthonously created, and to an extent transnational border identity did not prevent the Borderland from being an important site for both English and Scottish national identities; indeed its very pluralism and liminal status may have helped cause it to be so. In the Borderland as elsewhere, national identities were not simply constructed from free-floating discourse, but were rooted in the physical realities of lived experience.

The Borderland and National Identities

The significance of the Borderland to modern Scottish national identities has been overlooked by historians, who have emphasized the claimed importance of the Highlands as a repository of unpolluted national virtue in the context of advancing modernity. According to R.J. Finlay,

> By the early nineteenth century Scottish national identity increasingly focused on an exclusive set of Scottish symbols. The Highlandization of Scottish culture and the celebration of rural values was largely a middle-class response to the demand for nostalgia in an increasingly urbanized and industrialized society. The key components of British identity in Scotland had a distinctive tartan complexion.[76]

There is some truth in this: the centrality of the Highlands to modern Scottish nationalism cannot be gainsaid. But the Scottish nationalist imaginary had other storehouses too, and the Borderland was one of them.

For a start, the border landscape was subject to nationalist reading. Despite the fact that it followed the national boundary line for much of its course, the Tweed was widely understood to be a Scottish river, figuratively if not geographically part of the heartland landscape of Scotland.[77] And what was true on the grand scale was true in the particular, districts such as the Yarrow Valley being strongly associated with the Covenanting struggle, the wars of Scottish independence, the genius of Scott, and much else besides.[78] In this way, the geographically and culturally liminal landscape of the border was made central to narratives of the Scottish nation.

That being said, the Scottish Borders—as with the Highlands—could also be put to the service of unionism. The ideological freight carried by the landscape was compatible with what has been termed unionist-nationalism, a distinct sense of Scottish identity coexisting with a wider Britishness, the United Kingdom being after all a union of multiple identities.[79] Scott's role in the construction of this unionist-nationalist discourse is well known, and perhaps notorious for those who see him as recasting Scottishness in politically neutered, romance-suffused terms—a Scotland of Rob Roy, kilts, and Balmoralism, unthreatening to Sassenach sensibilities and the constitutional status quo.[80] But negative caricature is best avoided here: unionist-nationalism was a powerful means of articulating Scottish identities within the Union, while at the same time acknowledging Scotland's centrality to Britishness. As Graeme Morton has shown, the nineteenth-century cult of the medieval hero William Wallace emphasized how centuries of resistance to the English had made possible a union of equals in 1707.[81] *Mutatis mutandis*, the history of border strife performed the same function; indeed as reified in the Borderland landscape—all those battlefields, bleak moorlands, castles, and pele towers—it provided especially secure evidence of this resistance, and of Scotland's valorous part in it. In his *Border Raids and Reivers*, the Rev. Robert Borland, minister of Yarrow, stressed Scottish refusal to buckle before English efforts to reduce the nation to vassalage. In this account, the raiding and plundering of the moss-troopers is reimagined in terms of patriotic revenge: 'It was right to rob the English; it was disgraceful to turn your hand against anyone belonging to your own country. Here we have the ethical system of the Border reiver in a nutshell.'[82] Yet at the same time, the Union that ultimately resulted from the centuries of conflict caused Borland no regrets, being a felicitous terminus for the patriotic resistance of the Scots, and the Border Scots in particular.[83]

This resistance was also subject to organized commemoration, one example being the regular ceremony of the Hawick Common Riding in Roxburghshire. A relic of a practice originally legal-administrative in function, it had from the eighteenth century onward morphed into more of a civic celebration.[84] Held every May, the event featured a procession that rode the Burgh bounds, carrying a flag modeled on an English pennon captured in a skirmish after Flodden, in this way commemorating the valiant defense of the Borders even after crushing defeat. To underline the point, a stirringly patriotic song was sung during the festivities, one stanza of which went (as reported in 1839), 'We'll a' hie to the muir a-riding, / Drumlanrig gave it for providing / Our ancestors of martial order, / To drive the English off our border.'[85] Hawick was not alone; a similar ceremony was established in Selkirk in the late nineteenth century, that town's common riding having become 'the chief event of the whole year' by the twentieth century—bigger even than Hogmanay or Burns Night. Held in mid-June, the ceremony again involved a mounted procession around the bounds of the Burgh, and again the theme of continued resistance after Flodden loomed large.[86]

The prominence of Flodden is unsurprising. By the late nineteenth century it was a key element of the landscape of unionist-nationalism, the battlefield being something of a tourist attraction in its own right, drawing visitors from both sides of the border.[87] Its status in this respect was given concrete form by the Berwickshire Naturalists' Club, which erected a large cross of Aberdeen granite where King James was said to have fallen, the monument being unveiled on September 27, 1910 before a sizeable crowd. A plaque affixed to the monument proclaimed, 'Flodden / 1513 / To the brave of both nations.'[88] This inscription was a succinct expression of the organizers' unionist agenda: the battle had been a clash of national equals, and the soldiers on each side had displayed equal valor in the heat of the fight. The memorial was intended to stand as a reminder of the ancient enmity between England and Scotland, an enmity between two great nations, but also one that had passed. By all accounts it achieved its objective. One 1916 guidebook recommended it to tourists as 'a fitting memorial to splendid past bravery and present unity.'[89] In this way the inhabitants of the Borders reconceptualized Flodden for unionist ends without sacrificing its nationalist connotations and place in the narratives of two separate national histories.

Of course, Flodden featured far less in English than Scottish nationalistic discourse, but the Borderland was also important to unionist-national Englishness. As with its Scottish equivalent, this Englishness was suffused with patriotic feeling for the history of the local environment and its landscape, concentrated engagement with which had been widespread in the English Borders since the late eighteenth- and early nineteenth-century explosion in antiquarianism.[90] This historical-antiquarian dispensation grew stronger in the later nineteenth century, crystallizing into something akin to an organized movement—what Robert Colls has called the 'New Northumbrians.' These writers and other intellectuals comprised one manifestation of a regional expression of English patriotism. The *fons et origo* of the movement was arguably Wallington Court, ancestral home of the Trevelyan family, the central courtyard of which was—at John Ruskin's suggestion—covered with a roof and thereby transformed into a hall in 1855. This hall was decorated with murals by the pre-Raphaelite artist William Bell Scott, in which the landscape and history of the Northumbrian border loomed large. In addition to portraits of the Trevelyan family and other county luminaries,[91] Bell Scott's decorations featured a series of larger paintings, each depicting a scene from the history of Northumberland. Taken together, the story told was one of continuity, from the time of the Roman Wall through the descent of the Danes, the age of Bede, the raiding and reiving of moss-trooping times, all the way down to the present day—as illustrated by a scene of bustling activity on the quayside at Newcastle-Upon-Tyne. The prominence of the border was emphasized by the *Border Ballad of Chevy Chase*, which was depicted on the upper spandrils all around the room.[92]

As represented by the Wallington murals, New Northumbrians celebrated what the Liberal politician Robert Spence Watson called 'this wild, free northern land of ours.'[93] It found expression in locally produced guidebooks, flourishing naturalist and historical societies (not least the Berwickshire Naturalists' Club), and the pages of the county press. In the assessment of the antiquary Cadwallader Bates, 'the unique devotion of the people of Northumberland to the history of their country is brought out in every local newspaper you take up.'[94] One such organ, the *Newcastle Weekly Chronicle*, was an especially important outlet, publishing numerous articles on the history and folklore of the border—so many, indeed, the paper even brought out its own antiquarian journal in 1887, *The Monthly Chronicle of North-Country Lore and Legend*. Poets such as James Armstrong achieved some fame for their evocations of the scenery and associations of their 'muirland hame,' often—as in Armstrong's case—rendering their verse in the dialect of the Borderland.[95]

As in the discourse of the Berwickshire Naturalists' Club, now stronger than ever, New Northumbrianism was predicated on the reality of the Borderland as a transnational zone, involving but transcending English and Scots loyalties yet drawing on a common British heritage, formerly conflictual but now reimagined to serve the ends of unity. At the same time, however, the Unionist dimension to the New Northumbrian movement was compatible with a robust, regional Englishness, one quite different—in part because of its Scottish inflections and connections—from its equivalents elsewhere. New Northumbrians asserted the importance of their region by linking it with the themes and events of national history, which, given the proximity of the border with Scotland, they were well placed to do. It became, in some accounts, the cockpit of English history. Writing of Alnwick in 1888, W.W. Tomlinson remarked how

> The capture of William the Lion, the death of Malcolm Canmore, the march of the English armies northward and of the Scottish armies southward ... the movements of troops in the civil wars, Yorkists and Lancastrians, Royalists and Roundheads—these were the spectacles witnessed by the ancient burghers [of the town].[96]

As Colls has noted, New Northumbrians sought to preserve continuity with their regional past. In doing so, they did not seek to repudiate industrial modernity, which after all formed the inescapable context of the daily lives of many of them (not least those who lived in Newcastle-Upon-Tyne), but to 'affirm the modern world by re-charging it with historic meaning.'[97] This intention is evident in the paintings at Wallington, with their story of progress through the ages from Roman antiquity to contemporary Tyneside. It was also evident in the activities of Louisa Anne Beresford, marchioness of Waterford and chatelaine of Ford Castle from 1859 until her death in 1891. Ford was a historic site of some importance, having been visited by James IV before Flodden, and Lady Waterford was very sensible of its historical associations.[98] Seeking to better preserve these associations Waterford had the architect James Bryce restore the castle to something more suggestive of a Border fortress, stripping away what she called 'the trumpery Gothic style of a hundred years [ago].'[99] In this project she was motivated by a Ruskinian desire for 'authenticity' in architecture, while simultaneously seeking to recapture and preserve some of the romance attaching to border myth-history as imagined in Scott and ballad culture. Hers was a project which aimed at the preservation of continuities between the castle and its historical hinterland; the castle's restoration reconnected it to Flodden Field (a frequent conversation piece and day-trip destination for the marchioness's house guests), and also the historical-literary associations of the border landscape more generally.

In her desire to preserve these continuities, Lady Waterford—like other New Northumbrians—combined enthusiasm for medieval romance with a reluctance to turn back the clock. Frustrated by what she saw as the reactionary sensibilities of Augustus Hare, a regular visitor, Waterford wrote of how

> He cares for everything that belongs to other times ... but this, I think, is a taste that wants mixing up with a more onward march. I love old things too, but I rejoice in the *providence* of progress, without which England would be such a country as Spain—a blank among nations—and I can see a desolate waste made frightful (its beauty lost) with a most utilitarian delight. I love *heads* that have done such great things for England as her engineers, and think that that romance of their useful lives greater than that of a knight-errant; but then ... I am not a Conservative.[100]

The patriotic valorizing of a progressive sense of continuity on the part of New Northumbrians was in line with the nationalistic uses to which history was put in England generally at the *fin de siècle*. Maintaining links with the past helped maintain regional and national identities, the dissolution of which would render the transformations of modernity more difficult to negotiate.[101] The landscape of the Northumbrian Border, with its historical associations, provided a particular sort of connection with a particular sort of past, and supported a regional northern Englishness. For Colls, the articulation of this Englishness—which reached its apogee in New Northumbrianism—was a defensive reaction to the dominance of conceptualizations of nationhood focused on 'college cloisters and south country lanes.'[102] If this is right, it can be viewed as indicative of the marginal, secondary, and oppositional status of all versions of northern Englishness—a point that Dave Russell has made in a recent book-length study of the subject.[103] Yet while the Englishness of the inhabitants of the Northumbrian Border was based on a distinctive sense of regional identity, it cast neither the Scots nor the southern English as the 'other' in any very antagonistic sense. It was a form of unionist-nationalism, suffused with an intensely English pride of place while at the same time acknowledging its close connections with the Scottish past, now reimagined—as at Flodden—in ways calculated to express amity rather than enmity. In this way, the physical landscape of the English Border provides, a fortiori, an illustration of the diversity of nineteenth- and twentieth-century English identities, and the role of landscape in the construction of those identities. As the lived experience of the Borderland showed, the Scots had no monopoly on ideologies of unionist-nationalism.

Thus the physical reality of the nineteenth-century Borderland not only shaped the region's society and economy in distinctive ways, it was also central to the construction of overlapping languages of patriotism and nationalism. The inhabitants of the Borderland identified closely with their region, its history, topography, and landscape. This supported a powerful *regional* patriotism, despite—or perhaps because of—the relatively weak ties of *local* attachment, as indicated by the linked phenomena of flitting and low rates of religious observance. Based on mutual affinities and commonalities (not least in agricultural organization), and a shared if turbulent history, the Borderland identity was British; indeed it was emphatically unionist in ideological complexion. Yet withal it could accommodate Scottishness and Englishness too, being an important site for the commemoration, preservation, and celebration of different—but now happily compatible—narratives of nationhood north and south. This illustrates the capaciousness of unionist-nationalism, its rootedness in the realities of lived experience, and its utility as a concept for the understanding of English as well as Scottish history and culture. Employed heuristically in this way, it calls into question interpretations that emphasize the predominance of a sequestered 'south country' in constructions of Englishness, and a primeval-Celtic Highland in constructions of Scottishness. British identities were more plural and complexly imbricated than sometimes assumed.

Notes

1. *The Times*, Aug. 31, 1850; *Newcastle Courant*, Aug. 30, 1850; S. Smiles, *Lives of the Engineers. The Locomotive: George and Robert Stephenson* (new ed., London, 1879), pp. 310–12.
2. A.G. Bradley, *When Squires and Farmers Thrived* (London, 1927), pp. 63–4.
3. S. Oliver [W.A. Chatto], *Rambles in Northumberland, and on the Scottish Border* (London, 1835), p. 163; A.G. Bradley, *The Romance of Northumberland* (London, 1908), p. 293.
4. W. Howitt, *Visits to Remarkable Places, Old Halls, Battle-fields, and Scenes Illustrative of Striking Passages in English History and Poetry* (2 vols, London, 1840), I, pp. 189–90.
5. 'From the Tyne to the Tweed,' *Gentleman's Magazine*, July 1861, 19–20.
6. P. Bishop, *An Archetypal Constable: National Identity and the Geography of Nostalgia* (London, 1995); J. Taylor, *A Dream of England: Landscape, Photography, and the Tourist's Imagination* (Manchester, 1994).
7. See, for example, Taylor, *Dream of England*; N.J. Watson, *The Literary Tourist: Readers and Places in Romantic and Victorian Britain* (Basingstoke, 2006); P. Mandler, *The Fall and Rise of the Stately Home* (New Haven and London, 1997).
8. For some useful observations on this trend, see A. Jones, 'Word and Deed: Why a Post-Poststructuralist History Is Needed, and How It Might Look,' *Historical Journal*, 43 (2000), 517–41.
9. See, for a recent example, A. Groundwater, *The Scottish Middle March 1573–1625: Power, Kinship, Allegiance* (Woodbridge, 2010).
10. For example, *A Historical and Descriptive View of the County of Northumberland* (Newcastle-Upon-Tyne, 1811), pp. 79–82.
11. J. Hardy (ed.), *The Denham Tracts*, vol. I (London, 1892 [1846–59]), pp. 27–8.
12. P.A. Graham, *Highways and Byways in Northumbria* (London, 1920), p. 324; H. Pease, *Borderland Studies* (Newcastle-Upon-Tyne, 1893), p. 12.
13. *New Statistical Account of Scotland*, vol. III (Edinburgh, 1845), p. 389.
14. J. Christie, *Northumberland* (Carlisle, 1893), pp. 89–90; Graham, *Highways*, p. 294.
15. N. Pearson, 'The Scotch Borderland,' *Gentleman's Magazine*, Feb. 1886, 194; A.H. Japp, 'A Northumbrian Valley,' *Gentleman's Magazine*, March 1894, 285–7. Staying

at Coldstream in 1894, Beatrix Potter encountered an elderly salmon poacher who told her that 'If she met the "bailie" and he asked her, she always told him how many she had caught, and if she had one with her he took it … and [she] suggested he ate it himself, to which he cordially agreed, and with a little encouragement told me some amazing stories of this "bad man" with whom, however, she seemed on good terms': L. Linder (ed.), *The Journal of Beatrix Potter 1881–1897* (new ed., London, 1989), p. 361.

16. E. Bogg, *Two Thousand Miles of Wandering in the Border Country* (Leeds, 1898), p. 96.

17. *New Statistical Account*, III, pp. 166–8.

18. J.S., 'Border Gipsies,' *Once a Week*, 6 (Apr. 1862), 431.

19. J. Philipson, 'Whisky Smuggling on the Border in the Early Nineteenth Century,' *Archaeologia Aeliana*, 4th series, 39 (1961), 152–3.

20. J. Russell, *Reminiscences of Yarrow* (2nd ed., Selkirk, 1894), pp. 81–3; *New Statistical Account*, III, pp. 165, 176; Sir G. Douglas, *A History of the Border Counties* ([Edinburgh], 1899), p. 425.

21. J.D. Burn, *The Autobiography of a Beggar Boy* (London, 1978 [1855, 1882]), pp. 88–91, 106–7.

22. *New Statistical Account*, IV, p. 273.

23. A. Brack, *Marriages at Lamberton Toll 1833–1849* (n.p., 1995), p. vii; Graham, *Highways*, p. 17.

24. Brack, *Marriages*, p. vii.

25. K.D.M. Snell and P.S. Ell, *Rival Jerusalems: The Geography of Victorian Religion* (Cambridge, 2008), pp. 229–31.

26. Snell and Ell, *Rival Jerusalems*, pp. 291, 415–16.

27. *Seventh Report of the Commissioners of Inquiry into the Excise Establishment*, part I (1834), pp. 63, 301, 313, 315; Oliver, *Rambles*, pp. 89–90.

28. A. Somerville, *The Autobiography of a Working Man* (London, 1848), p. 101.

29. H. Pease, *Tales of Northumbria* (London, 1899), p. 5.

30. For example, *Historical and Descriptive View*, p. 78; J. Grey, *A View of the Past and Present State of Agriculture in Northumberland* (Berwick-on-Tweed, 1841), pp. 3–5.

31. R. Heath, 'Northumbrian Hinds and Cheviot Shepherds' (1871), in R. Heath, *The English Peasant* (London, 1893), pp. 207–15.

32. F.G. Heath, *British Rural Life and Labour* (London, 1911), pp. 54, 154; P.A. Graham, *The Rural Exodus* ([London], 1892), p. 109; Heath, 'Northumbrian Hinds,' pp. 207–15.

33. Russell, *Reminiscences of Yarrow*, pp. 64–5.

34. W.S. Gilly, *The Peasantry of the Border* (Edinburgh, 1973 [1841]), pp. 14–15.

35. J. Caird, *English Agriculture in 1850–51* (London, 1852), pp. 389–90.

36. M. Creighton, *The Story of Some English Shires* (London, 1897), pp. 24–5, 37–8.

37. B.W. Robertson, 'Family Life: Border Farm Workers in the Early Decades of the Twentieth Century,' in J. Beech et al (eds), *Scottish Life and Society, vol. IX: The Individual and Community Life* (Edinburgh, 2005), p. 421.

38. Somerville, *Autobiography*, p. 102.

39. Christie, *Northumberland*, pp. 95–6.

40. Graham, *Rural Exodus*, p. 100.

41. H.M. Neville, *A Corner in the North* (Newcastle-Upon-Tyne, 1909), p. 10.

42. Robertson, 'Family Life,' p. 422

43. Robertson, 'Family Life,' p. 423.

44. W. Howitt, *The Rural Life of England* (2 vols, London, 1838), I, pp. 165–84, at 165, 166, 184.

45. Heath, *Rural Life*, pp. 15–16, 94; Neville, *A Corner in the North*, pp. 12–13; J.P.D. Dunbabin, *Rural Discontent in Nineteenth-Century Britain* (London, 1974), pp. 139–44, 164, 172.

46. M. Robson, 'The Border Farm Worker,' in T.M. Devine (ed.), *Farm Servants and Labour in Lowland Scotland 1790–1914* (Edinburgh, 1984), pp. 71–4

47. Sir J. Sinclair (ed.), *Statistical Account of Scotland*, vol. III (Wakefield, 1979 [1791–99]), pp. xxvi–xxvii.

48. D.J. Rowe, 'The Culleys, Northumberland Farmers, 1767–1813,' *Agricultural History Review*, 19 (1971), 156–74; Sinclair, *Statistical Account*, III, pp. 101, 124–5, 217–18.

49. Evidence of Ralph Carr, a magistrate and poor law guardian: *Sixth Report from the Select Committee on Settlement, and Poor Removal*, Parliamentary Papers (1847), XI, p. 319.

50. See for example, Gilly, *Peasantry*, pp. 7–8.

51. L. Creighton, *Life and Letters of Mandell Creighton* (2 vols, London, 1904), I, pp. 155–7, at 155.

52. Neville, *A Corner in the North*, p. 29.

53. Snell and Ell, *Rival Jerusalems*, pp. 229–31, 291, 415–16.

54. J. Bossy, *The English Catholic Community 1570–1850* (London, 1975), pp. 80–5, 408.

55. M.R. Watts, *The Dissenters, Volume I: From the Reformation to the French Revolution* (Oxford, 1978), p. 271.

56. J.D. Gay, *The Geography of Religion in England* (London, 1971), pp. 128–9; Snell and Ell, *Rival Jerusalems*, pp. 97–8, 221.

57. Snell and Ell, *Rival Jerusalems*, p. 221.

58. Heath, 'Northumbrian Hinds,' pp. 216–17; Christie, *Northumberland*, pp. 96–7; H.M. Neville, *Under a Border Tower* (Newcastle-Upon-Tyne, 1896), p. 121.

59. Cf., for example, H. Trevor-Roper, 'The Invention of Tradition: The Highland Tradition of Scotland,' in E. Hobsbawm and T. Ranger (eds), *The Invention of Tradition* (Cambridge, 1992 [1983]), pp. 15–41.

60. Scott's Border 'raids', as he termed them, commenced in 1792, and were made annually until 1799. See D. Hewitt, 'Scott, Sir Walter (1771–1832)', *Oxford Dictionary of National Biography* (Oxford, 2004), online edn, May 2008; and J. Sutherland, *The Life of Sir Walter Scott* (Oxford, 1995), pp. 46–7.

61. J. Bathgate, *Aunt Janet's Legacy to Her Nieces* (2nd ed., Selkirk, 1894), pp. 4–5.

62. See J. Veitch, *The History and Poetry of the Scottish Border* (2 vols, Edinburgh, 1893 [1878]), II, esp. pp. 52, 71–4, 366.

63. In the Northumbrian village of Ford at the turn of the twentieth century, some of the harvest songs were Scottish; see Neville, *A Corner in the North*, pp. 53, 88–9.

64. G.M. Trevelyan, *A Layman's Love of Letters: Being the Clarke Lectures Delivered at Cambridge, October–November 1953* (London, 1954), p. 85.

65. G.M. Trevelyan, 'The Middle Marches,' *Independent Review*, 5 (1905), 340–1.

66. See for example, 'Border Thieves,' *Monthly Chronicle of North-Country Lore and Legend*, 1 (Nov. 1887), 405–6.

67. *History of the Berwickshire Naturalists' Club*, sesquicentenary volume, p. 2.

68. *History of the Berwickshire Naturalists' Club*, 1–2 (1832–41), 179.

69. G. Tate, 'On the Stature, Bulk and Colour of the Eyes and Hair of Native Northumbrians,' *History of the Berwickshire Naturalists' Club*, 6 (1870), 133–40.

70. R. Carr, 'The Northumbrians between Tyne and Tweed,' *History of the Berwickshire Naturalists' Club*, 6 (1870), 141–2.

71. Veitch, *History and Poetry*, pp. 56–9. For a general discussion, see C. Kidd, 'Race, Empire, and the Limits of Nineteenth-Century Scottish Nationhood,' *Historical Journal*, 46 (2003), 873–92.

72. J.A.H. Murray, 'The Dialect of the Southern Counties of Scotland,' *Transactions of the Philological Society* (1870–72), part II, pp. 4–5.

73. C. Påhlsson, *The Northumbrian Burr: A Sociolinguistic Study* (Lund, 1972).

74. Murray, 'Dialect,' pp. 4–5. See also K.M.E. Murray, *Caught in the Web of Words: James A.H. Murray and the Oxford English Dictionary* (New Haven and London, 1973), esp. pp. 12–13, 51–2, 81–3.
75. Kidd, 'Race, Empire,' 892.
76. R.J. Finlay, 'Caledonia or North Britain? Scottish Identity in the Eighteenth Century,' in D. Broun, et al. (eds), *Image and Identity: The Making and Re-Making of Scotland through the Ages* (Edinburgh, 1988), p. 153.
77. For example, Veitch, *History and Poetry*, I, pp. 1–3; Sir H. Maxwell, *The Story of the Tweed* (London, 1909).
78. See for example, G. Eyre-Todd, *Byways of the Scottish Border* (Selkirk, 1893), pp. 11–13, 46–93.
79. See G. Morton, *Unionist-Nationalism: Governing Urban Scotland, 1830–1860* (East Linton, 1999); C. Kidd, *Union and Unionisms: Political Thought in Scotland 1500–2000* (Cambridge, 2008). Cf. L. Brockliss and D. Eastwood (eds), *A Union of Multiple Identities: The British Isles c.1750–c.1850* (Manchester, 1997).
80. C. Kidd, 'The Canon of Patriotic Landmarks in Scottish History,' *Scotlands*, 1 (1994), 1–17.
81. G. Morton, *William Wallace: Man and Myth* (Stroud, 2001); G. Morton, 'The Most Efficacious Patriot: The Heritage of William Wallace in Nineteenth-Century Scotland,' *Scottish Historical Review*, 77 (1998), 224–51.
82. Rev. R. Borland, *Border Raids and Reivers* (2nd ed., Dalbeattie, 1910), pp. 4–13, 301.
83. Borland, *Border Raids*, pp. 5–6, 282–8.
84. Maxwell, *Tweed*, pp. 172–3; R.S. Craid and A. Laing, *The Hawick Tradition of 1514* (Hawick, 1898); J.C. Goodfellow, 'Hawick Common-Riding,' *Monthly Chronicle of North-Country Lore and Legend*, 1 (1887), 207–11.
85. *New Statistical Account*, III, p. 399.
86. J.S. Muir, 'Selkirk Common Riding,' *History of the Berwickshire Naturalists' Club*, 27 (1929–31), 372–8.
87. The young Beatrix Potter was among those fascinated by Flodden, making several trips to the battlefield while on holiday near Coldstream in summer 1894: *Journal*, pp. 329, 334, 341, 344, 349–50, 352, 355, 360.
88. P. Usherwood et al., *Public Sculptures of North-East England* (Liverpool, 2000), pp. 21–2; *History of the Berwickshire Naturalists' Club*, 20 (1906–8), 273–4, 307; *History of the Berwickshire Naturalists' Club*, 21 (1909–11), 165–8.
89. E. Morris, *Northumberland* (London, 1916), p. 162.
90. R. Sweet, '"Truly Historical Ground"': Antiquarianism in the North,' in R. Colls (ed.), *Northumbria: History and Identity 547–2000* (Chichester, 2007), pp. 104–25.
91. Such as the engraver Thomas Bewick, in some ways a New Northumbrian *avant la lettre*: see J. Uglow, *Nature's Engraver: A Life of Thomas Bewick* (London, 2006), p. 398 *et passim*.
92. Sir C. Trevelyan, *Wallington* (Pelaw-on-Tyne, 1935), pp. 30–6.
93. R.S. Watson, 'Northumbrian Story and Song,' in T. Hodgkin et al, *Lectures Delivered to the Literary and Philosophical Society, Newcastle-Upon-Tyne* (Newcastle-Upon-Tyne, 1898), pp. 25–172, at 26.
94. C.J. Bates, *The Border Holds of Northumberland*, vol. I (Newcastle-Upon-Tyne, 1891), p. vii.
95. J. Armstrong, *Wanny Blossoms* (Carlisle, 1876).
96. W.W. Tomlinson, 'Views in North Northumberland,' *Monthly Chronicle of North-Country Lore and Legend*, 2 (March 1888), 128.
97. R. Colls, 'The New Northumbrians,' in Colls, *Northumbria*, p. 151.

98. Letters from Lady Waterford to Rev. Canon T.F. Parker and Mrs Osborne, July 30, 1859 and Sept. 21, 1859: A.J.C. Hare, *The Story of Two Noble Lives* (3 vols, London, 1893), III, pp. 69, 74.

99. Lady Waterford to Mrs Bernal Osborne, Jan. 30, 1865: Hare, *Story of Two Noble Lives*, III, p. 257.

100. Waterford to Mrs Bernal Osborne, Aug. 28, 1865: Hare, *Story of Two Noble Lives*, III, pp. 277–8.

101. For this theme, see P. Readman, 'The Place of the Past in English Culture, *c.*1890–1914,' *Past and Present*, 186 (2005), 147–99.

102. Colls, 'New Northumbrians,' pp. 175–7.

103. D. Russell, *Looking North: Northern England and the National Imagination* (Manchester, 2004), pp. 268–9.

9

Church Fights: Nationality, Class, and the Politics of Church-Building in a German–Polish Borderland, 1890–1914

Jim Bjork

The American humorist Robert Benchley once observed that there are two kinds of people in the world: those who think that there are two kinds of people, and those who don't. The joke has particular poignancy for the student of borderlands. As discussed in the introduction to this volume, borderlands fascinate scholars because they can be viewed in two ways: as showcases for the power of nation-states and empires to sort out territories and populations, thereby establishing exclusive control over those under their jurisdiction; and as liminal zones of cultural contact and mixing, in which local inhabitants can defy and subvert state-driven agendas. The attention given to the latter dimensions of borderland experience in recent scholarship has provided a welcome counterpoint to more teleological stories focused exclusively on the self-reinforcing dynamics of border-drawing. But we should be careful not to imagine this tension as a straightforward conflict between border-makers (agents of state power, ideological entrepreneurs, national activists) and border-subverters (locals, ordinary people, borderlanders). In practice, individuals, groups, and institutions have played both roles, articulating and policing some boundaries, inscribing others with new meanings, dismissing still others as meaningless or illegitimate. The following micro-level case study is intended to illustrate the complicated and dynamic interplay created by different actors simultaneously engaging a variety of boundaries: those of empires and nation-states; of religious jurisdiction and devotional practice; of linguistic usage and ethnic identification; of class and status. The attempt to channel these disparate interests and identities into a single, quotidian conflict—in this case, over the subdivision of a Roman Catholic parish—proved endlessly confusing, to contemporaries no less than to readers today. But it is by tracking the uncertainty and volatility of these debates that we can best get a sense of what different boundaries meant to various actors in their everyday lives and how this changed over time.

Map 9.1   Central Europe, c. 1905

194

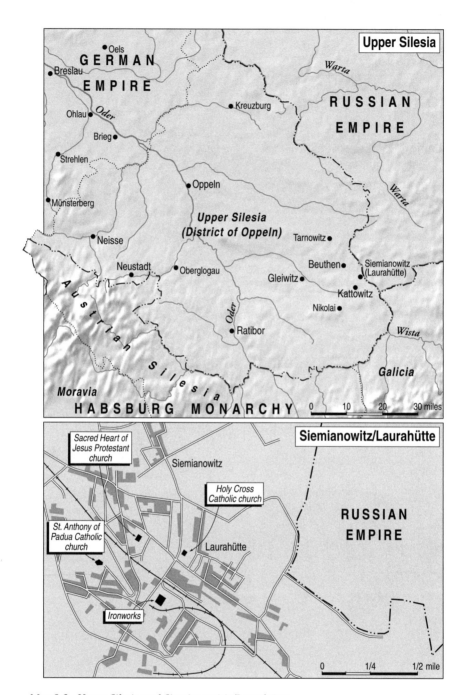

Map 9.2   Upper Silesia and Siemianowitiz/Laurahütte

The 4th of November 1913 was election day in the town of Siemianowitz,[1] a densely populated metallurgical center in Upper Silesia's industrial district, situated in the far southeastern corner of the German Empire. The results of previous elections had been thrown out due to various irregularities, so partisans on each side of the electoral struggle spared no effort to turn out their voters for what they hoped to be a final and decisive round of voting. Industrial employers, lobbying heavily for one slate of candidates, were accused of dragooning sick or injured miners to the polls to support their cause.[2] Catholic priests, in turn, reportedly urged the wives of local coal miners to douse their husbands with hot water after work and send them on to cast ballots in favor of the opposing slate.[3] Each side distributed fliers, put up posters, called public meetings, and published appeals in the local press. Already during a previous round of voting in the spring of 1912, voter interest had been described as unprecedented, with reports of residents waiting in line for hours to cast their ballots.[4]

Hotly contested elections, mass mobilization, and melodramatic rhetoric were not, in themselves, anything unusual in imperial Germany. Indeed, in the face of longstanding stereotypes of passive and apolitical Germans, recent studies of German elections have portrayed a voting culture that was far more vigorous, sophisticated, and complex than had previously been appreciated. Most of this scholarship has understandably been focused on the marquee level of national politics—in particular, elections to the Reichstag, in which the introduction of universal male suffrage, the secret ballot, and high turnout created a vibrant arena for mass politics.[5] But the election-day scene in Siemianowitz described above did not involve a Reichstag election, or even a state-level (Landtag) or municipal election. It was an election for a local church council.[6] What could possibly have been at stake in such an election? At the most basic level, the church fight in Siemianowitz was about something disarmingly pragmatic: the subdivision of the existing parish and the construction of a new church building in the 'daughter' parish. And yet this discussion and the prosaic questions that one might expect to accompany it—How much will the new church cost? How can we be sure that the building will not collapse?—came to be described in terms of epochal ideological struggles: supporters vs. critics of the Roman Catholic church; workers vs. employers; and, most explosively, Germans vs. Poles.

The framing of the Siemianowitz church fight as a *national* struggle seemed intuitively plausible to many observers.[7] The region in which the town was located—Upper Silesia, which in this essay refers to the Prussian Regierungsbezirk of Oppeln—was, after all, a linguistic borderland, an area where census figures suggested a fairly even balance between those claiming German and those claiming Polish as their mother tongue, with many residents bilingual to varying degrees.[8] Over the past decade, the region had also become a nationalist battleground. Previously a bastion of the Center, a party that defended the rights of the Catholic Church and that had routinely won lopsided majorities among the 90 percent of the local population who were Roman Catholic, Upper Silesia had more recently proved fertile ground for the Polish national party, which won five of the district's twelve Reichstag seats in the 1907 elections.[9] The success of the Polish national movement did not, to be sure, transform the region into a genuine geopolitical borderland. There was no immediate prospect of Upper Silesia becoming part of Poland; Poland, after all, had ceased to exist as an independent state after the partitions of the late eighteenth century, and the pre-partition Polish state had not even included Upper Silesia. But residents were well aware that the border between the German and Russian empires, a boundary that ran just about a mile to the east of Siemianowitz, had served for centuries as the border between Silesia and the Kingdom of Poland.[10] If a Polish state were ever to be revived, in other words, it would be right next door. And each local electoral success by the Polish national movement held the promise (in the view of Polish activists) or the threat (in the view of German state officials) of stretching the claims of such a hypothetical 'Poland' just a bit further to the west, to include the majority-Polish-speaking lands of Upper Silesia.

Claiming Roman Catholic parishes as 'Polish' was arguably an even more important stepping stone in this process than claiming Reichstag districts. The latter, after all, were simply pragmatic containers for holding legislative elections; a voter only needed to worry about the specific contours of these districts once every few years. Parish boundaries, by contrast, shaped residents' everyday lives, determining where—and under whose jurisdiction—they were supposed to worship, confess their sins and receive absolution, marry, bring their children for baptism, and be buried. If Catholic parishes were becoming battlegrounds and, ultimately, prizes in a zero-sum nationalist competition, it makes sense to view parochial politics as not only a faint echo of national polarization but one of its primary locations, a micro-level harbinger of the national sorting out that Upper Silesians would face less than a decade later, when they were asked to declare their preferences for union either with Germany or with the newly constituted Polish state in the frontier plebiscite of 1921.

This essay argues that parish-level politics on the one hand and Upper Silesia's fate as a national battleground on the other are, indeed, closely intertwined. But this is not because narratives of inexorable national polarization played out in miniature in localities such as Siemianowitz. Instead, scrutinizing attempts to make nationality tangible and meaningful in the everyday life of the parish—to mobilize people behind Germanness and Polishness by linking these concepts to disputes about bricks and iron, run-off ponds and property speculation—reveals not only the ubiquity but also the ambiguity, fragility, and mutability of national affiliations in the region, and hence the fundamental flaws in the narrative of national polarization. This is, in short, a story that tells us as much about the resilient dynamics of borderlands as it does about the inexorable logic of the nation-state.

'Betraying the Parish, Betraying Yourselves':
What Was at Stake in a Church Fight?

The industrial district in which the town of Siemianowitz is situated is often referred to as 'Black Silesia.' It is a straightforward, literal characterization, contrasting the coalfields situated across eastern Upper Silesia and the blackened facades of the region's built environment with the relatively greener landscapes of Middle and Lower Silesia. The region thus defied the stereotype of linguistic borderlands as 'an overwhelmingly rural phenomenon,'[11] but it certainly did conform to the perception of linguistic borderlands as places of exceptional religious piety—providing an alternative, parallel meaning to the moniker 'Black Silesia.' Upper Silesia's huge Roman Catholic majority was renowned for its high levels of religious observance, and it had a tradition of support for pro-clerical (that is, 'black') political parties, first the German Center party, later Polish Christian Democrats. The foil for *this* 'Black Silesia' was not 'Green Silesia' but rather 'Red Silesia' or 'Brown Silesia,' the areas of Lower and Middle Silesia that tended to favor Social Democracy around the turn of the century, the National Socialists by the 1930s, and—after the complete demographic rupture at the end of World War II—the Polish post-Communist party in the mid-1990s.[12] These two meanings of 'Black Silesia' were often fused into a portrait of a landscape defined by the juxtaposition of industry and piety. In a speech to a local Catholic journeymen's association in 1909, Father Jan Kapica, a local pastor and political and social activist, asked rhetorically,

> And has modern Man only built factories and chimneys? Take a trip through Upper Silesia and you will see the most beautiful houses of God—as large and numerous as in few places on earth. Who has built these houses of God? The faith, the piety of the Upper Silesian people ... The proud masters of the powerful steam engines, the strong conquerors of the raw forces of nature bend their knees before their God, humble and quiet, the evidence of work visible on their hands and faces, the sign of faith confessing before the entire world. Looking at such churches, looking at these faithful masses, what priest could condemn the machine, the progress of industry?[13]

The perception of industrial expansion and church-building as proceeding symbiotically was enhanced by the success of parish subdivision and church-building in the final years of the nineteenth and first years of the twentieth century. Dozens of new church buildings arose across the region at this time, making the pairing of steeples and smokestacks, crucifixes and colliery wheels, a genuinely ubiquitous sight. The fact that one author, writing during World War I, could refer derisively to the 'clichéd character' of the region's neo-Gothic brick churches offered backhanded testimony to the success of the building campaign.[14]

The 'churching' of industrializing and urbanizing landscapes in the late nineteenth century was often viewed as a top-down technocratic challenge: providing sufficient square footage of worship space to accommodate a growing number of bodies. In situations where state officials tended to view popular piety as unambiguously desirable (among Protestants in nineteenth-century Prussia; among Catholics in Habsburg Austria), the financing and management of new church construction was duly left to municipal or diocesan authorities or to royal patronage.[15] Prussia's heavily Protestant political establishment, however, viewed the promotion of *Catholic* piety with much greater ambivalence and was reluctant either to provide subsidies to Catholic church construction or to give the Catholic hierarchy direct access to the resources for doing so. Indeed, one of the least discussed but most durable and far-reaching pieces of legislation enacted in the context of the Prussian Kulturkampf—the 'struggle for civilization' waged by Chancellor Otto von Bismarck and his National Liberal allies against the Catholic Church after German unification—was an 1875 law on the financial administration of Catholic parishes, which sought to use democratization as an indirect check on clerical overreach. Financial oversight, control of real property, and the power to set church tax rates within individual parishes were placed in the hands of democratically elected governing bodies—the Kirchenvorstand, a board of between four and ten members which served as a kind of executive committee for the parish, and the Kirchengemeindevertretung, which was to have a membership three times the size of that of the Kirchenvorstand and had to approve all of its major decisions. The two bodies were elected by parishioners under a franchise that was closer to the universal male suffrage of Reichstag elections than to the restrictive, income-weighted system of Landtag and municipal elections. Any man eligible to vote in a Reichstag election was also eligible to vote for a local church council as long as he had his own household (rented or owned), had lived in the same location for at least a year, and was eligible to pay church taxes.[16]

This formal democratization of control over a considerable portion of church affairs provoked predictable alarm among Prussia's Roman Catholic episcopate. Fears that this could trigger an uncontrollable secularization from within were not entirely unfounded. In one dramatic case in *fin de siècle* Munich, where a similar system of elected church councils prevailed, lay discontent with increased church taxes and cost overruns from the construction of a new church spurred local Social Democrats in the working-class parish of Sendling to put forward their own slate of candidates for the church council elections. The socialist candidates emerged victorious and eventually pressured the disgraced parish pastor to resign.[17] This was, however, the exception rather than rule. In Prussia especially, where the plebiscitary rallying around the clergy during the Kulturkampf became and remained an extraordinarily powerful folk memory, church council elections generally served as vehicles for affirming

rather than challenging pastoral authority. Just how sleepy and noncompeti-
tive church council elections could be was evident just a few miles south of
Siemianowitz, in the densely populated pit village of Zalenze. In 1912, even as
control of the church council in Siemianowitz was being so hotly contested,
the Zalenze church council election drew a whopping turnout of 13 parishioners; candidates, apparently, outnumbered voters.[18]

So what was different in Siemianowitz? The prospect of the subdivision of
the parish, inhabited by an estimated 35,000 people by 1912,[19] was hardly controversial in itself. No one quibbled with the judgment of the local dean that
the current church of the Holy Cross was insufficient and that the creation of
a new parish and construction of a new church in such an overcrowded community was 'a necessity.'[20] But beneath this thin consensus a feud was brew-
ing over where and how a new church should be constructed. Ewald Hilger,
the general director of the company (the Vereinigte Königs- und Laurahütte
Aktiengesellschaft für Bergbau und Hüttenbetrieb) that operated the mam-
moth Laurahütte ironworks and served as the formal patron of the existing
parish, offered to donate a piece of land for the building site. But he stipulated
that the church be constructed out of iron rather than brick to reduce the structure's weight and allow for the continued exploitation of adjacent coal fields.[21]
Hilger assured the local pastor, Franz Kunze, that such a building could be a
'beautiful, worthy church, fit for purpose.' Kunze, however, complained to the
bishop that it would be a 'monstrosity ... less like a church than a workshop.'[22]
He advocated an alternative site, close to the existing Protestant church, and
for emulation of the familiar brick neo-Gothic style of other local churches.

The key to pushing through one of these plans was winning the support
of the local church council. Here is where a conflict that might otherwise
have remained a spat between willful local notables intersected fatefully
with the explosive world of mass politics in early twentieth-century Upper
Silesia. Siemianowitz happened to be the hometown of the region's most
famous politician, the Polish national activist Wojciech Korfanty. Over the
previous decade, Korfanty had spearheaded a series of breakthrough victories
for the Polish party in Upper Silesia, persuading the majority of the local
working-class, Polish-speaking population to abandon their Kulturkampf-era
allegiance to the Catholic Center and articulate their grievances in national
terms. This nationalist mobilization had even spilled over into elections to the
local church council. By the spring of 1911, as the construction controversy
was coming to a head, Polish activists had won control of the Siemianowitz
Kirchenvorstand and Kirchengemeindevertretung.[23] This Polish majority had
rejected the Hilger plan and pushed ahead with plans to build a church on the
alternative site favored by Father Kunze. By approving an order for 1.5 million
bricks, the council made abundantly clear what style it thought appropriate for
the new church.[24]

The next year and a half would witness a near-constant campaign for control of the council, as successful complaints about electoral irregularities triggered new elections every few months. In September 1911, the 'Polish party' was heavily defeated, bringing in a new majority more favorable to the Hilger plan for construction of the new church on the Laurahütte site. The new slate maintained control following a subsequent round of voting in the spring of 1912, but a subsequent election in the autumn resulted in a narrow majority for the 'Polish' candidates. This party's grip on the council was reaffirmed in a further round of voting a year later.

To generate enthusiasm throughout this electoral marathon, opponents of the 'iron church' framed the struggle in terms of populist resistance to the region's overbearing 'bread lords.' Given that 'the people,' in this instance, were overwhelmingly proletarian, and their opponent was the dominant employer in what was—for all intents and purposes—a company town, it is tempting to describe the mobilization in class terms. Indeed, the local Polish-national press often referred to supporters of Hilger's plan as the 'industrialist party' (Hüttenpartei or Hütten- und Grubenpartei),[25] and the more populist-minded among German-language Catholic newspapers also portrayed the conflict in terms of defiance of the quasi-military discipline imposed by employers both in and out of the workplace. Following one of the victories of the 'industrialist party' slate, the *Oberschlesische Kurier* complained that 'in the foundries and mines and on the estates, columns were formed and marched out [to the polls] ... the election proceeded under the full economic pressure of the company.'[26]

Where socioeconomic oppression was often left implicit, confessional cleavages were placed front and center. In a stylized dialogue between the characters Wojciech and Jakob, printed in a local Polish daily, Wojciech wonders aloud why the 'lords' were so concerned about church affairs when so many of them were 'of a different faith and liberals.' It was a shame, he continued, that so many ordinary workers 'are afraid to vote Catholic and [instead] vote Protestant.' He and Jakob agreed that workers' wives were sometimes part of the problem, encouraging their husbands to vote according to the wishes of employers. Good Catholics should be listening to their priest on such matters, not women.[27] The message that only one side in this church fight actually cared about the church at all was reiterated in fliers distributed in the run-up to council elections. To vote for the 'industrialist party,' one flier insisted, 'means betraying the Catholic Church, betraying the parish, betraying yourself.' The only reason that this party could be 'ordering us to build a house of God on a pond' (a reference to the building site having been used as a run-off pond for the ironworks) was that they wanted to build the church on the cheap rather than well.[28]

The framing of the conflict in national terms was, in a sense, another natural extension of the language of populism. Just as 'the people' were overwhelmingly working class and Roman Catholic, they were also predominantly Polish-speaking. Both the Polish-national press and local government officials often reverted to national shorthand in characterizing the electoral contests. Rather than a 'citizens party' vs. an 'industrialist party' or a 'Siemianowitz party' vs. a 'Laurahütte party' (reflecting the locations of the rival building sites), the church fight was most pithily recorded as a struggle between a 'Polish party' and a 'German party.' And yet a closer look at the discursive nationalization of the church council elections in Siemianowitz reveals not only the power but also the limits of this language of national polarization.

One problem was the awkward fit between nationalist terminology and the individuals involved in the conflict. Common markers of ethnicity, such as sur-names and language of daily use, failed to correspond to national labels in any intuitive way. The slate for the 'Polish' party included a candidate named Röther, while the 'German' slate included a Bujotzek and a Kolodziej.[29] One of the lead-ers of the 'German' party reportedly tried to address a rally in Polish.[30] Among the marquee figures of the church fight, Herr Hilger was, to be sure, a German-Protestant industrial bread lord straight from central casting. The hero of the 'Polish party,' however, was not nearly as convincing in his assigned national role. Indeed, as late as the spring of 1911, even as Father Kunze was starting to express his opposition to the plans for an iron church near Laurahütte, he was simultaneously complaining to the bishop about the very church council members who would come to champion his position in this church-building struggle. The Polish party that had recently taken over the council, he lamented, 'treats interference in church affairs as its main activity' and was responsible for agitation against him in the local Marian congregation and for rumors of finan-cial irregularities in the parish accounts. A Polish daily had also just included Kunze on a list of 'clergy of German heritage with German names' who had regrettably been sent to minister to the predominantly Polish-speaking parishes of the industrial region.[31] Kunze's sudden reinvention over subsequent months as the leader of a 'Polish party' is therefore a bit of a mystery. Most attempts to explain the conversion focused on an increasingly close friendship with his personal physician, Dr Jan Stęślicki, one of the small number of middle-class Polish national activists active in the region and one of the 'Polish party' candidates for the church council. Stęślicki, as well as other council members, reportedly owned property opposite the alternative site for the new church and thus had a financial interest in seeing construction proceed at that location.[32] It is not clear how Father Kunze was to benefit from such machinations by 'Polish property speculators,'[33] though the paper trail does suggest that the pastor may have been trying to drag out the subdivision of the parish in order to delay the loss of half of his parishioners and the resulting diminution in surplice fees.[34]

The point of exploring these more cynical interpretations of the dynamics behind the Siemianowitz church fight is not to debunk but to understand more fully the grander, more ideological framings of what this conflict was all about. In Upper Silesia, the Polish national cause was still a relative novelty, and its promoters were keenly aware that filling it with meaning—associating Polishness with the right bundle of values and interests and with appealing personalities—required constant hard labor and careful strategizing. As Adam Napieralski, the editor of the newspaper *Katolik*, had noted back in the 1890s, soon after his arrival in the region, 'Workers' wages, worker protection laws, the tax system, railroad rates, tariff politics—all of these are national issues for us in Silesia.'[35] Repackaging such desiderata as those of a 'Polish' people—rather than the 'Catholic' people mobilized during the Kulturkampf or the working class promoted by Social Democratic activists—had proved remarkably success-ful around the turn of the century. This was due, in part, to the charisma of the national movement's Siemianowitz-born standard-bearer, Wojciech Korfanty, who had boldly insisted on campaigning under a Polish-national flag rather than as part of the Center—the region's most venerable political 'brand.' But such reinventions of 'the people' can be fragile constructs, as the subsequent rollercoaster of support for the Polish party attests. After achieving dramatic breakthroughs in the Reichstag election of 1903 (winning 43 percent of the vote in Siemianowitz, 42 percent in Laurahütte) and consolidating those gains further in a by-election in 1905 and the subsequent regular election in 1907 (52 percent and 60 percent in Siemianowitz; 55 percent and 50 percent in Laurahütte), the Polish party had suffered an electoral collapse in 1912, with showings of only 21 percent and 24 percent, respectively, as voters questioned the accomplishments of their Polish representatives and shifted to Social Democracy, back to the Center, or even to German-national parties as alterna-tive ways to voice their grievances.[36]

In the aftermath of this debacle, Polish activists reverted to the emphasis on *mała praca* or 'small work' that they had been fruitfully pursuing since the mid-nineteenth century.[37] Polishness would need to be rebuilt from the ground up again, through individual conversions and incremental transformation of lived experiences into national activity. The church fight in Siemianowitz provided one occasion for such activity. Tellingly, it was as much about elite recruitment as mass mobilization. Convincing individual pastors such as Father Kunze that the Polish-national movement was a reliable ally in buttressing their personal authority within their parishes was as important as convincing ordinary parish-ioners that the 'Polish' party spoke for their interests. In one article, published when the 'industrialist party' had control of the church council, a leading Polish daily begged its readers to demonstrate the value of such an alliance: 'Is it not a shame when, through our behavior, we betray for a Judas-dime a priest who wants to defend us? Is it therefore any wonder if priests who are well-disposed to us have nothing good to say about us?'[38] The eventual triumph of the slate supporting the pastor's position on church construction showed that the 'Polish Catholic people' could be put back together again as an electoral force—but dangers lurked even within such successes. By this time, after all, the German-language and pro-Center-party Oberschlesische Kurier had joined the attack on the 'industrialist party' and thrown its weight behind Father Kunze's position. Was this a matter of some germanophone fellow travelers riding the coattails of a Polish-national triumph? Or did the *Kurier*'s involvement risk changing the narrative altogether, transforming the victors into a nationally indeterminate 'citizens' party'? Such anxieties haunted national activists, whether the contest at hand was for the German parliament, a municipal election, or voting for the Siemianowitz church council.

Church Buildings, Parish Boundaries, and the Limits of National Legibility

One of the things that made the church council election distinctive, of course, was the prominence of aesthetic issues. The debate over church construction, while sometimes shorthanded in terms of the rival sites, was more evocatively summarized by references to how the proposed building would *look*—in particular, how this 'house of God' would relate to the imposing industrial workplaces that dominated the region's built environment. Father Kunze and his allies emphasized the need for stylistic distinctiveness and a certain physical distance from the ironworks. As the members of a building commission set up by the 'Polish' church council argued to the bishop, 'we do not want "iron Baroque," but rather a house of God that takes into account our religious feelings.'[39] Kunze warned that proceeding with construction at the Laurahütte site would leave the church too literally in the shadow of the local ironworks: 'Aside from the aesthetic contrast between the new, colossal factory and the projected church building, the direct proximity of the ironworks is—due to the unavoidable noise and emission of smoke and gasses—extremely dangerous and disruptive.'[40] The bishop of the diocese, Georg Cardinal Kopp, who deplored Kunze's stirring of populist opposition to the local employer, sought to allay these concerns. He sent the pastor a copy of the magazine *Die Christliche Kunst* (*Christian Art*) featuring examples of iron church construction in France and the United States, which, he argued, showed that an iron church could avoid 'the odious appearance of a factory.'[41] But such appeals to key notables to stretch their aesthetic imaginations in more avant-garde directions never seem to have been translated into effective mass-level promotion of the look of the new church. Opponents of the iron church, by contrast, could rely on brief, evocative phrases to conjure a mental contrast between a church-that-looked-like-a-church (that is, like the brick neo-Gothic structures they were used to) and one that threatened to look like their workplaces. It was an unequal contest.

While these aesthetic contrasts resonated at the level of populist mobilization—simple, plebeian religious sensibility vs. the encroachments of soulless corporate managers—they fit only awkwardly with the nationalist terminology of the Siemianowitz church fight. In Upper Silesia, as noted above, national divisions *were* often understood as class divisions—not Germans here, Poles there, but rather Germans on top, Poles on the bottom—and so discourses apparently appealing to class-based solidarities could, in context, seem quite plausibly national. But this was a profoundly localized, contingent, and conditional—and thus, for national activists, insufficient—understanding of the nation. What was absent here was any suggestion that the church favored by the 'Polish' party represented a typically 'Polish' aesthetic (one favored by Polish architects or to be found disproportionately in Polish-speaking lands) or that the church favored by the 'German' party represented a typically 'German' style. Any such framing would have been a tough sell. Gothic revival, after all, was an international style most famously pioneered in Great Britain, and the brick-Gothic version was ubiquitous across northern Europe. The specific inspirations for some of Kunze's ideas reportedly came from his holidays in Austria.[42] Iron-and-stone church designs were an equally cosmopolitan phenomenon; as just noted, the examples invoked came from as far afield as France and the United States. So whatever other benefits might have been expected from using the Siemianowitz church fight as an opportunity for nationalist mobilization, there would not seem to have been any expectation among nationalists that the resulting edifice would facilitate the reading of the local landscape as either typically 'Polish' or typically 'German.'[43]

A similar kind of disconnect emerges when we look at the territorial dimension of the Siemianowitz church fight. As noted at the beginning of this essay, there would seem to have been an obvious resonance to a 'Polish' church council taking over a parish situated on the German–Russian frontier—in other words, directly adjacent to 'Poland.' And yet a reader of the extensive paper trail churned up by this controversy would be unlikely to come across any reference to this fact. The conflict between German and Polish parties over control of the local parish council played out in ways that were oddly indistinguishable from similar conflicts in ethnically mixed neighborhoods in the Ruhr (the heavy-industrial region of western Germany that drew substantial Polish in-migration) or even in comparable immigrant neighborhoods in the United States, areas that were clearly not German–Polish borderlands in the geopolitical sense.[44]

One could go further still. Although Polish activists in, say, Bottrop (in the Ruhr) knew that winning control of a local church council would not amount to staking out the territory of a future Polish nation-state, they would certainly have had a strong sense of their campaign having a territorial anchor: those neighborhoods with concentrated settlement by Polish-speaking immigrants.

In Upper Silesia, by contrast, the fact that the 'Polish party' doubled as the 'Siemianowitz party' and the 'German party' as the 'Laurahütte party' did not really translate into a coding of the former neighborhood as 'Polish' or the latter as 'German.' As noted earlier, levels of support for the Polish party in the two parts of the parish had been virtually indistinguishable over the previous decade. And although both census figures and the results of the later (1921) plebiscite do suggest a stronger German element in Laurahütte than in Siemianowitz, the differences were relatively modest.[45] Indeed, in the sprawling network of industrial suburbs and pit villages that made up the Upper Silesian industrial conurbation, short-range moves were common and identification with specific neighborhoods correspondingly weak. The advocates of a Siemianowitz site plausibly argued that 'members of the communities of Laurahütte and Siemianowitz will visit the new church together without regard to whether the new church happens to stand on Laurahütte or Siemianowitz territory.'[46] It was, in other words, understood that the division of the parish was not really a process of secession, in which the new parish could take on a radically different identity than its 'parent,' but rather of mitosis, in which both the 'new' and the 'old' parishes would carry more or less identical cultural DNA. Like every other large parish in the region, each would offer worship services and other devotional activities in both German and Polish, so there would be no pressure for German-speakers to gravitate toward one church and Polish-speakers toward the other.

The severe constraints on the ability of a 'Polish' parish council to 'Polonize' a parish were traceable to the 1875 Prussian laws on parochial governance. Whereas the Kirchenvorstand and Gemeindevertretung had extensive authority over finances, they had no authority over pastoral matters, including the use of specific languages in worship and pastoral care. The ironic result was a sphere of political expression wide open to nationalist mobilization but without any potential to impact the policies most central to nationalist concerns. This was, in one sense, a contingent and fragile constraint. Changes in regime and certainly changes in state sovereignty could, and eventually did, overthrow many of these particular limitations on how politics could shape everyday religious activity. But when we track the outcome of the Siemianowitz church fight and the later evolution of these parishes, what is striking is how durable those constraints ended up being, as structures that started out as impositions of the Prussian state ended up being recast as internal traditions of local Roman Catholic Church and markers of local identity.

In December 1913, exasperated by the standoff over a site for the new parish church, Herr Hilger offered the use of a planned market hall in Laurahütte (down the road from the run-off pond originally offered as a site for the new parish church) as an 'emergency' church (Notkirche). Cardinal Kopp happily agreed, and the building became the temporary meeting place for a newly established parish in Laurahütte—in the teeth of continued opposition from Father Kunze.[47] Within a few months, World War I had broken out, and raising funds for a new permanent church became more daunting than ever—even leaving aside the ongoing standoff over where the church might be built. Residents of the newly established parish (named after St Anthony of Padua) apparently got used to worshipping in the would-be market hall and continued to do so through the plebiscite campaign of 1919–21. In the aftermath of the plebiscite, the German–Polish frontier was shifted several miles to the east, placing Siemianowitz/Laurahütte in Poland. Officials of the new (Polish) diocese of Katowice decided to adapt rather than completely rebuild or replace the 'emergency church': a pair of towers was added to the front to give it a somewhat more traditional look. The would-be market hall remains one of the Siemianowice Śląskie's two Roman Catholic parish churches.

The Slow Death and the Afterlife of the Bilingual Parish

The absorption of parishes such as St Antoni and its 'parent,' Holy Cross, into the new Polish state might have been expected to signal a decisive Polonization of their internal activities. But the results were, in fact, spectacularly anti-climactic. From 1922, local pastors duly switched from German to Polish in writing correspondence to their new diocesan superiors, but the bilingual nature of local religious life was barely disturbed. In most parishes, at least one and sometimes two Sunday services continued to be offered in German (that is, with a German homily and German singing), and the clergy continued to offer German religious instruction and sponsor an extensive array of German-language parish associations. Indeed, the regional norm of having the second Sunday mass in German remained so entrenched that one would have been hard pressed to hear a Polish homily in the industrial parishes of interwar Polish Silesia between the hours of 7:30 and 8:30 a.m.[48] Whether this represented a minimal provision of pastoral care in the mother tongue of large German minorities or gratuitous Germanization of local Poles depended on what view one took on how many 'Germans' lived in the parish. According to one estimate from 1928 (apparently by the local pastor, Father Scholz, who had succeeded Father Kunze in 1924), fully 50 percent of the parish was composed of Germans, suggesting that they might actually be somewhat underserved by pastoral provision.[49] Five years later, a visitation report characterized the parish as 40 percent German;[50] three years after that, another survey put the figure at 33 percent.[51] The head of the Social Circle of Polish Women in Siemianowice, complaining in 1934 that the pastor 'in a visible way favors Germans to the disadvantage of Poles' dismissed all such figures as gross exaggerations; Germans, she insisted, did not make up more than 15 percent of the parish.[52] As this wild variation suggests, shifting figures had less to do with migration—though out-migration by German speakers certainly played some role in decline over time—than with slippery categorizations. Considering that 95 percent of all 'Germans' in the region were deemed by the local pastor to be fluent in Polish, it is easy to understand the difficulty of pinning down who was a German and who was a Pole.[53]

This resilient model of the bilingual parish came under much more sustained pressure after the German reconquest of the region in the autumn of 1939. Just how shocking the change was for the region's parish clergy can be seen in a letter sent to the curia in Katowice by Josef Kubis, the pastor of the parish of Załęże, located just to the southeast of Siemianowice. Although Kubis himself identified with the German minority, he was horrified that the demands of local police—expressed, he noted, in a 'very unseemly manner'—not only included the addition of a German service but the outright cancellation of a Polish service.[54] Despite near-universal misgivings about such ruptures with tradition, the local clergy ultimately accommodated the regime's escalating demands for the Germanization of parish governance and of devotional life. Already in late November 1939, the recently appointed pastor in St Antoni's in Siemianowice, Father Lubina, reported to diocesan authorities in Katowice that a new church council, consisting exclusively of 'German national comrades' (*Volksgenossen*), had been installed[55]—though one of these 'national comrades,' Jozef Grabowski, had been described a couple years earlier as 'a Pole and ardent Catholic.'[56] By 1941, Stanisław Adamski, the Polish bishop of the diocese of Katowice, was in exile, and a germanophile vicar-general, administering the region on the absent bishop's behalf, had accepted a near-total ban on the public use of Polish in Upper Silesia's churches. The increasingly emphatic suppression of Polish was based on the regime's understanding of Upper Silesia's national make-up: whereas the last interwar Polish census had recorded a German minority of only 6 percent in the former Prussian portion of Polish Silesia, a police survey conducted in December 1939 found 95 percent of the local population to be 'German' by nationality.[57] In the eyes of Nazi officials, it was scandalous for good Germans to be listening to Polish homilies or singing Polish hymns.

With the return of Polish rule in 1945, these national categorizations and corresponding language bans were turned upside-down. Local inhabitants who had 'masqueraded' as Germans were now 'rehabilitated' as Poles. They were expected to abandon any connection to German culture, and pastoral care in local churches was to be conducted exclusively in Polish. As Bishop Adamski emphasized in a pastoral letter read in the parishes of the dioceses in February 1945, 'only one thing has changed [compared to the prewar status quo]. Other than Latin, the only language that will be heard in church is the Polish language.'[58] In the St Antoni parish in Siemianowice, a new church council, reflecting 'new conditions,' was duly put in place in late 1945, and Father

Lubina switched back to writing to his diocesan superiors in Polish.[59] This new understanding of Upper Silesia's Catholic parishes as monolingual and mononational remained dominant throughout the Communist era. Indeed, even as some allowances for German-language pastoral care have reemerged in the region after 1989, parishes such as St Antoni's continue to define themselves as exclusively Polish. If one looks today at the St Antoni parish website—even at its section on the 'history of the parish'—one cannot find any mention of Germans or the German language.[60]

This seemingly effective erasure of the national/linguistic controversies that shaped the founding of the parish should not, however, obscure the ongoing importance of the local Catholic Church's borderland traditions. For although the boundary between the German and Russian empires, which had long run just to the east of Siemianowice, was erased after the 1921 plebiscite and has never again served as an international borderline, and although the Polish state has carefully ensured that internal administrative units (*województwa*) straddle the former frontier, that line has nonetheless remained salient in one crucial respect: it has continued to serve as the eastern boundary of the diocese of Katowice. For almost a century, in other words, residents of communities such as Siemianowice have known that however local parish boundaries might be redrawn, they would never result in their attending the same church as people residing in the Zagłębie industrial basin, on the 'Russian' side of the diocesan boundary. And whoever might be assigned to serve as their pastor or curate would almost certainly have been born and raised within the diocese—thus on the 'Prussian' side of that historic borderline. This is not to suggest that Poles inhabiting former Prussian territory and those inhabiting former Russian territories have remained completely isolated from one another; postwar migration across these old frontiers has been substantial, as has out-migration to Germany. But there has been enough demographic continuity to ensure Upper Silesia's continued—indeed, in some ways, even waxing—distinctiveness.

When the statistics on nationality from Poland's 2011 census were released in 2012, they revealed a startlingly large number of people identifying as 'Silesian': 847,000, with about half declaring this as their sole nationality, about half as a second nationality alongside Polish. Already in 2002, Silesians ranked as the largest national minority in Poland; over the next decade, their number increased five-fold. Although precise county-level breakdowns have not yet been released, it is clear that self-identification as 'Silesian' has little to do with living within the various communist- and post-communist-era administrative units that have been labeled as 'Silesia'—units that have deliberately straddled the old imperial frontier and have always included the Zagłębie region. 'Silesianness' instead remains the preserve of those with genealogical connections to the old German–Polish linguistic borderland to the west of the German–Russian state boundary. Indeed, a map showing counties where more than 10 percent of the population declared 'Silesian' as a nationality in 2011 is virtually indistinguishable from a map of the plebiscite zone of 1921.[61]

Such ghostly reappearances on twenty-first-century maps of an early twentieth-century plebiscite zone might be described as certain proof of the referendum's failure. The point of this one-off vote, after all, was to produce a clear-cut territorial demarcation, separating out which parts of the Upper Silesian borderland would be attached to the narratives of German history and which would be attached to those of Polish history. Instead, the subsequent history of the entire region became a shared narrative of serial nationalization. Regardless of where in the region they lived, and regardless of any national preferences that they or their ancestors expressed on a ballot cast in March 1921, the vast majority of Upper Silesians would go on to compile family histories involving several episodes of nation-switching—Polish to German in 1939; German to Polish in 1945; Polish to German for many postwar emigrants from the 1950s through the 1990s; and now Polish to Silesian for many, perhaps most, of those who remained in the region.

Turning from this longer-range trajectory back to the church fight that convulsed the communities of Siemianowitz and Laurahütte on the eve of the World War I, it becomes clear that it is precisely the difficulty of reading the episode in national terms that makes it such a useful case study of how nationality worked in a borderland such as Upper Silesia. Rather than being based on either a clearly delineated constituency or a discrete, constant set of linguistic desiderata, the nation drew strength from a grab-bag of socioeconomic resentments, personal affinities and animosities, aesthetic prejudices, religious sentiments, and civic values. This could generate temporary victories for national activists, but it left the national cause precariously dependent on contingent and contested—indeed, to an outside observer, seemingly quite arbitrary—bundles of issues. If Polishness or Germanness could be pieced together from so many disparate interests and local points of reference, it becomes easier to imagine how Upper Silesians moved between national categories so readily over the following century. Close scrutiny of these processes reminds us that mobilization along national lines—or for that matter, those of confession, class, or other putative cleavage—is not best understood as an inexorable, *longue durée* process, cumulatively generating 'identities' of architectural solidity and durability. It is often better seen, as Rogers Brubaker has argued, as an event: 'a contingent, conjuncturally fluctuating and precarious frame of vision and basis for individual and collective action.'[62] In this view, the protracted Siemianowice/Laurahütte church fight is best symbolized not by either of the proposed church buildings—neither of which was ever actually built—but by a more appropriately ephemeral image: the crowds of voters that periodically coalesced to vote in the 'industrialist party' or the 'citizens party,' the 'Siemianowice party' or the 'Laurahütte party,' the 'German party' or the 'Polish party.' If we are to imagine nations as 'daily plebiscites,' as Ernest

Renan famously invited us to do,[63] we would do well to imagine these kinds of voting experiences—not predictable, ritual mass affirmations, but confused, provisional, endless wranglings, in which both the outcome and the question being posed are far from clear, either to those being asked to vote at the time or to us today.

Notes

1. I use German spellings for place names when referring to the period before 1922, when the places under discussion were part of Imperial Germany. I switch to Polish spellings when these places came under Polish control.
2. *Dziennik Śląski*, Nov. 11, 1913, translated in a report from Grenzkommissar Mädler to Landrat Kattowitz, Nov. 11, 1913, Sygnatura 462, Archiwum Państwowe w Katowicach (hereafter APK).
3. Testimony from Berufung gegen den ablehnenden Bescheid des katholischen Kirchenvorstandes zu Siemianowitz (dated Dec. 27, 1913), Die K.V. (Kirchenvorstande) im Arch. Myslowitz, Archiwum Archidiecezjalne w Katowicach (AAK).
4. *Oberschlesische Kurier*, May 30, 1912, Nr. 121, 2. Beiblatt, p. 2.
5. M.L. Anderson, *Practicing Democracy: Elections and Political Culture in Imperial Germany* (Princeton, 2000); S. Suval, *Electoral Politics in Wilhelmine Germany* (Chapel Hill, 1985); J. Sperber, *The Kaiser's Voters: Electors and Elections in Imperial Germany* (Cambridge, 1997).
6. As will be discussed further, I use 'church council' as shorthand for the bicameral set of governing bodies that prevailed in Prussian parishes of the time.
7. In one early study of the development of a Polish national movement in Upper Silesia, the author included a paragraph on church elections, concluding that a 'general Polish electoral movement' could be seen here as well as in municipal politics by the eve of World War I: I. Schwidetzky, *Die Polnische Wahlbewegung in Oberschlesien* (Breslau, 1934), p. 105.
8. As explained later in this essay, the parish of Siemianowitz originally included the town of Laurahütte as well as the town of Siemianowitz. According the 1910 census, 52% of Laurahütte's Catholic population listed German as their mother tongue, 44% Polish, while Siemianowitz was 35% German-speaking and 59% Polish-speaking. The remainder listed both languages as mother tongues. See *Gemeindelexikon für die Regierungsbezirke Allenstein, Danzig, Marienwerder, Posen, Bromberg und Oppeln* (Berlin, 1912).
9. For a detailed account of these developments, see J. Bjork, *Neither German nor Pole: Catholicism and National Indifference in a Central European Borderland* (Ann Arbor, 2008), ch. 2.
10. From the fourteenth century until the early eighteenth century, Silesia was part of the Kingdom of Bohemia. In 1740, the province was famously seized by Frederick the Great (the 'rape of Silesia') and incorporated into the Kingdom of Prussia. The areas to the east of the Prussian/German–Russian frontier had constituted a formally distinct Kingdom of Poland, in personal union with the Russian Empire, between 1815 and 1867, when the distinction was eliminated in the aftermath of a failed Polish-national uprising. Even after its formal dissolution—indeed, up until the present day—the 'Congress Kingdom' (a reference to the Congress of Vienna, which had created the entity) remains a well-known geographic point of reference.
11. P. Judson, *Guardians of the Nation: Activists on the Language Frontiers of Imperial Austria* (Cambridge, MA, 2006), p. 33.

12. On the role of religiosity in Poland's electoral geography, see J. Bartkowski, *Tradycja i polityka: Wpływ tradycji kulturowych polskich regionów na współczesne zachowani społeczne i polityczne* (Warsaw, 2003), pp. 249–54.

13. J. Kapica, 'Kirche und Handwerker' (address to the Catholic apprentices' association in Kattowitz, July 2, 1905), *Mowy-Odezwy-Kazania* (Katowice, 1931), pp. 176–80.

14. P. Knötel, *Kattowitz 1865–1915: Eine Denkschrift zum fünfzigjährigen Bestehen der Stadt* (Kattowitz, 1915), p. 27.

15. On (generally unsuccessful) efforts to finance new church construction in Vienna through a centralized fund, see J. Weissensteiner, 'Erzdiözese Wien,' in E. Gatz (ed.), *Pfarr- und Gemeindeorganisation* (Paderborn, 1987), esp. pp. 34–9. On royal and aristocratic patronage of Protestant church-building in Berlin, see H. McLeod, *Piety and Poverty: Working-Class Religion in Berlin, London and New York, 1870–1914* (New York, 1995), pp. 20–1.

16. *Die preußische Gesetzgebung über die Vermögensverwaltung in den katholischen Kirchengemeinden und Diözesan mit Anmerkungen und Sachregister von Dr A. Förster* (Berlin, 1907): voting eligibility criteria on pp. 44–5.

17. K.H. Pohl, 'Katholische Sozialdemokraten oder sozialdemokratische Katholiken in München: ein Identitätskonflikt?,' in O. Blaschke and F.-M. Kuhlemann (eds), *Religion im Kaiserreich: Milieus–Mentalitäten–Krisen* (Gütersloh, 1996), pp. 233–53; H.-J. Nesner, 'Die katholische Kirche,' in F. Prinz and M. Krauss (eds), *München—Musenstadt mit Hinterhöfen: Die Prinzregentenzeit, 1886–1912* (München, 1988), p. 201.

18. *Katolik*, Feb. 8, 1912, p. 3.

19. Bernhard Stephan, et al. to General-Vikariats-Amt, May 2, 1912, Acta parafialne Laurahütte, vol. 1, AAK.

20. Victor Schmidt to Cardinal Kopp, Feb. 16, 1909, Acta parafialne Laurahütte, vol. 1, AAK.

21. Kunze to Kopp, Feb. 18, 1911, Acta parafialne Laurahütte, vol. 1, AAK.

22. Hilger to Kunze, Feb. 21, 1911, Kunze to Kopp, Feb. 28, 1911, Acta parafialne Laurahütte, vol. 1, AAK.

23. Kunze reported to the bishop that nine-tenths of the council was made up of 'Polish-national-oriented personalities': Apr. 25, 1911, Acta parafialne Laurahütte, vol. 1, AAK.

24. Mitglieder der Kirchenvorstand und Gemeindevertretung to Kopp, March 16, 1912; Anschrift signed by Stephan, Sauer, Widera, Mokrski, and Ramatschi, Apr. 21, 1912, Acta parafialne Laurahütte, vol. 1, AAK.

25. For example, in a poster prepared for (1912?) church council elections, where it is explained that the 'Hüttenpartei' was the same as the 'Deutschepartei': Sygnatura 462, Zespół Landratsamt Kattowitz, APK. In a Polish-national newspaper, the 'foundry- and mine-owners party'—Hütten und Grubanpartei—was preferred: *Gazeta Ludowa*, Nov. 15, 1912.

26. June 2, 1912, p. 3. See Anderson, *Practicing Democracy,* pp. 199–237 on the opportunities for industrial 'bread lords' to 'make' elections, but also the limits on this influence.

27. *Dziennik Śląski*, Nov. 3, 1912, translated in memo from Grenzkommisar Mädler to Landrat Kattowitz, Sygnatura 462, Zespół Landratsamt Kattowitz, APK. Hilger, like most higher officials in the state and private sector in Upper Silesia, was a Protestant and from western Germany.

28. Poster prepared for (1912?) church council election: Sygnatura 462, Zespół Landratsamt Kattowitz, APK.

29. 'German' slate taken from letter of Kirchenvorstand to Landrat, Dec. 12, 1911; 'Polish' slate from a poster prepared for (1912?) church council election: Sygnatura 462, Zespół Landratsamt Kattowitz, APK.

30. *Gazeta Ludowa*, Nov. 15, 1912.

31. 'Polnische Geistliche für polnische Parochie,' *Gazeta Ludowa*, March 15, 1911, translated in *Gesamtüberblick über die polnische Presse* (1911), 332–3.

32. Abschrift by Stephan, sent to Generaldirektor Hilger, Apr. 4, 1914, Acta parafialne Laurahütte, vol. 1, AAK.

33. *Laurahütter Zeitung*, Aug. 1, 1911, p. 2.

34. After the division of the parish went through, Kunze sought financial compensation from the Laurahütte joint-stock company (the old parish's patron), apparently for this loss of revenue linked to reduction in the number of parishioners. The company rejected the claim, noting that the subdivision was already planned at the time that Kunze became pastor (1909). See Aktien-Gesellschaft to Bertram, Nov. 28, 1914, Acta parafialne Laurahütte, vol. 1, AAK.

35. M. Czaplinski, *Adam Napieralsk, 1861–1928: Biografia polityczna* (Wrocław, 1974), p. 17.

36. Table of electoral results by community: Sygnatura 286, Zespól Landratsamt Kattowitz, APK.

37. On 'organic work' or 'small work,' see W. Hagen, 'National Solidarity and Organic Work in Prussian Poland, 1815–1914,' *Journal of Modern History*, 44 (1972), 38–64.

38. Translation of article from *Dziennik Śląski*, October 22, 1912, Mädler to Landrat Kattowitz, same date, Sygnatura 462, Zespół Landratsamt Kattowitz, APK.

39. Dembinski, et al. to Kopp, May 15, 1912, Acta parafialne Laurahütte, vol. 1, AAK.

40. Abschrift by Kunze, March 14, 1914, Acta parafialne Laurahütte, vol. 1, AAK.

41. Abschrift of Kopp, Aug. 14, 1912, Acta parafialne Laurahütte, vol. 1, AAK.

42. Abschrift by Stephan, sent to Generaldirektor Hilger, Apr. 4, 1914, Acta parafialne Laurahütte, vol. 1, AAK.

43. This is not to deny that architectural styles that are 'objectively' transnational can be read as nationally loaded in particular contexts. As Gregor Thum has shown in his study of post-World War II Breslau, traces and reconstructions of Gothic architecture were valorized by Polish authorities as offering a connection to the Piast period, when the region had robust political ties to the Polish state—ironically ignoring the widespread association of Gothic architecture with German romanticism. But this concerned Gothic rather than neo-Gothic architecture—in other words, its resonance hinged on archeological claims about early settlement patterns, not just suggestive stylistic appropriations in the present. See Gregor Thum, *Uprooted: How Breslau Became Wroclaw during the Century of Expulsions* (Princeton, 2011), pp. 329–47.

44. On comparable Polish–German church fights in the Ruhr, see R.C. Murphy, *Guestworkers in the German Reich: A Polish Community in Wilhelmine Germany* (New York, 1983), pp. 85–91; and C. Klessman, *Polnische Bergarbeiter im Ruhrgebiet, 1870–1945* (Göttingen, 1978), pp. 62–3. 'Church wars' in the United States had a somewhat different focus: the use of explicitly 'national' (i.e., non-territorial) parishes shifted national controversies to the diocesan level. See A.J. Kuzniewski, *Faith and Fatherland: The Polish Church War in Wisconsin, 1896–1918* (Notre Dame, 1980).

45. See census statistics in note 8. In the 1921 plebiscite, Siemianowitz residents voted for Germany by a margin of 56 to 44 percent; Laurahütte residents voted for Germany by a margin of 67 to 33 percent: *Journal officiel de Haute-Silesie*, May 7, 1921, p. 109.

46. Dembinski, et al. to Kopp, May 15, 1912, Acta parafialne Laurahütte, vol. 1, AAK.

47. Hilger to Kopp, Dec. 6, 1913; Kopp to Hilger, Feb. 19, 1914; Kunze to Kopp, March 14, 1914, Acta parafialne Laurahütte, vol. 1, AAK.

48. Ironically, some of the best evidence of continuity in bilingual pastoral care can be found in a publication by a German minority leader: see appendix 13 in T. Szczeponik, *Die Gewissensnot der Deutschen Katholiken in Polen* (Kattowitz, 1927), p. 172. This shows that in many parishes, such as St Antoni in Siemianowice, there

were no changes whatsoever to the linguistic balance of services. On provision of German-language associations, see K. Śmigiel, *Die statistischen Erhebungen über die deutschen Katholiken in den Bistümern Polens 1928 und 1936* (Marburg, 1992), pp. 48, 206.

49. Śmigiel, *Die statistischen Erhebungen*, p. 48.
50. Visitatio Pastoralis, May 9, 1933, Akta lokalne, sw. Antoni, Siemianowice (AL 2100), AAK.
51. Śmigiel, *Die statistischen Erhebungen*, p. 206.
52. Rezolucja, Zofia Gabzylylowa (sp?), on behalf of the Koło Towarzyskie Polek in Siemianowice, published in Polska Zachodnia, Oct. 15, 1934, AL 2100, AAK.
53. Śmigiel, *Die statistischen Erhebungen*, p. 206.
54. Kubis to Kuria, Oct. 27, 1939, Akta lokalne Załęże (AL 2521), AAK.
55. Lubina to Bischöfliche Amt, Nov. 4, 1939, AL 2100, AAK.
56. Scholz to Kuria, Feb. 3, 1936, AL 2100, AAK. On the Germanization of parish governance, see also J. Myszor, *Historia Diecezji Katowickiej* (Katowice, 1999), pp. 346–55.
57. R. Kaczmarek, *Gorny Śląsk podczas II wojny świątowej* (Katowice, 2006), pp. 164, 172–5. How much agency residents exercised in these categorizations and recategorizations is a complicated and contentious issue. Suffice it to say here that they were obviously conducted under duress—people would have been aware of the differential life chances of Poles and Germans in the Third Reich—but did nonetheless involve individualized participation and acquiescence.
58. List pasterski, to be read on Feb. 18, 1945, in Syg295/VII/243, Zespół 1400 (Polska Partia Robotnicza, Komitet Centralny), Archiwum Akt Nowych. On the interpretation of mass adoption of German nationality as a 'masquerade,' see S. Adamski, *Pogląd na rozwój sprawy narodowościowej w województwem śląskim w czasie okupacji niemieckiej* (Katowice, 1946).
59. Lubina to Kuria, Dec. 7, 1945, AL 2100, AAK.
60. http://www.swantoni.siemianowice.wiara.pl (accessed Feb. 20, 2013).
61. 'Przynależność narodowo-etniczna ludności—wyniki spisu ludności i mieszkabń 2011,' Notatka informacyjna, Wyniki Głównego Urzędu Statystycznego, Materiał na konferencję prasową w dniu 29.01.2013 r.
62. R. Brubaker, *Nationalism Reframed: Nationhood and Nationalism in the New Europe* (Cambridge, 1996), p. 19.
63. E. Renan, 'What Is a Nation?,' in G. Eley and R. Suny (eds), *Becoming National: A Reader* (New York, 1996), pp. 53–4.

Part V
Labor and Social Experience

10

'Frontier Indians': 'Indios Mansos,' 'Indios Bravos,' and the Layers of Indigenous Existence in the Caribbean Borderlands

Jason M. Yaremko

This study of borderland experiences in eighteenth- and nineteenth-century Cuba encompasses several case studies in borderlands as lived, physical realities for three large groups of Amerindian immigrant peoples: southeastern cultures such as the Calusa, Creek, and other indigenous peoples of Florida, the Apache and other southwestern peoples, and the Mayas of the Yucatán peninsula. The essay will focus on the varied forms of struggle, resistance, adaptation, and persistence of various Amerindian individuals and communities in eighteenth- and nineteenth-century Cuba, working toward an understanding of borderlands as multileveled, multifaceted, and fluid experiences in a place where cultures from Europe, the continents, and the Caribbean met, and where 'subaltern' indigenous peoples sometimes influenced empires.[1]

This perspective has often been overlooked for Cuba, in part because of the historical endurance of the extinction trope accepted by most Cuban and foreign scholars concerning both indigenous and immigrant Amerindians. This has been facilitated by an emphasis in Cuban historiography on a national history based on unity, in turn inspired by a theoretical racial integration ('Cubanidad' or 'Cubanía') that has historically narrowed the space for debate about other cultures in Cuba outside the undeniable one of Africa-Cuba, and, to a limited extent, Chinese indentured labor. Amerindian passages to Cuba— voluntary and involuntary—predated these other diasporas, eventually intersecting with them through transculturation. What follows focuses on the 'other'—Amerindian—diaspora, through the conceptual framework of borderlands as zones of conflict, negotiation, and diverse adaptation. It highlights the role of indigenous geopolitics, in contrast to the conventional emphasis on European imperial powers in Amerindian migration and intercultural relations.

218

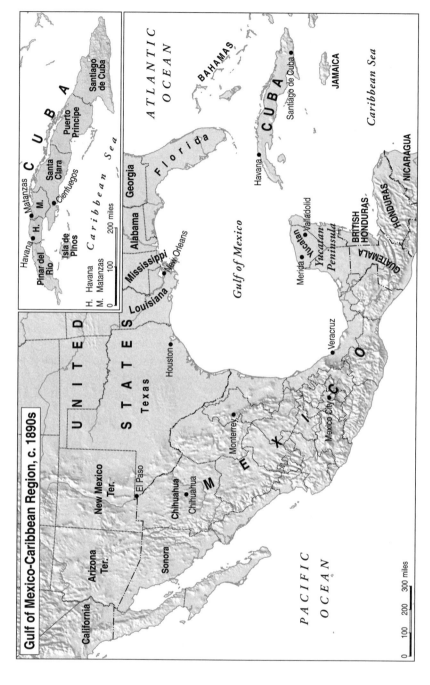

Map 10.1 Gulf of Mexico and Caribbean Region, c. 1890s

Florida: Indigenous Geopolitics and the Utility of Havana

The eighteenth century ushered in an era of change in the long-term history of Caribbean imperial rivalries. For the Spanish Empire in the Americas, it meant a series of growing challenges from competing empires in the autumn of Spain's hemispheric hegemony. Imperial European geopolitics took its regional toll on both Spain and indigenous allies like the Calusa of south Florida, for whom the century was fraught with colonial wars that brought hardships but simultaneously offered the opportunity to salvage, adapt to, and even influence the changes often violently imposed by competing colonialisms. This was the case in the southeastern peninsula of Florida, where the autochthonous communities of indigenous peoples had both succumbed to and confounded two of the most powerful empires in the region, Spain and Britain. While Spaniards used the largest island in the Caribbean as a base for expeditions and missions into southeastern North America, many southeastern indigenous peoples, like the Calusa, Ais, Tequesta, Timucua, and others, journeyed to Cuba and its primate city of Havana as traders, diplomats, refugees, and immigrants, to treat, trade, and work. The frequency of indigenous visits and stays in Havana was a function of several factors. In part a perpetuation of precolonial practices and indigenous geopolitics, indigenous forays into Spanish Cuba were also a consequence of Spanish adaptation to a failed policy of conquest in Florida. This failure was followed in the late sixteenth and seventeenth centuries by a new policy emphasizing diplomacy, the instruments of which were the presidio (garrison) and Christian mission.[2]

At the dawn of the eighteenth century violent change came to the Caribbean basin in the form of colonial wars, punctuated by the War of the Spanish Succession and the Seven Years' War with their regional and local manifestations. British alliances with the Yamasee and then Uchisi facilitated intensive campaigns from the north into Florida, forcing Spain's indigenous allies to fight or take flight, generating a series of indigenous migrations to Cuba. Throughout the eighteenth century hundreds fled. In February 1711 alone some 280 of Florida's indigenous people arrived in Havana.[3] By 1763, when Spain succumbed to British naval power, losing Florida in exchange for English-occupied Cuba, contemporary reports observed that Spanish evacuation of the peninsula included the flight of over two hundred Amerindians, including Calusa, Apalachee,Timucua families and others, most of whom went to Cuba.[4]

Many of these indigenous refugees became residents in Cuba. According to the royal Protector of Indians, Cristóbal de Sayas Bazán, more than two hundred of the Amerindians who had fled to Cuba in 1711 still resided in a community near Havana, likely the settlement of Guanabacoa, in the 1720s.[5] Furthermore, parish records for Havana and environs describe new generations of Calusa and other Amerindians, the sons and daughters of indigenous immigrants, born in Cuba in the 1720s, 1730s, and later.[6] More refugees arrived in 1763 and were eventually settled in various parts of the island colony.[7] Some appear to have married among themselves or intermarried with others, adding to the mestizo population. Occupationally, fragmentary evidence suggests that, as 'Indios,' they lived at the lower socioeconomic rungs of colonial Cuban society, working as fishing guides, stevedores, domestic servants, and laborers. Some worked within the European and indigenous geopolitics that endured, as official interpreters for Amerindian individuals and delegations who continued journeys to Cuba into the late eighteenth and early nineteenth centuries.[8]

The endurance of relations with Spanish authorities in Cuba, demonstrated in the countless trips to Havana during the British occupation of Florida, was no mere matter of loyalty. The Lower Creeks discerned their interests as logically and strategically as did their European counterparts: Lower Creek leaders, their headmen, families, and other relations continued their trek across the straits to receive gifts, trade with, and work among the Spanish in Cuba. Creek leaders like Thlawhulgee, Escuchape (Escuchabe, or the Young Lieutenant), Estimslayche, and Tunape, met with the governor and captain general and other colonial officials in Havana, a place for negotiation and the venue for Amerindian leaders and other representatives to assert their interests, not infrequently at Spanish material (and British political) expense. Throughout the late eighteenth century, Creek leaders held extended meetings with colonial officials in Cuba, communicating their need for support to oust the British and facilitate the return of the Spanish. In a May 1775 conference with the governor in Havana, Escuchape reported on rising clashes with the British, and reminded colonial officials of the Spanish crown's claims of allegiance to its indigenous allies.[9] Likewise, Estimslayche reassured the colonial government of Creek fealty while reminding them of the Spanish promise that the Lower Creek 'would not be abandoned.'[10] Both vowed to honor the alliance and to 'maintain with this island [Cuba] a reciprocal trade of pelts, horses, and other fruits.'[11]

The responses of the government in Havana reflect the necessary paradoxes of geopolitics: governors complained of the frequency of the Amerindian visits, their repeated requests for aid and resolution, and for material support when in Havana and for the return trip—costs that in one year ran to thousands of pesos.[12] The Governor Marquis de la Torre conveyed the crown's gratitude to Amerindian delegates, at the same time attempting to subtly discourage the frequent visits. Privately, the governor protested about the 'continuous troupes of men, women, and children.'[13] Spanish officials were divided between those like the Marquis and the Indian agent Juan Josef Eligio de la Puente who supported such expenses as worthwhile, even vital. Yet both recognized the Creeks as significant political and economically strategic allies, both to regional imperial geopolitics and the economy of Cuba. The Marquis de la Torre conceded that he could take no immediate strong measures, 'due to the necessity of placating them so that they will not impede our fishing on their coasts, which would cause great harm to this city.'[14]

The colonial government in Havana continued to receive indigenous delegations, accommodate them, whether for a few days or several months, and gift and supply them on their return. As the Marquis de la Torre realized, accommodating Creeks in Havana encompassed more than sensitivity to the role of the visiting indigenous in regional geopolitics. Local needs became intertwined with European and indigenous geopolitics in a way that, to a significant degree, made Havana dependent on good relations with the Creeks and other indigenous allies in the peninsula. Put another way, the indigenous peoples of Florida played direct and indirect roles in the economy, livelihood, and sustenance of the primate city of Havana.

Spanish relations with the Creeks, therefore, were also a matter of economic necessity. The Lower Creeks worked in Havana and environs as traders. The most voluminous trade was in the form of deer hides, furs, ambergris, horses, bee's wax, dried fish, and other goods, not a few of which colonial Cuban society depended upon. As one contemporary observer noted in the mid-1770s, the indigenous trade with Havana was considerable, and 'the Spaniards received them very friendly.'[15] At another level, by the late eighteenth and early nineteenth centuries, the many ranchos or small farms and fisheries secured by Cuban fishermen on Florida's eastern shores also employed indigenous labor, especially women. In many cases, these female employees became sexual and marital partners, some receiving the sacrament in Havana, all such unions regarded as legally binding in Cuba. Evidence strongly suggests that a considerable number of these unions endured into the nineteenth century, as Creeks and Seminoles immigrated to, were baptized, educated, and remained in Cuba.[16]

Though the numbers and longevity of the children, wives, and mothers of Cuban-Amerindian fishing families who resided in Cuba are less clear, there is evidence to show that they were not the only indigenous of Florida who chose to stay. Spanish government reports provide glimpses of life for those who came to Cuba as refugees, spouses, and immigrants, and struggled to make a living in Havana and its environs. Reports in the early 1770s described the diverse circumstances and conditions of Florida Amerindians resident in Cuba. According to one report in the summer of 1771, some Amerindians endured destitution, while others, especially the men, were characterized as 'robust' workers, employed as fishermen, wood cutters, and in various trades.[17] Others, particularly indigenous women, intermarried with 'naturals' of Cuba, and 'they have other means and resources on which they can live.'[18] The growth of settlements like Guanabacoa suggests that autochthonous adaptation and resourcefulness would serve indigenous immigration in Cuba well into the nineteenth century.

'Indios Barbaros' in Cuba

Unlike the indigenous migrants from Florida, the various Apaches, Chichimecas, and other 'indios barbaros' or 'wild Indians' of northwestern New Spain shared a radically different relationship with the Spaniards, influenced in no small way by their persistent resistance to Spanish colonization. The Provincias Internas del Norte, the frontier provinces of the northern interior, arguably resembled the region of Florida during the sixteenth century: in these lands indigenous peoples almost uniformly refused to be colonized and assimilated. They fought aggressively against Spanish military expeditions, slavers, and mining prospectors who aimed to control the native population in order to be able to proceed with colonization, resource exploitation, and evangelization. As noted above, generations passed before the Spanish reverted to diplomacy in Florida and began to make inroads into their relations with the region's indigenous peoples.

In northwestern New Spain Spanish colonization advanced along the corridors of the Sierra Madre Occidental through missions established among sedentary farming peoples and the labor drafts that brought mixed populations to the mining centers, despite concerted indigenous resistance to enslavement. During the seventeenth and eighteenth centuries Spanish expeditions, garrisons (presidios), and mining and ranching settlements pushed into the northernmost reaches of their American empire, sparking increased confrontation with the similarly expanding 'raid and trade' networks of the Apaches, Comanches, Kiowas, and Navajo equestrian and nomadic groups for the acquisition and trading of various commodities—including livestock and human captives—that stretched from California to Louisiana, Mexico City, and Cuba.[19] Spanish Apache policy relied on the precedent of the earlier, grueling war with the Chichimecas, which had ground to a halt by the end of the sixteenth century through Spanish inducements of material (agricultural) aid, government protection, religious instruction, and Chichimec willingness to make peace. Colonial officials hoped to repeat such success with the Apaches. To a significant degree, they did: throughout much of the eighteenth century alliances were secured with Lipan and other bands of Apaches.[20] Many, however, remained defiant toward Spanish colonization. For these Apaches the Spanish resorted to war, imprisonment, and forced relocation or deportation.

The practice of forced, physical relocation of recalcitrant Apaches was given royal sanction through the Reglamento of 1729. By the 1730s, the first *colleras* or convoys of Apaches captured as prisoners of war—men, women, and children—were conveyed by Spanish military escort to remote locations such as Mexico City. By the mid- to late eighteenth century, the rising rate of escape, and, in the eyes of the Spanish, repeat offending, among Apache prisoners, generated calls for deportation to insular sites, places where, colonial officials believed, escape was far less likely and prospects for pacification and assimilation far more promising. Some, like the military commander Jacobo Ugarte, insisted on Cuba for the permanent banishment of defiant Apache leaders and their followers. The Spanish crown eventually concurred, sanctioning, in 1783, the deportation of about one hundred Apache prisoners of war south 'to a place from which they could never return.'[21] This was followed by a series of *colleras* and deportations involving hundreds of Apache and other 'indios barbaros.' Their journeys began in regions like Chihuahua in Nueva Vizcaya, Nueva Santander, or as far west as Sonora province in the Intendancy of Arizpe, from which they were marched to Alta Mira or Veracruz, and from there to the principal destination, Cuba.[22]

Spanish colonial officials believed Cuba an obvious solution to the escape-plagued prison of San Juan de Ullua in Veracruz, and a more practical place of exile for 'los indios infieles.'[23] The island colony was an ideal site, officials argued: both distant and insular, it would force the 'indios barbaros' to become resigned to their fates, increasing the probability of 'taming' defiant indigenous like the Apaches, putting them to 'good use' in royal works of road and fortification construction and repair, field work, and domestic service, and facilitating the eventual Christianization and assimilation of men, women, and children.[24] Though not all colonial officials in Cuba supported the argument,[25] dissenting officials appeared to be in a minority and were otherwise overruled by the crown: the new deportation policy was officially decreed in April 1799, reinforced two years later, and insisted on thereafter.[26]

For the duration of the eighteenth and early nineteenth centuries, at least until the last known convoy in 1816, a stream of indigenous deportees followed, ebbing and flowing relative to clashes with Spanish military forces in the provinces. Apache and other indigenous men, women, and children suffered the harsh trek across difficult terrain, bound and under guard, were sequestered in Veracruz, and then delivered in Spanish warships to Havana. Surviving arrivals in Cuba fulfilled a dual purpose: to rid the colonial government in New Spain of parties disruptive to the expansion of Spanish colonialism, and, at the same time, help remedy an endemic labor shortage in the island colony.

The dilemma born of a paradox for dissenting Cuban colonial officials was that, although some complained of the 'dumping' of Apache and other prisoners from the mainland, they required them to fill chronic labor shortages in Cuba. The colonial government needed laborers to work in the crown's arsenals and in the construction and repair of fortifications, roads, and other infrastructural works. Furthermore, if certain captains general of Cuba opposed Apache deportations, *Habaneros* proved very receptive: demand became substantial among private citizens and institutions on the island whose apparent need for laborers and domestic servants was considerable. All of this was entirely consistent with the objectives of the crown. As a royal order reasserted in November 1799, in spite of the 'inconveniences' of transporting the 'Indios barbaros prisioneros,' once in Cuba, Amerindian men were put to work, and the women and children distributed among the families of the elite classes of the island, to be maintained and 'educated in the religion,' trusting that, with catechization, Apache exiles would acquire 'some docility.'[27]

Cuba's historic labor shortage played an overriding role in the government's intentions for Apache males. Amerindian laborers like the Apaches were viewed as significant additions to the labor system, contributing to a relatively diverse system of forced or enslaved labor that included African, Chinese, and Yucatecan Maya workers. This acceptance was also based, in part, on the enduring concern to maintain an appropriate racial balance between white and non-white populations in the colony, a sentiment voiced by a number of colonial political and economic elites on the island.[28]

'Orders' for the distribution ('repartimiento') of individual Amerindian men and women were directed to individuals and institutions in Havana at the will of the crown and/or through requests (peticiones). Interested citizens and institutions included the small and incipient middle classes of colonial Cuba who, in exchange for Amerindian labor, extended their hand to provide education and also instruction in the tenets of Christianity. Corporate interests weighed in too, for the colonial Catholic Church and religious orders like the Jesuits and Bethlemites were major patrons who were in need of laborers as well as souls to save.[29] Indeed, a range of institutions and individuals in and around Havana petitioned the colonial government for indigenous slave labor. In an order dated February 9, 1802, 'Mecos' or indigenous men were variously distributed to the royal intendent, the rector of the University of Havana, military officers, and various others among colonial Cuba's military, economic, and political elites—among them administrators of several hospitals (including Paula and the Casa de Beneficencia), ministry officials, and the owners of a tobacco factory. Families constituting colonial Cuba's embryonic middle classes were also counted among those in line to receive 'Mecas' maids, among them lawyers, teachers, and shop owners.[30]

While evidence for the quotidian realities of Apache and other Amerindian exiles in Cuba during this period is fragmentary, we can glean some sense of occupation, social place, working conditions, and Amerindian responses to forced exile. Apaches and other Amerindians arrived in bondage, most knowing little or no Spanish. It is clear that many if not most Apache and other 'Mecos' Indians who arrived in Cuba were destined for employment as domestic servants or laborers on fortifications and road works. Petitions to the colonial government clearly indicate, moreover, that at least some Apache and other 'Mecos' prisoners were directed to labor in one of various sectors in the colonial economy, as evidenced in the request from the tobacco factory, to give one example. Apache and other Amerindian workers represented small but significant groups that reinforced the diversity of forced or enslaved labor in colonial Cuba. At the same time, while African slave laborers and Chinese indentured workers outnumbered them, indigenous laborers predated both in their immigration to the island colony, even as they shared the realities of forced relocation with African and creole slaves.

Apache men often did the heavy work of building or repairing roads and/or fortifications like el Morro in Havana, while others were assigned agricultural work like that in the tobacco vegas. Like indigenous *forzados* from other regions who were transported to Cuba during this period—rebellious Yucatecan Mayas, for example—Apache men likely found themselves in a number of worksites, such as Havana's shipyards, which, by the late eighteenth century, 'were building some of the largest and best warships in the world.'[31] Nor were children ineligible for work. Indigenous children and youth were generally favored by the Catholic Church and religious orders both as servants and students, initiates into Spanish Christianity and culture. In the interests of 'religion, the State and humanity,' the 'Indios Mecos of minority age' were sent to religious orders like the Bethlemites to learn letters and trades and then placed in the employment of local shops in and around Havana, measures that were justified in order to transform the indigenous youths into both Christians and useful adult members of Havana society.[32]

Overwhelmingly, the Apache women and other 'Mecas' coveted by *Habanero* and other elite families found themselves in the homes and estates of these recipients of the crown's latent *repartimiento*. Amerindian women, and even some men, were chosen to keep house, mind aged relatives, and care for the children. In one such instance, a 'Meca' servant was directed to work for a regimental officer as a nanny (*una criada*) for his child.[33] Others were assigned to the homes of widows, others still in accordance with the requests of military wives, their husbands absent in defense of the realm.[34] Yet, contrary to the assertions of the proponents of Amerindian deportation, Apache and other indigenous exiles did not uniformly resign themselves to their fates, but actively sought freedom no less so than their African and Chinese counterparts. Of the many incidents reported, typical was the escape in August 1802 of six Amerindians, *cimarrones* (runaways) who had worked in a munitions depot in Havana, and who survived through raids on local farms in the district of Tapaste, southeast of Havana, stealing livestock for food and evading capture.[35] Reports circulated of a fugitive slave settlement that they had formed in the woods near the town of San José de Las Lajas. Amid the pursuit and violent struggles that ensued, several people, including two of the Apaches, were killed.[36] Following the capture and eventual conviction of the surviving Apache fugitives Rafael Vitaque and Oste, the Captain General Marquis de Someruelos attempted to send one back to prison in Veracruz; Iturrigaray, the viceroy of New Spain, sternly reminded him that Amerindian prisoners 'were never to be returned to New Spain' regardless of the grounds.[37]

Fragmentary evidence suggests further that when Apache and 'Mecos' laborers resisted violently they did not do so indiscriminately or without cause;
nor were they averse to forming alliances with other fugitives. In a number of
cases documented by Cuban anthropologist Gabino La Rosa Corzo, reports of
'rebellious Indians'—variously described as 'Mecas,' 'Indios Bravos,' or 'Indios
Feroces'—roaming and raiding in western and central Cuba, abounded during
the early nineteenth century, which was consistent with a period of stepped-up
Amerindian deportations to the island.[38] Further evidence speaks to another of
the important dynamics of Amerindian existence in nineteenth-century colonial Cuba. In October 1802, for instance, two 'rebellious Indians' reportedly
roamed the Filipinas, an area west of Havana, raiding plantations. While one
was killed by local authorities, the second was later joined by other 'Mexican
Indians' as well as eight African slave fugitives. Still at large two years later, the
reward for the capture of this group exceeded 2000 pesos; the outcome of the
pursuit is unclear.[39] Similar incidents involving 'Mexican' and/or 'rebellious
Indians' were reported throughout western and eastern Cuba; La Rosa Corzo
suggests that the protagonists were 'possibly Apache.'[40]

Crucial for long- and short-term survival, Amerindians like the Apache exiles
in Cuba, therefore, did not always shun alliances with African slaves or with
other Amerindians. More research needs to be done on these alliances in part
because of their significant implications for a thorough understanding of the
history of intercultural relations, race, and *mestizaje*—the 'middle ground'—in
Cuba, which is a recurring theme in Cuban history. If Spanish deportation
of Apache and other 'Mecos' Amerindians ceased with the independence of
Mexico in 1821, by no means did this mark the end of the dislocation and
forced exile of indigenous peoples altogether or their forced relocation to the
largest remaining insular Spanish colony in the region.

'Guachinangos':
The Maya Middle Ground in Nineteenth-Century Cuba

Maya journeys to Cuba have endured for more than half a millennium. Since
at least the mid-sixteenth century many had come involuntarily, captured and
forcibly relocated as slaves destined for the labors of early colonial society.
These early Amerindian exiles, more than a thousand of them, helped found
one of the greatest cities in the world. The Barrio de Campeche, aptly and
deliberately named, was the 'Indian' barrio, one of two founding barrios in the
village of la Habana (the other, la Punta, was inhabited by Spanish colonists).
These Mayas or 'Indios campechanos' fulfilled some of the daily labor needs of
a growing colonial town, building and repairing churches, fortifications, and
roads in and around Havana. In their barrio in what is now Habana Vieja, they
and their children played pelota in the plaza for distraction.

Two centuries later, the Mayas continued to come to Cuba, some voluntarily as individuals seeking a new life or escape from harsh conditions in Mexico, others involuntarily as individuals and groups. Forced Maya migration to Cuba increased during periods of Yucateco unrest and rebellion in New Spain, like that led by Jacinto Canek/Uc in 1761. Sparked by festering tensions between Spanish and indigenous communities, these conflicts often resulted in captive Mayas, not unlike resistant Apaches, being banished to Cuba for ten years or more to labor in Havana's shipyards and fortifications.[41] Evidence also points to other dimensions in the lives of Mayas in Cuba, as in the case of the Maya slave owner or the Maya *curandero*, or healer.[42] While they were principally laborers, the Mayas' role in Cuba's economy and society was not limited to forced labor nor was it significant solely for the early colonial period.

The period during which perhaps the greatest concentration of Mayas descended on the island colony was the mid-nineteenth century, a definitive crossroads for both Cuba and Mexico. Cuba had become the world's foremost producer of sugar, having inherited the mantle from a revolutionary Haiti whose slaves had rebelled against and destroyed the system of labor that had powered a former world-class sugar-producing colony. Cuba reaped the rewards: from the late 1790s to the 1840s sugar production advanced rapidly, mills multiplied and expanded, and huge tracts of land were absorbed for cane cultivation. The demand for slave labor skyrocketed, generating a massive importation of African slaves, their numbers soon rivaling the white populace. To facilitate production and counterbalance the problems associated with chattel slavery, namely the rising slave population, British abolitionism, rising slave prices, escalating slave rebellions, and an increasing proportion of slaves who were becoming freed persons, indentured Chinese laborers were brought to Cuba. But by mid-century, dissatisfaction with Chinese workers prompted proposals from Cuban planters for the introduction of Mexicans, particularly Yucatecans deemed more culturally affinitive, acclimatized, and therefore more fit for work in a Caribbean climate. The timing was propitious.

Mexico was ripe with rebellion at various levels, especially between the center in Mexico City and the states, and between the center and periphery within the states themselves. Defiant Yucatán exemplified Mexico's regionalist, politico-economic, and cultural struggles. The rebellion of 1847, known but misnamed as the Caste War, was the culmination of various factors, prominent among them being growing burdens of civil and church taxation, the loss of communal lands to sugar latifundism (especially in eastern Yucatán), the deterioration of living standards, removal of social compacts like the traditional Maya right to present complaints and petitions (*petición* or *okotba than*), and the general reversal of rising expectations. By the 1840s, all of this converged into rebellion among the Mayas. The imposition of more burdensome taxes and the reimposition of the authority of the state and landowning elites, coupled with the loss of traditional Maya negotiating mechanisms, left Mayas with little recourse than rebellion. Hit hardest, the Maya peasantry of eastern Yucatán provided the bulk of the resistance forces that ignited the rebellion, which went on to expand, spreading westward.

The rebellious Mayas of eastern Yucatán also constituted the majority of prisoners and other dissidents transported to Cuba. The evidence here indicates that most of the laborers shipped to Cuba during the mid-nineteenth century were Mayas, evident both in the Maya surnames in the documentation and through direct reference to the terms 'indio' or 'indio yucateco' in cases of either Maya surnames or Spanish *apellidos*. (The etymology of the umbrella term 'Maya' is understood here; the difficulty of pinpointing the exact regions or villages of origin of most of the indigenous under study here forces the use of the name.) Quantitative estimations of Yucatecan Maya in Cuba during this period remain in dispute, ranging from approximately one thousand to several thousand. As González Navarro noted, a widespread contraband trade in Yucatecan Maya migrant workers makes an accurate estimate difficult to establish. Arguably, existing estimates may be conservative.[43]

Both the colonial government and creole planters in Cuba craved a stable, cheaper, even whiter, form of labor to offset the rising costs and liabilities of a slave system that appeared less congruent and economically viable in a plantation system moving toward a more technologically modern agrarian capitalism. Immigration, specifically imported Maya labor, offered a solution. For the Yucatecan state, the shipment of Maya prisoners of war removed sources of instability and facilitated prosecution of the war. Despite some initial resistance from Mexico City, both the federal and state governments cooperated in facilitating and formalizing the process for transporting Mayas to Cuba, including labor contracts and legislation passed to ensure the well-being of workers, the voluntary and temporary nature of the contracts, and the retention of Mexican citizenship. In practice, such contracts, when they existed at all, appear to have been honored more often in the breach.

Those Maya men, women, and children who disembarked at the harbors of Havana or Matanzas (most were concentrated in western Cuba) entered the ranks of a labor system almost wholly subordinated to Cuba's growing sugar industry. Though many worked as mill workers and cane cutters, others worked in diverse sectors of the economy directly or indirectly linked to the sugar industry: on *estancias*, tobacco vegas, *cafetales* or coffee plantations, cattle ranches, and cotton fields, on roads and railroads, in manufacturing, and in various other sectors, including domestic service.[44] Mayas in Cuba also entered into an indentured labor system rife with confusion, corruption, and abuse. Confusion stemmed from, on one hand, the bureaucratic labyrinths of Mexican and Spanish immigration policies and, on the other, the absence of any effective consolidation of such policies. At the same time, state officials, employers, and others often abused, ignored, or evaded importation policies, engaging in circumscription and contraband, thus deceiving the Mayas and the Cuban and Mexican governments. In numerous cases (those who were caught), for example, the entry of Maya individuals and groups into Cuba went unrecorded in the register books.[45] In one typical case that is representative of both the systemic abuse and its duration, the patrón Carlos López was charged in the summer of 1860 with illegally importing various Mayas or *colonos yucatecos*, that is, 'without proper license' or government authorization of any kind.[46]

Furthermore, if the Cuban indentured labor system was not as brutal as the African slave system, this remained a question more of degree than of kind. Forced labor was the reality for deported Mayas during this early period, and, not unlike the old encomienda system, patronos, while they tended to adhere strictly to the provisions allowing for corporal punishment of their charges, they often ignored or neglected the regulations regarding the well-being of Maya workers and their families. By the early 1850s, reports of the maltreatment (*maltrato*) and abuse of Maya workers in Cuba were legion. Brought to the attention of Mexican and colonial Spanish authorities in Cuba, the charges included slave-like working conditions, excessive corporal punishment, lack of payment, separation of families, refusal to terminate fulfilled contracts, and unauthorized transfer of a contract to another patrón. Transgressions by employers were both numerous and extended, persisting through the 1850s and 1860s and belying the effectiveness of protective legislation. Some of the sources of these charges included associates of the offending patronos themselves. As late as 1861, on the point of the official cessation of the traffic by Mexico, one associate acknowledged the 'numerous complaints' of abuse that persisted.[47] Many, if not most, were submitted by the Mayas themselves.

Maya workers were a substantial source of revelation and resistance, as many of the documented reports of abuse arose from their complaints.[48] Though the magnitude of Maya grievances is unclear, that it was significant is suggested in a number of letters to the Captain General of Cuba from 1853 through 1860, in which the Spanish government responded to reports of Mayas reduced to slavery with calls for investigation.[49] Mayas defended themselves and their families with a range of responses from the legal (*peticiones*) to the extralegal (flight, arson). The available evidence suggests that legal means of redress were more often utilized than illegal or violent means. In April 1860, Marcelino Peche filed a claim against his patrón for having been forced to work without a contract for eight years; another complained against his patrón for denying him the right to work for another employer at the end of his contract.[50] José María Chan and his wife María Antonia Pérez filed their claim against the patrón for illegally extending their contracts.[51] Felipe Cap asserted his claim against his patrón with charges of 'abuse, excess of punishment, lack of payment,' and other violations.[52]

The cases of Yucatecan Mayas Victoriana Acosta, Juana Alcántara, and Juana Poot illustrate the various charges and complaints asserted by Maya colonos in Cuba and the role played by Yucatecan Maya women in the island colony. All three women filed claims against their respective employers for unpaid labor and other contract violations, including physical abuse and family separation.[53] Many Maya women accompanied their husbands contracted as colonos; other women were contracted individually or in groups, domestic service the principal occupation. Maya women asserted their own grievances, and played a substantive part in household and labor politics in Cuba. Matthew Restall's analysis of Maya women in post-conquest Yucatán holds true for Cuba: 'The unfavorable colonial and gender structures of Yucatán elicited more than the silent resistance of anonymous Maya women'; in Cuba, alongside or independent of Maya men, 'Maya women also actively engaged the individuals and institutions that represented those structures.'[54]

Maya workers and their families employed an array of defenses ranging from the subtler forms of everyday resistance to more overt forms of defiance. Though the extent to which the colonial government addressed Maya claims and, also importantly, decided in favor of the Maya claimants is unclear, the evidence suggests that the frequency of such cases was considerable, from the first decree of 1853 that sanctioned the investigation of abuses to government reports that confirmed such investigations in the 1860s.[55] In instances where conditions militated against successful filing and prosecution of legal claims or other forms of negotiation, Maya workers resorted to several types of drastic actions: some nonviolent, others not.

In some cases, when Maya laborers in Cuba failed in their bid to ameliorate their conditions and defend their interests through legal means, they registered their grievances with their feet: they ran away. Some, for example, simply left the plantation and secured employment elsewhere on the island; others sought refuge in the Mexican embassy or in *el monte* (the woods), joining the ranks of *cimarrones*, or escaped African slaves.[56] In some cases, workers were falsely reported as deserters, as in the instance of one Maya family in early 1860. Upon apprehension and investigation, the police determined that the family had earlier fulfilled their contract and they were freed.[57] Flight from a patrón also served as a means to the end of ensuring justice, as the Maya escapee returned to pursue legal charges or exact revenge. While evidence for the latter point is more limited, there were a few cases in which Mayas responded to abuses with violence directed at the patrón and plantation property. Among the more prominent are acts of arson and other attacks on property. In Cienfuegos in 1861, colono Juan Oy, upon requesting discharge from his contract for having fulfilled his obligations, was found by colonial officials to have previously fled from another estate (*finca*), where he was suspected of other offenses against the owner.[58] Such reports, however, are few in number compared to those of Mayas in Cuba who asserted their rights through legal claims against the transgressions of the owners of plantations, farms, ranches, and other enterprises. Mayas sought resolution to their struggles with a patrón through negotiation or legal redress considerably more frequently than they did through violence. That at least some of these struggles were resolved for Maya claimants may account in part for the considerable number of Maya families and individuals who chose to renew their contracts or to find other employment and remain in Cuba.

A year after the Mexican government's official cessation of indentured Maya labor traffic to Cuba in 1861, the colonial government in Cuba initiated a new policy under the Consejo de Administración de la Isla de Cuba for introducing twenty thousand more Yucatecan colonos as free laborers, to be administered by the Sección de Agricultura, Industria, y Comercio.[59] While it is unclear how many more Mayas came to Cuba after 1861 under the new policy or returned to Mexico after fulfilling their contracts, documented court cases show that they continued to come. Furthermore, at the end of their contracts, many Mayas chose to either renew their contracts (*contratas renovadas*) or seek employment elsewhere in the colonial economy.

Whether after 1861 or earlier, Maya workers sometimes struggled to secure release from fulfilled contracts. The complaints of Felipe Cap and Feliciana Poot, filed in the autumn of October 1864 against their respective employers for refusing to discharge them (among other abuses), are typical.[60] Many if not most of these cases appear to have been decided in favor of the Maya claimants. At the same time, just as many colono contracts appear to have been settled amicably, other Mayas chose to renew their agreements or work for new employers. In one of a number of such instances, Güines, José Hu(h) renewed his contract with the patrona Rosa Figueroa de la Torre.[61] Others like Faustino Atum completed their pre-1861 contracts and signed new ones as free laborers who continued to work in Cuba; many of them worked in agriculture as farmers or ranchers, others were employed in and around Havana.[62]

The desire of Mayas like Hilaria Iba to work beyond their original contracts and 'remain on the island' was not uncommon, nor particularly surprising.[63] By the late nineteenth century, great change had overtaken the peninsula: land concentration in eastern Yucatán, the origin of most Mayas in Cuba, facilitated rapid expansion of agrarian capitalism, violently displacing Maya small farmers. Opportunity, meanwhile, appeared to beckon in Cuba. The combined effect of a booming sugar economy, enduring labor shortages, elite concerns about racial balance amid growing populations of African slaves and freed persons, and the threat of their involvement together with Chinese contracted immigrants in looming independence movements, encouraged a relatively more open attitude—if not always practiced—toward peoples characterized as docile, hard-working, and racially, culturally, and climatically appropriate. For the duration of the nineteenth century and the struggles of a lingering war in the Yucatán, Mayas continued to come, work, and live in Cuba as individuals, forming families and communities.

Conclusion

The eighteenth and nineteenth centuries marked epochs of distinctive sea changes in the Americas; social, economic, and cultural transformations were wrought by imperialism, nationalism, war, and political revolution. Colonial wars, wars for independence, and civil wars in newly independent American republics replaced their respective *ancien régimes*, laying the foundations for a new order and giving rise to new conflicts. Caught, literally and figuratively, in the middle, indigenous peoples in the affected regions responded in defense of their communities, their families, and their interests. Whether in resolute defiance of the expansionist tendencies of Spain or Great Britain, in alliances with the agents of these imperial powers (or both at different times), or through negotiation, indigenous nations and their leaders employed a diverse array of mechanisms that were both reactive and adaptive in the borderlands between European and indigenous realities. The historiography that is witness to this dynamic in the continental Americas is comprehensive, rich, and relatively well established. Considerably less developed and seriously understudied is the examination of the other 'middle ground' or borderlands between the continents and the first theater of sustained encounters between European and indigenous peoples: the Caribbean.

One crucial outcome of both European and indigenous geopolitics in the continents was the creation or generation of another borderlands region; or, to put it another way, the addition of a new dimension to the frontier zone, extending into the Circum-Caribbean. Historically, Cuba figured prominently and strategically as Spain's 'Pearl of the Antilles' and 'key' or 'gateway' to the Americas. At the same time, as this history demonstrates, Cuba lay both within and at the interstices of empire, not merely as a stage for imperial geopolitics but, through the actions and interventions of indigenous actors like the Creeks and Mayas, as both a dynamic middle ground where different cultures of the Caribbean and the continents met and from which indigenous geopolitics sometimes influenced powerful empires as well as rising republics. As the largest island and principal destination, Cuba is geographically and conceptually a microcosm in a new study of borderlands as zones of conflict, negotiation, and dynamic adaptation, as is illustrated in both the diverse layers of indigenous experience there and in the projection of indigenous influence beyond the island: in the intervening spaces between 'dominant' and 'subaltern' cultures, and between islands and continents. As the preceding case studies demonstrate, Cuba—including the straits between the island, Florida, and mainland Mexico—served as an important borderland site. Contrary to the perceptions of both contemporary opponents and proponents of the Amerindians in Cuba, moreover, agency rested not only with the era's imperial or national powers like Spain or Mexico, but also with 'los Indios infieles.'

Notes

1. This research was funded by the Social Sciences and Humanities Research Council of Canada (SSHRC). I also wish to thank the Canadian Association of Latin American and Caribbean Studies (CALACS) for permission to reprint portions of my article 'Colonial Wars and Indigenous Geopolitics' published in the *Canadian Journal of Latin American and Caribbean Studies*, 35 (2010), 165–96.
2. See Pedro Ménendez de Avilés to Francisco Borgia, Jan. 18, 1568, in F. Zubillaga (ed.), *Monumenta Antiquae Floridae, 1566–1572* (Rome, 1946), pp. 228–32, and Diego Ebelino de Compostela, Bishop of Santiago de Cuba to the Dean and Chapter of the Holy Cathedral Church of Santiago de Cuba, Jan. 2, 1690 (Archivo General de Indias [AGI], Santo Domingo, legajo 154).
3. Bishop Geronimo de Valdes to the King, Dec. 9, 1711, Santo Domingo, legajo 860, AGI.
4. See, for example, 'Listas de familias de Indios de Florida alojadas en Guanabacoa,' 1764, Reales cedulas y ordenes de Florida, Cuba, legajo 416, folios 755–70, AGI. See also B. Romans, *A Concise Natural History of East and West Florida* (New York, 1775), pp. 69–70.
5. Cristóbal de Sayas Bazán, Letter and record of service, Aug. 17, 1727, Santo Domingo, legajo 860, fol. 38–9, AGI.
6. See Libros Registros, Archivo Parroquial, Church of Nuestra Señora de la Asunción (Guanabacoa), Cuba, c. 1700, and also J. Worth, 'A History of Southeastern Indians in Cuba, 1513–1823,' paper presented at the 61st Annual Meeting of the Southeastern Archaeological Conference, St. Louis, Missouri, Oct. 21–3, 2004, p. 7.
7. J. Landers, 'Africans and Native Americans on the Spanish Florida Frontier,' in M. Restall (ed.), *Beyond Black and Red: African–Native Relations in Colonial Latin America* (Albuquerque, 2005), p. 63.
8. See, for example, Spanish reports on indigenous relations in Cuba, legajos 1220–2, AGI.
9. Don Rafael de la Luz, Interim Senior Assistant to the Plaza of Havana, May 2, 1775: Cuba, legajo 1220, AGI.
10. Juan Josef Eligio de la Puente to the Governor of Havana, the Marquis de la Torre; Havana, March 6, 1773, Cuba, legajo 1164, AGI.
11. Juan Josef Eligio de la Puente to the Governor of Havana, the Marquis de la Torre; Havana, March 6, 1773, Cuba, legajo 1164, AGI.
12. See report of Juan Josef Eligio de la Puente, Havana, May 16, 1777, Cuba, legajo 1222, fol. 748–9, AGI.
13. Governor of Havana, Marquis de la Torre, to don Julián de Arriaga; Havana, May 4, 1775, Cuba, legajo 1220, AGI.
14. Governor of Havana, Marquis de la Torre, to don Julián de Arriaga; Havana, May 4, 1775, Cuba, legajo 1220, AGI.
15. W. Bartram, *Travels through North and South Carolina, Georgia, and East and West Florida* (Philadelphia, 1791), pp. 227–8.
16. J. Yaremko, 'Colonial Wars and Indigenous Geopolitics: Aboriginal Agency, the Cuba–Florida–Mexico Nexus, and the Other Diaspora,' *Canadian Journal of Latin American and Caribbean Studies*, 35 (2010), 178.
17. Juan Josef Eligio de la Puente to Julián de Arriaga, Havana, July 2, 1771, Cuba, legajo 1211, AGI.

18. Juan Josef Eligio de la Puente to Julián de Arriaga, Havana, July 2, 1771, Cuba, legajo 1211, AGI.

19. J.F. Brooks, *Captives and Cousins: Slavery, Kinship, and Community in the Southwest Borderlands* (Chapel Hill, 2002), p. 33.

20. M. Moorhead, *The Apache Frontier: Jacobo Ugarte and Spanish–Indian Relations in Northern New Spain, 1769–1791* (Norman, 1968), pp. 11–14.

21. Cited in C. Archer, 'The Deportation of Barbarian Indians from the Internal Provinces of New Spain, 1789–1810,' *Americas*, 29 (1973), 377. See also M. Moorhead, 'Spanish Deportation of Hostile Apaches: The Policy and the Practice,' *Journal of the Southwest*, 17 (1975), 205–20, and M. Santiago, *The Jar of Severed Hands: Spanish Deportation of Apache Prisoners of War, 1770–1810* (Norman, 2011).

22. See colonial government correspondence and reports, 1797–98, in expediente 13, caja 099, vol. 208, Archivo General de la Nación, Mexico (AGN); report, Apr. 11, 1799, exp. 233, vol. 172, AGN; reports, 1800–02, exp. 14, caja 113, vol. 238, AGN.

23. Letter, Diego Josef Navarro, Governor of Cuba, to Viceroy Martín de Mayorga, Jan. 22, 1781; Diego Josef Navarro to Viceroy Martín de Mayorga, Jan. 23, 1781, Havana, exp. 5, vol. 21, AGN.

24. Diego Josef Navarro to Viceroy Martín de Mayorga, Jan. 23, 1781, Havana, exp. 5, vol. 21, AGN. Luis de Unzaga y Amezaga to Matías de Gálvez, Havana, Sept. 20, 1783, expediente 38, vol. 1083, AGN.

25. Luis de Unzaga y Amezaga, Governor of Cuba, to Matías de Gálvez y Gallardo, Viceroy, Sept. 20, 1783, Havana, exp. 38, vol. 1083, AGN.

26. Viceroy Miguel José de Azanza to Captain General of Cuba, Nov. 17, 1799, San Lorenzo, Archivo General de Simancas, Guerra Moderna, Simancas (AGS), legajo 7029.

27. Viceroy Miguel José de Azanza to Captain General of Cuba, Nov. 17, 1799, San Lorenzo, Archivo General de Simancas, Guerra Moderna, Simancas (AGS), legajo 7029.

28. See, for example, D.C. Corbitt, 'Immigration in Cuba,' *Hispanic American Historical Review*, 22 (1942), 283–4.

29. Antonio Cornel to Governor Captain General, Havana, Jan. 28, 1800, legajo 37, no. 3, Reales, Cedulas y Ordenes, Archivo Nacional de Cuba (ANC).

30. Noticia del Repartimiento de Mecos y Mecas, Feb. 9, 1802, legajo 1716, AGI.

31. R.W. Patch, *Maya Revolt and Revolution in the Eighteenth Century* (New York, 2002), p. 177.

32. Antonio Cornel to Governor Captain General, Havana, Jan. 28, 1800, legajo 37, no. 3, Reales, Cedulas y Ordenes, ANC.

33. Francisco Mendieta to Governor Captain General, Havana, Feb. 11, 1802, legajo 1716, AGI.

34. Maria del Rosario de Acosta to Governor Captain General, Havana, Feb. [?], 1802, legajo 1716, AGI.

35. Statement, Jan. 18, 1804, legajo 1716, Cuba, AGI. See also Archer, 'Deportation of Barbarian Indians,' 383–4, and G. La Rosa Corzo, *Runaway Slave Settlements in Cuba: Resistance and Repression* (Chapel Hill, 1988), p. 88.

36. Statement, Jan. 18, 1804, legajo 1716, Cuba, AGI.

37. Archer, 'Deportation of Barbarian Indians,' 384.

38. La Rosa Corzo, *Runaway Slave Settlements*, pp. 88–9.

39. La Rosa Corzo, *Runaway Slave Settlements*, pp. 88–9.

40. La Rosa Corzo, *Runaway Slave Settlements*, pp. 89–90.

41. See, for example, Autos criminals seguidos de oficio de la Real Justicia sobre la sublevacion que los Yndios del Pueblo de Cisteil y los demas que convocaron hicieron

contra Ambas Magistrades el de 19 de Noviembre de 1761. Mexico, legajo 3050, AGI, Microfilm reels C-7595–7. Also cited in Patch, *Maya Revolt and Revolution*, pp. 178–9.

42. See, for example, Jan. 5, 1750, Libros de Bautismos de los Indios, Negros y Pardos, Libro 4, 1749–1755, Iglesia Santo Cristo del Buen Viaje, Havana, Cuba; Actas de Cabildo, Matanzas, 1773, Archivo Histórico Provincial Matanzas (AHPM), Cuba.

43. See M. González Navarro, *Raza y Tierra: La Guerra de Castas y el Henequen* (México, 1970), and T. Rugeley, *Rebellion Now and Forever: Mayas, Hispanics, and Caste War Violence in Yucatan, 1800–1880* (Stanford, 2009).

44. See, for example, the labor obligations in a 'Contrata de Hombres Solos,' 1859, legajo 640, Gobierno Superior Civil, ANC.

45. See, for example, Manuel Arroyas to Governor Captain General of Cuba, Aug. 20, 1859, legajo 640, no. 20225, Gobierno Superior Civil, ANC. See also González Navarro, *Raza y Tierra*, pp. 127–9.

46. F. Fernández del Pino to Governor Captain General of Cuba, July 28, 1860, legajo 640, no. 20222, Gobierno Superior Civil, ANC.

47. Declaración de D. Pedro José Crescencio Martínez, Havana, July 14, 1861, legajo 640, Gobierno Superior Civil, ANC.

48. Ultramar, Presidencia del Consejo de Ministros to Governor Captain General, November 11, 1853, legajo 172, no. 327, Reales Cédulas y Ordenes; Sección de Fomento [?] to Gobierno Político, [1860], legajo 641, no. 20249, Gobierno Superior Civil, ANC.

49. Ultramar, Presidencia del Consejo de Ministros to Governor Captain General, November 11, 1853, legajo 172, no. 327, Reales Cédulas y Ordenes; Sección de Fomento [?] to Gobierno Político, [1860], legajo 641, no. 20249, Gobierno Superior Civil, ANC.

50. Sección de Fomento to Governor Captain General of Cuba, Apr. 23, 1860, legajo 640, no. 20220, Gobierno Superior Civil, ANC. F. Fernández del Pino to Governor Captain General, June 19, 1860; Sección de Fomento to Governor Captain General, September 6, 1860, legajo 640, no. 20247, Gobierno Superior Civil, ANC.

51. Consejo de Administración de la Isla de Cuba, la Habana, to Gobierno Superior Civil, Sept. 12, 1863, legajo 641, no. 20248, Gobierno Superior Civil, ANC.

52. Juan G. [?] to Gobierno Superior Civil, Oct. 28, 1864, legajo 642, no. 20297, Gobierno Superior Civil, ANC.

53. Victoriana Acosta to Governor Captain General, Oct. 21, 1859; Victoriana Acosta to Governor Captain General, Jan. 27, 1860, legajo 640, no. 20215, Gobierno Superior Civil, ANC; F. Fernández del Pino to Governor Captain General, Sept. 13, 1860, legajo 640, no. 20225, Gobierno Superior Civil, ANC; [illegible] to Governor Captain General, Jan. 20, 1862; Fernando de Levanco to Governor Captain General, Nov. 2, 1862, legajo 643, no. 20318, Gobierno Superior Civil, ANC; Sección de Fomento to Governor Captain General, June 2, 1862, legajo 641, no. 20248, Gobierno Superior Civil, ANC.

54. M. Restall, '"He Wished It in Vain": Subordination and Resistance among Maya Women in Post-Conquest Yucatan,' *Ethnohistory*, 42 (1995), 580.

55. See, for example, case reports in legajos 638, 640, and 641, Gobierno Superior Civil, ANC.

56. Secretaría Política to Governor Captain General, Matanzas, July 20, 1859; Jefatura Superior de Policía to Governor Captain General, Nov. 28, 1859, legajo 640, no. 20225; Letter to Manuel [?], Comisario de Policía, Distrito Regla, Sept. 10, 1859, legajo 640, no. 20215; Sección de Fomento to Gobierno Superior Civil, Oct. 12, 1861, legajo 641, no. 20248, Gobierno Superior Civil, ANC.

57. F. Fernando del Pino to the Governor Captain General of Cuba, Jan. 4, 1860; Felipe Arango to Governor Captain General of Cuba, Feb. 3, 1860, legajo 640, no. 20225, Gobierno Superior Civil, ANC.

58. Ramón Garbaly to Governor Captain General of Cuba, Feb. 24, 1861; F. Fernando del Pino to Governor Captain General, March 23, 1861; Lieutenant-Governor, Cienfuegos, to Governor Captain General of Cuba, May 18, 1861, legajo 640, no. 20215, Gobierno Superior Civil, ANC.

59. Consejo de Administración de la Isla de Cuba to Gobierno Superior Civil, March 21, 1862, legajo 641, no. 20248, Gobierno Superior Civil, ANC.

60. Juan G. [?] to Gobierno Superior Civil, Oct. 28, 1864, legajo 642, no. 20297; Juan Gerez, Secretaría de Agricultura, to Gobierno Superior Civil, Sept. 9, 1864, legajo 642, no. 20296, Gobierno Superior Civil, ANC.

61. Tenencia de Gobierno to Governor Captain General, Güines, April 13, 1861, legajo 641, no. 20248, Gobierno Superior Civil, ANC.

62. Tenencia de Gobierno, Sección de Fomento, to Gobierno Superior Civil, Sept. 6, 1864, legajo 642, no. 20296, Gobierno Superior Civil, ANC. See also Relación de los colonos yucatecos que pasan a la capital para ingresar en el Depósito de su clases, Cardenas, Sept. 25, 1863, legajo 641, no. 20248, Gobierno Superior Civil, ANC.

63. Sección de Fomento to Governor Captain General, July 28, 1860, legajo 640, no. 20215, Gobierno Superior Civil, ANC.

11
The Twisted Logic of the Ohio River Borderland

Matthew Salafia

In 1802, Thomas Worthington, Virginia slaveholder turned antislavery Republican leader of the Ohio statehood movement, said of his move across the Ohio River, 'I was decidedly opposed to slavery long before I removed to the territory—the prohibition of slavery in the territory, was one cause of my removal to it.' In contrast, formerly enslaved Kentuckian Richard Daly recalled of his situation in the 1850s, 'I worked on the farm and attended market at Madison, IN across the river, and never thought I would run away ... I knew that I could be free whenever I wanted to.' These statements run contrary to conventional logic. The white slaveholder leaves a slave state to advance his freedom, whereas the enslaved African American seemingly temporarily forgoes his freedom and chooses to remain in bondage in a slave state.[1]

How can we reconcile this apparent contradiction? First, the issue of choice is perhaps overstated for Daly. Richard Daly lived in Trimble County, Kentucky, on a plantation along the Ohio River. He married Kitty, a house servant from a neighboring plantation, and they had four children before Kitty died in childbirth. Daly protected his family as best he could and visited his children nightly. He did not accept the legitimacy of slavery as, by his own estimate, he helped thirty slaves escape from bondage. Daly did not believe, however, that the uncertain status he would hold in the 'free' states was necessarily better than his present situation. More important, his affection for his family overshadowed the advantages of freedom. Bondage conditioned his life, but love motivated him. Then Daly learned that his daughter was to be given away to her master's own daughter in Louisville, roughly fifty miles away. In 1857 the devoted father escaped to Canada with his four children.

Thomas Worthington, on the other hand, did choose to live in the nominally free state of Ohio, but his move was perhaps less dramatic than he made it sound. Worthington acquired vast landholdings in Ohio as a speculator and freed his slaves before making the move. His former slaves labored on Adena, his new estate in Ohio, and Worthington promised each family 'a freehold ... whenever

238

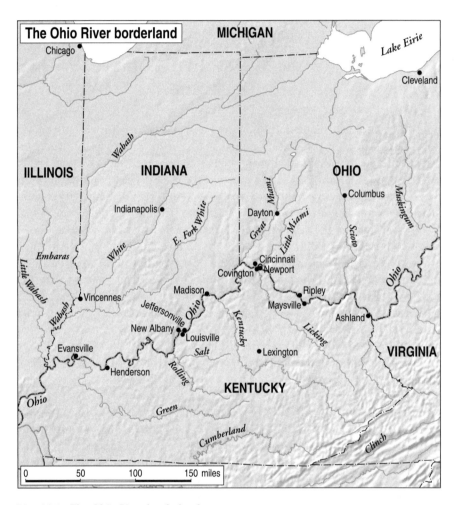

Map 11.1 The Ohio River borderland

he should judge them capable of preserving the cabin and adjacent acres, which he allowed them by way of probation.' None of the families ever received their own land. Thomas Worthington spoke truthfully when he said that he moved to Ohio because it was free from slavery. Worthington, like many other Americans, disliked chattel slavery in principle, but his devotion to liberty did not translate into color-blind equality. While Worthington's decision to keep African Americans as lifetime servants appears contradictory to his self-espoused devotion to antislavery principles, it is entirely possible that Worthington did not see this contradiction in his Adena estate. He feared the potential growth of the free black population and the social upheaval that might accompany immediate emancipation. So, in his mind, by providing employment and limiting the freedom of his freed slaves he was safeguarding the republic (and profiting from it).[2]

In antebellum America, apparent contradictions such as these were common along the Ohio River. In 1787, the Northwest Ordinance made the Ohio River the border between free and slave states in the early West, providing the necessary conditions for the development of the borderland. While the United States Constitution protected the right of fugitive reclamation, the existence of free territory complicated the issue because it put state laws into conflict. Did slaveholders have the right to enforce enslavement in a state that guaranteed the freedom of its residents? When, exactly, did an enslaved person become free? These questions spoke directly to the issue of state sovereignty. When crafting the Constitution, James Madison had imagined the equal sovereignty between independent states as central to the stability of the young republic. Therefore only a federal authority had the power to resolve interstate disputes such as fugitive reclamation. Lawmakers passed a Federal Fugitive Slave Law in 1793 to clarify the issue. While the law guaranteed slaveowners the right to retrieve their human property across state lines, it failed to give responsibility for enforcement to any one person or official; nor was the requirement of proof precise. As a result, when state officials appealed to the federal government to resolve interstate disputes, federal officials were reluctant to step in. Instead, state officials and local residents were often on their own to accommodate this federally created divide.[3]

But slavery and freedom represented more than legal statuses; they represented distinct cultural identities. Even in regions with fewer slaves, like Kentucky, race and enslavement were inextricably intertwined. In fact, the Kentucky courts defined color as presumptive evidence of status. So powerful were these perceived differences that the European observer Alexis de Tocqueville wrote in his *Democracy in America* that from the Ohio River one had only to 'cast glances around himself to judge in an instant' the differences between slave and free territories. Tocqueville wrote that in Kentucky 'society is asleep' and in Ohio there is 'a confused noise that proclaims from afar the presence of industry.' In Tocqueville's understanding, the Ohio River was a clear border dividing two distinct worlds.[4]

Commentary by local white and black residents, however, suggested that the line defined by the river was far more permeable than Tocqueville suggested. As one abolitionist newspaper from Cincinnati remarked, 'Our situation peculiarly exposes us to the insidious and incessant inroads of the spirit and practices of slavery ... slavery itself, as she passes through our midst, is suffered to rattle her chains, unrebuked.' Not all lamented the permeable border, however. Across the border, Kentuckian Cassius Clay argued that the border made the region unique, explaining that radicals in the Deep South 'do not understand the position and feelings of the people of Missouri, Kentucky, Virginia, and large portions of them in North Carolina and Tennessee.' While the Northwest Ordinance defined the Ohio River as a border, local residents made the region a borderland.[5]

The Ohio River borderland, however, was an inherent contradiction because the river served as both a unifying economic force and a symbol of division. As residents on both sides struggled to accommodate the river border, they came to understand the region as a borderland defined by the existence of both conflict and accommodation. Thus, I argue that while the Ohio River borderland appeared contradictory, it operated through an internal logic of flexibility that held the region together against the pull of sectional animosity. Indeed, it is suggestive that in 1861, when the Civil War tore the nation apart, the region failed to split at this seam.

This essay takes the analysis of the Ohio River borderland, and all of its apparent contradictions, in three parts. First, I identify the factors that prevented the Ohio River border from becoming a strict divide. The Ohio River was a highway drawing people and goods into its flow and dispersing them throughout the country and its various sections. Also, the division between slave and free labor was anything but clear in antebellum America, and the border highlighted the ambiguities. Second, I trace how the liminality of the Ohio River border led those in power to craft limits to contain the corrosive power of human movement. These limits, in turn, allowed the oppressed to open up spaces that defied the clarity of the border. Finally, as these limits and the spaces within came to define the borderland, I examine how residents understood the Ohio River borderland region as distinct from both the North and the South.

Life and Labor on the Ohio River

The Ohio River's centrality to the region undermined efforts to use it as a divide. Along its thousand-mile course to the Mississippi River, 18 major tributaries empty into the Ohio from both northern and southern sources. The drainage basin of the Ohio River covers over 200,000 sq miles and reaches 14 states. The enormous reach of the Ohio River's drainage basin made it a critical artery of trade and movement. During the colonial period, Ohio Indians capitalized on the Ohio River to trade with the French in the Illinois country on the Mississippi as well as the British in Virginia and Pennsylvania. Indeed, the Ohio River marked the meeting place of the French and British Empires: a middle ground occupied by Ohio Indians. But rather than a divide, imperial rivalries and hostilities made the Ohio River an artery for an extensive captive exchange system. The French engaged in the trade in captive bodies to both facilitate relations and breed rivalries among their Native American allies and enemies. South of the river, Euro-Americans used bound labor to facilitate expansion and settlement. Thus, through agricultural production and imperial warfare and politics, by the 1780s a thriving slave trade developed that stretched from the Ohio River Valley to New Orleans.[6]

Thus, when Americans made the Ohio River the border between slavery and freedom in 1787, the river already functioned as a western highway. Often beginning in Pittsburgh, Pennsylvania and ending in New Orleans, Louisiana, Americans loaded their goods on flatboats and traveled down the Ohio and Mississippi Rivers. In this early system, Americans had to rely on the southerly flow of the river, disassemble their flatboats at the end of their journey and return by foot. While all could not undertake this three-month trek, the river still connected northern farmers with national and potentially international markets. The improvement of the steamboat for travel in shallow waters in 1817 cut the voyage down to a three-week excursion and allowed for travel and trade in both directions. Steamboats revolutionized life along the Ohio River by drawing farmers into larger networks of market exchange. Kentuckian John Corlis wrote to his wife,

> I am still of the opinion that the vicinity of the river is the best place for farming, as there will be always a growing and unceasing market. The communication will become so easy and cheap that much of the produce of Kentucky and Ohio will reach the Atlantic market.[7]

Because the Ohio River served as a central highway, it widened spheres of interaction and demanded cross-river associations. First, as they wound their way along the river's path, steamboats brought a diverse population into close contact. The casualty reports from steamboat crashes give us a window into the heterogeneity of life and work along the Ohio. One report from June 1828 listed among the dead 'Collins, carpenter; Bradley, white fireman, Peter, a white French boy; Hunstman and Ferral, sailors; five Negroes, four of them firemen, and a sailor.' The injured included 'the mate of the boat, a black fireman, and a sailor.' Second, steamboats made several stops along their journey. Thus hotels, bars, stores, and brothels sprung up in river cities, fed by the constant flow of people up and down the river. When slaveholders stopped in Cincinnati hotels some were brazen enough to bring their personal slaves with them, while others placed their slaves in holding pens across the river in Newport, Kentucky. Finally, the steamboat industry developed in support of the Ohio River economy. Free and enslaved African Americans labored along the river, helping build steamboats, unloading cargo, building docks, working as ferrymen, as draymen, and even sometimes as musicians to entertain travelers.[8]

After decades of interaction, violence, and sectional conflict, instead of dividing the region, by 1860 the Ohio River still allowed residents to forge distinct cross-river economic zones. These economic zones were shaped by the presence or absence of slavery, but they were not determined by it. Major urban manufacturing centers on one side of the river typically faced smaller manufacturing centers on the other bank. This cross-river relationship in manufacturing production was related to the free black population. For example, Jefferson County, Kentucky, which included the city of Louisville, was the manufacturing center of the region, but across the river Floyd County, Indiana, including the city of New Albany, also had a high manufacturing output. Both counties held a relatively large free black population. Attracted by labor opportunities and the presence of a free black nucleus, African Americans traveled to the area, settled on both sides of the river, and developed a cross-river community. In Clark County, Indiana, near Louisville, Kentucky, only 58 per cent of the residents were born in Indiana and Kentuckians made up nearly 15 per cent of the population in 1850. This suggests that even if they chose to live in Ohio and Indiana to be free from slavery, these white Americans perched themselves on a river that connected the region with the rest of the South. Thus, even by 1860 the economic development of the region and Americans' interaction with the river undermined a clear separation between free and slave states.[9]

The unique mixture of enslaved and free labor along the Ohio River made these economic connections possible. In theory, the division between freedom and slavery was a sharp dichotomy representing antithetical forms of labor. On the one hand, historians have argued that the decline of indentured servitude and bound labor in general in America furthered the separation of freedom and slavery in Americans' minds. They suggest that, over the course of the antebellum period, sectional animosity and the rise of free-labor ideology in the northern states erased the connections between slavery and freedom so that by 1860 white Americans understood the commodification of labor power and the commodification of laborers as two entirely opposite things: freedom and slavery. However, historians of American slavery have also suggested that in a national market economy that increasingly demanded the commodification of labor power *and* laborers, African Americans could be commodified as capital, labor power, and laborers.[10]

The hiring out of enslaved African Americans exemplified the combination of commodified labor and commodified people. Slave-hiring followed principles similar to free labor, especially if the slave was self-hired and living on his or her own. Essentially, in the case of self-hiring, the slave was selling his labor, not his person, on the market. Even when slaveholders were responsible for negotiating the contract, they sold the labor on the market while maintaining their claim to ownership of the person. Slave-hiring complicated the distinction between free and slave labor because the commodified labor was a product of a commodified laborer.[11]

While hiring out was present throughout the South, the presence of the Ohio River and its symbol as a border gave it unique features in this region. First, both free and enslaved African Americans worked along the docks and in shipyards, thus potentially blending free workers from free states with enslaved workers from a slave state. Hired Kentucky slaves performed virtually every type of labor from field work, to factory work, to work on a steamboat, to work in a hospital; they could be barbers, musicians, draymen, or domestic servants. Across the river, the 1842 Cincinnati directory reveals that the occupations of free blacks included barber, laborer, cook, river workers, domestics, and washerwomen, which were the same types of jobs held by enslaved people from Kentucky. Certain jobs, such as menial labor along the docks, work in the service industry, and barbering, were the domain of African Americans, which made it difficult for free blacks to find higher-paying jobs.[12]

While the commodification of labor trapped enslaved and free African American laborers in the same economic web, the commodification of human capital further undermined the security that freedom supposedly guaranteed. The growth of the slave trade throughout the 1830s and 1840s made Kentucky a point of departure for enslaved blacks. The Ohio River facilitated the transfer of mobile capital from Kentucky southward to Natchez and New Orleans, which contributed to the commodification of enslaved Kentuckians. While slaveholders found endless ways to differentiate their human capital, skin color was the ultimate representational quality that white Americans used to commodify African Americans. Overall, skin color overshadowed distinctions between free and enslaved African Americans, which meant that either could be sold 'down the river' at any time. As Mrs Colman Freeman, a free black woman, explained, 'I lived in Ohio ten years, as I was married there,—but I would about as lief live in the slave States as in Ohio. In the slave States I had protection sometimes, from people that knew me—none in Ohio.'[13]

The Extent of Slavery and the Limits of Freedom

The Ohio River's role as an economic conduit undermined distinctions between free and slave states, between free and slave labor, and between the lived experience of slavery and freedom for African Americans. In addition, the success of the Ohio River economy demanded interaction both along and across its banks. Thus in a variety of ways, Americans' interaction with the Ohio River undermined its role as a border. In an effort to outline the functionality of the border without impeding economic development, white Americans established limits to African Americans' mobility. These limits did not appear all at once, nor were they part of some grand design. Instead, they were established piece-meal in response to specific circumstances.[14]

First, in the Northwest Territory white leaders determined that slavery could not be imposed on a person who had a legal right to freedom. When white settlers moved north of the Ohio River many brought their freed slaves with them. Once in free territory, they imposed indentured servitude on their for-mer slaves. While they argued that both parties entered these contracts in a state of perfect freedom, one side clearly had more 'freedom' than the other. Nonetheless, once an African American legally transitioned from a slave into a servant she or he could not become a slave again. In addition, servitude had a time limit. In Indiana, some indenture contracts were for 99 years, essentially making bondage a lifetime condition. However, the children of indentured servants would become free at a set age. Kentucky courts also supported the legal distinction between servant and slave, and in several cases judges granted freedom to someone who had lived either as a free person or as a servant in a free state. The movement of African Americans forced white Americans in power to make the Ohio River the northern limit of the chattel principle.[15]

While slavery could not be imposed on a person traveling south across the river, enslaved African Americans could not escape from slavery by fleeing north of the river. First established by the United States Constitution in 1787, and reinforced by the Federal Fugitive Slave Laws of 1793 and 1850, slavehold-ers had the right to retrieve escaped slaves anywhere in the country. Notably, neither Ohioans nor Indianans ever granted freedom to fugitives seeking refuge in their state borders, thus tacitly recognizing the authority of the national law. Initially, free states found means of legally supporting the federal law. Ohio established a precedent in 1803 by requiring African Americans to register with local courts upon entry into the state, thus making the assumption that all African Americans were enslaved until proven free. Indiana passed a similar law in 1831. These laws made it impossible for fugitive slaves to feel secure by forcing all African Americans to carry proof of their freedom. In 1850, the federal government strengthened the Fugitive Slave Law by requiring free state residents to aid in the reclamation of alleged fugitives. Ohioans repealed their laws requiring registration in 1849, but could never grant immediate freedom to escaped slaves. The Ohio courts ruled that any enslaved African American who had entered Ohio with permission (perhaps on an errand) had a right to freedom upon reaching Ohio soil. Indiana, in contrast, did not follow the same pattern, but granted tacit approval for the federal law by never passing any personal liberty laws. Overall, these national and state laws defined escape as illegal and closed it off as a legal avenue to freedom.[16]

In addition to defining fugitive slaves as perennial outsiders, Indianans and Ohioans also meant their laws to restrict the immigration of free African Americans by forcing them to enter into a $500 security bond. Although free blacks did not have to pay $500 upon entry into the state, they were required to find a sponsor willing to swear to pay $500 if the black immigrant ever became a charge on the state. Historians have gone back and forth on whether or not these laws actually restricted the migration of African Americans. While the effectiveness of these laws is debatable, they served as evidence of white Americans' efforts to define the limits of black movement. In 1849, after considerable political wrangling, Ohioans repealed their black laws, arguably opening the doors to black immigration. Indiana, on the other hand, completely banned the immigration of African Americans in 1851. Similarly, Kentucky banned the immigration of free African Americans. Thus it appears that by the 1850s the borderland had an internal contradiction. Ohio's move to repeal the black laws had much to do with the unique development of the state. Ohio had a stronger antislavery movement than did Indiana, a more outspoken free black community, and a politician with enough wherewithal to broker a difficult political deal in Salmon P. Chase. In southwestern Ohio, white residents despised abolitionists and slavery equally, and Chase had to make concessions with southern Democrats to secure the repeal of the black laws. Thus, despite the state's repeal of the laws, in many ways southwestern Ohio remained hostile to the immigration of African Americans.[17]

In sum, in their effort to define the meaning of the Ohio River border, white Americans established limits on the movement of African Americans. These limits controlled the volatility of the borderland in two ways. First, the law protected free African Americans from kidnapping by making the Ohio River the northern limit of the chattel principle. Second, making fugitive status cross the border prevented enslaved African Americans from streaming across the river in pursuit of their freedom. While restricting immigration did not stop slaves from fleeing altogether, white Americans in southern Indiana and Ohio were 'protected' from a flood of black migrants. These laws added stability, but free and enslaved African Americans created spaces of interaction within the established limits to make their lives more tolerable.[18]

Finding Space within the Limits

African Americans took advantage of the unique labor system along the river to create conditions and associations that flouted the slave/free dichotomy. The combination of prevalent slave-hiring and the presence of the river offered some enslaved African Americans a wide range of mobility and access to free states. As laborers on steamboats, enslaved and free African Americans plied western rivers from Pittsburgh to New Orleans. Others traveled with slaveholders up and down the river, ran errands across the river, or lived and worked with free blacks along the river. Three enslaved musicians lived in Louisville and played at parties across the river in Madison, Indiana, and Cincinnati, Ohio with the written permission of their owner, who actually lived in Lexington, Kentucky! In his narrative Josiah Henson recalled that he traveled well into Ohio preaching with the full permission of his owner. Henson wrote of a trip he took in fall 1828:

> The annual Methodist Conference was about to be held at Chillicothe [Ohio], to which my kind friend accompanied me, and by his influence and exertions I succeeded well there also. By his advice I then purchased a suit of respectable clothes, and an excellent horse, and travelled leisurely from town to town, preaching as I went, and, wherever circumstances were favorable, soliciting aid in my great object.[19]

In fact, the mixture of slave and free labor and the mobility of some allowed for the development of illicit fugitive labor. Working toward freedom became a viable means of escape for the enslaved, and some white Americans north of the river seemed to have looked the other way. Slaveholders commonly listed the skills of runaways in advertisements, suggesting they might try to hire themselves as free persons. During his escape, Henry Bibb got a job as a cook in Ohio, and was so successful the landlady wanted to hire him permanently. He declined, but he 'got a job of chopping wood during that winter which enabled me to purchase myself a suit, and after paying my board the next spring.' Business owners in Cincinnati hired slaves and sometimes escaped slaves from Kentucky. Certainly many African Americans in Cincinnati were free, but many were just as certainly fugitive slaves. While advertisements suggested the interchangeability of free and slave labor, they also made distinctions between free and enslaved people. The runaway adverts listing occupations suggested ownership of the person. Even if runaway slaves attempted to use their labor, they could not escape from the slaveholder's claim to ownership of their persons.[20]

To be certain, Henson and the traveling musicians did not represent the majority of the enslaved population. Indeed many enslaved African Americans spent their lives contained on the small farms of their owners. The significance is not in the frequency of these cases, but in their implications for such liberality of movement. This space exploited by African American laborers heightened white Americans' suspicions. In their advertisements for runaways, slaveholders warned their neighbors that enslaved African Americans would have written proof of their freedom. For example, an advertisement for 'Jacob Herral' read that 'he may procure forged papers and attempt to pass as a free man.' Clearly Herral had enough contact and interaction with the free black community along the river to know how to obtain some papers, and his owner feared the possibility of a market for free papers. Some mobile black laborers used their autonomy to learn how to read and write. In another example, Frank's owner wrote that he could 'read and write, and it is likely that he has got a pass of his own writing.' Adverts such as these meant that African Americans could be jailed or at least questioned on the mere suspicion of being runaways. Gilbert Dickey explained, 'I was making money in Indiana, and had plenty of friends. I liked there, all but one thing—slavery was there, as it is all over the United States.' David Grier made his point even more clearly: 'From Ohio, I came here [Canada] on account of the oppressive laws demanding security for good behavior.'[21]

Enslaved African Americans attempted to exploit the blended labor system to gain legal freedom. Some used their wages earned from hired-out work to purchase themselves. This was a tacit recognition that they were commodified laborers, but also an attempt to capitalize on their commodified labor power to overcome their situation. An agreement for self-purchase put a time limit on servitude and ideally created a mutual obligation between slave and slaveholder. Once they had entered into an agreement, slaves strove to earn and expressed great pride in their ability to purchase their freedom. 'If a slaveholder offers his servant freedom, on condition that he will earn and pay a certain sum, and the slave accepts freedom on that condition,' explained ex-slave Henry Blue, 'he is bound in honor to pay the sum promised.' Richard Keys worked in Cincinnati and paid $20 per month to his owner in Kentucky for 12 years and then paid an additional $850 for his freedom. But, if men like Henry Blue felt honor-bound to earn their freedom, slaveholders did not always share the same sense of obligation. Slaveholders' willingness to break these contracts by refusing freedom, raising the price of purchase, or selling the slave down the river frequently triggered escape. Alfred Jones explained that he made an arrangement to purchase his freedom for $350, but 'before the business was completed, I learned that my master was negotiating with another party to sell me for $400.' Upon learning of this betrayal, Jones wrote himself a pass and left for Canada.[22]

In addition to using the river economy, African Americans forged ties to their families and to the larger black community to improve their condition. In their devotion to family members and safety found in the protection of the community, African Americans used factors outside the slave/free dichotomy to structure their lives. Historians have demonstrated that family provided the first line of defense against the isolation of bondage throughout the antebellum South. In the borderland familial affection both eased the suffering under bondage and created ties that made the enslaved think twice about escape. Some African Americans sacrificed freedom to save their families. Mrs L. Strawthor recalled that her husband had earned enough to purchase his family before he was sold to the Deep South. Strawthor never saw her husband again and was unsure if he remained alive. William Wells Brown explained that when he thought of escaping to Canada his 'resolution would soon be shaken by the remembrance that my dear mother was a slave in St. Louis, and I could not bear leaving her in that condition.'[23]

Enslaved women in particular seemed to have felt duty-bound to protect their children. Slave women in the borderland often lived apart from husbands, many of whom resided on other farms or were hired out to work elsewhere. As a result the responsibilities of childcare fell largely to women. Hired female slaves had to earn enough to pay their owners and support themselves and their children. The Kentucky slave Charlotte, for example, worked as a washerwoman to support her family. She took pride in her ability to fend for her family, even covering the difference when her employed children failed to make their contractual earnings. 'I get along very well,' she stated; 'you couldn't pay me to live at home, if I could help myself. My master doesn't supply me with anything ... no more than if I didn't belong to him.' Charlotte tolerated her bondage because she lived on her own and could provide for her children. If given the option, slave women such as Charlotte undoubtedly would have purchased their freedom, but few had that choice.[24]

For slaves who lacked nuclear families or whose families had been torn apart by sale, the African American community provided emotional and sometimes physical support. In response to the isolation, prejudice, and instability that were part of life in the Ohio River Valley, mobile slaves mingled with free African Americans and built communities that extended across the slave/ free border. This process took place most often in urban places because they afforded greater opportunities for interaction between free and enslaved African Americans. Enslaved Louisville barber Isaac Throgmorton said, 'I lived with free people, and it was just as though I was free.' But whether blacks were enslaved on isolated hemp plantations in the Kentucky interior, hired out in Louisville, or lived as free people in Cincinnati, the African American community provided better protection against the hazards of white racism than did the law. In Cincinnati the African Methodist Episcopal Church served as a safe house for fugitives, and church members often harbored and aided fugitives. As former Kentucky slave Lewis Clarke wrote after he escaped his bondage,

> I am yet accounted a slave, and no spot in the United States affords an asylum for the wanderer. True, I feel protected in the hearts of the many warm friends of the slave by whom I am surrounded; but this protection does not come from the LAWS of any one of the United States.

Clarke suggested that the law offered him no protection. Instead the antislavery community kept him out of bondage.[25]

In forming active black communities, African Americans also unnerved white residents. White residents and visitors looked with trepidation on any group of African Americans gathered in city streets. These fears often translated into frequent complaints in the press identifying African Americans as the source of vice and basically all conflict and catastrophes that arose. In Cincinnati the editor of *Liberty Hall* warned 'that the rapid increase of our black population, to say nothing of slavery, is of itself a great evil.' Across the river in Louisville, one writer complained that

> We are overrun with free negroes. In certain parts of our town throngs of them may be seen at any time—and most of them have no ostensible means of obtaining a living. They lounge about through the day, and most subsist by stealing, or receiving stolen articles from slaves at night ...

The editor ultimately called for 'prompt measures to drive the vagrant negroes from among us.'[26]

Duality of Mobility in the Borderland

Liminality certainly characterized life in the Ohio River borderland. However, movement added volatility to the mix. The insatiable demand for bondspeople in the 'cotton kingdom' of the Deep South transformed the border South into a slave-exporting region. This domestic slave trade along western rivers represented a material link between the Ohio River borderlands and the Deep South, one that placed Kentucky slaves in a precarious position. Most slaves either experienced sale personally or witnessed the sale of family members, friends, and fellow slaves, often at public auctions. Historians have estimated that roughly one out of every three enslaved families were torn apart by sale.[27]

Sale introduced volatility to the borderland, because the threat of sale left borderland slaves with two options: take a chance at freedom by heading north or endure a lifetime of servitude and die a slave in the Deep South. As Kentucky slave Harry Smith recalled, 'going to New Orleans was called the Nigger Hell, few ever returning who went there.' When local enslaved people became 'aware of the presence of ... slave buyers,' he noted, 'a number of them would run away to the hills and remain often a year before they returned. Some would reach Canada for fear of being sold.' Likewise, when Louisville slave Henry Morehead learned of his family's potential sale to the Deep South, he decided it was time to act. 'I knew,' he remembered, 'it was death or victory.'[28]

We may ask, if the threat of sale was constant and some enslaved blacks had a considerable amount of mobility and even the opportunity to escape, then why was there not a torrent of enslaved blacks crossing the Ohio River? Those African Americans who escaped and wrote narratives left us clues. Often borderland fugitives described moments when they had opportunities to escape but ultimately decided to remain where they were. Their specific reasons varied, and in some cases they may have overstated their chances to escape. This recurring theme of standing at the precipice of freedom tells us that enslaved Kentuckians believed that the border offered the hope of obtaining liberation. This, in turn, explains why enslaved blacks actively worked to protect their situation to maintain this hope. In order to truly escape from slavery, however, enslaved African Americans had to leave the borderland. Crossing the Ohio River did not confer freedom; instead one had to flee farther north.[29]

As a slave in Madison County, Kentucky, south of Lexington, Lewis Clarke hired his own time, provided for his own room and board, and enjoyed considerable geographic mobility. In order to retain his liberties as a hired slave, Clarke denied his desire for freedom. As he later explained,

> Now if some Yankee had come along and said 'Do you want to be free?' What do you suppose I'd have told him? ... Why, I'd tell him to be sure that I didn't want to be free; that I was very well off as I was. If I didn't, it's precious few contracts I should be allowed to make.

Clarke certainly wished for freedom, but he also wanted to remain in Kentucky because his close proximity to the border made gaining freedom a tangible possibility. So he feigned contentment in order to protect his current situation. Only the threat of sale to the Deep South prompted Clarke to make his escape in 1841.[30]

During his flight, Clarke encountered a Baptist minister who suspected that he was a runaway and, according to Clarke, attempted to 'read [his] thoughts.' In order to allay the minister's suspicions, Clarke emphasized his favorable situation as a slave, noting, 'I wondered what in the world *slaves could* run away for, especially if they had such a chance as I had had for the last few years.' This apparently satisfied the minister, who believed that a slave who enjoyed so many privileges would not run away. Clarke closed this conversation by adding, 'I do very well, very well, sir. If you should ever hear that I had run away, be certain it must be because there is some great change in my treatment.' With these words Clarke actually explained to the minister why he was fleeing. Clarke had long entertained the idea of escape, but sale was the 'great change' that convinced him to run away. On a Deep South cotton plantation Clarke would have few or no opportunities to hire out his time and live independently. More important, sale away from the border extinguished his hope for eventual freedom.[31]

Breaking the chains of bondage was particularly difficult because enslaved African Americans knew that the only way to truly escape from slavery was to leave the Ohio River borderland altogether. This meant breaking all ties to the community and even to family. Henry Bibb's devotion to his family prevented him from leaving the borderland even after escaping from his Kentucky owner.

He lived as a slave in Shelby County, Kentucky, near Louisville, and escaped north on three separate occasions. Each time Bibb placed himself in danger and returned to Kentucky because he wanted to save the wife and child he had left behind. 'I felt,' he wrote, 'as if love, duty, humanity, and justice, required that I should go back.' His devotion to his wife meant that while he could escape the grasp of his master in Kentucky, he could never become truly free in southern Ohio because of the constant risk of recapture. Bibb made his final escape after his owner sold him and his wife and they were taken to the Deep South. With permanent slavery looming, Bibb's commitment to freedom became equally permanent. While in Louisiana, sale separated him from his wife and so Bibb decided to make his final escape and did not stop until he reached Detroit, far enough north to secure his freedom once and for all. In short he had to leave the borderland, and with it his family, to experience freedom.[32]

Conclusion: Twisted Logic or Contradiction?

As this suggests, African Americans in the borderland understood the differences between slavery and freedom because they experienced characteristics of both. Their understanding of this borderland reveals how contradictions allowed it to function. First, African Americans understood the Ohio River borderland as distinct from the rest of the country. Taken as a whole, stories like Lewis Clarke's demonstrated that former slaves weighed the benefits of freedom against its limitations before they fled. They perceived the practical differences between slavery and freedom, in the knowledge that freedom in the borderland was limited and not always worth the risk. Furthermore, the insecurity of freedom in the borderland meant that they could not leave the chains of bondage on the Kentucky bank. At the same time, sale to the Deep South promised a lifetime of brutal labor and likely death as a slave. Thus they were trapped between hope and death, and in order to escape they had to escape the borderland.

Second, the geography of the border mattered. The fact that these states shared access to the Ohio River and with it a lane of commerce that connected the region with markets as far south as New Orleans made residents more likely, and perhaps more willing, to foster harmonious relations with their neighbors across the slave/free border. Not only that, however, but the Ohio River economy necessitated mobile labor which linked borderland slavery with borderland free labor. African Americans' ability to cross the border as runaways and travel along the river as workers highlighted the similarity between bound labor and wage labor along the river. In fact, along the Ohio River wage labor and chattel slavery became points on a capitalist continuum rather than mutually exclusive categories for African Americans. Ironically, the Ohio River border, as the representation of the slave/free divide, was a place where dichotomies could not apply.[33]

Finally, gradual development of this borderland highlights the centrality of border-crossing in the definition of a border. Both free and slave state residents shared a desire to limit the movement of African Americans. Kentucky slave-holders demanded the right to reclaim escaped slaves from free states. They also demanded vigilance on the part of their free-soil neighbors because they wanted to stem the flow of slaves fleeing across the border. African Americans' depiction of freedom in the borderland as insecure suggests they were at least partly successful. Free-state residents defended their state from invasion by slaveholders by establishing the precedent that a free person could not become a slave. Thus battles over reclamation were equivalent to battles over state sovereignty. At the same time, white free-state residents also wanted to limit the flow of free blacks into their state and did so with immigration restric-tions. These efforts represented a shared desire to establish racial barriers to freedom. In other words, the very method that white residents used to empha-size their differences became the common ground on which they negotiated. The American poet Walt Whitman may as well have been speaking specifically about the Ohio River Valley when he said of America in 1855, 'Do I contradict myself? Very well then … I contradict myself.'[34]

Notes

1. *Chillicothe Scioto Gazette*, Aug. 28, 1802, Nov. 18, 1802; J.W. Blassingame, *Slave Testimony: Two Centuries of Letters, Speeches, Interviews, and Autobiographies* (Baton Rouge, 1977), pp. 519–21.
2. S.W.K. Peter, *The Private Memoir of Thomas Worthington* (Cincinnati, 1882), pp. 29–31; M. Mangin, 'Freemen in Theory: Race, Society and Politics in Ross County, Ohio, 1796–1850,' PhD dissertation, University of California San Diego, 2002, pp. 33, 114–18; A.R.L. Cayton, *Ohio: The History of a People* (Columbus, 2002), p. 35.
3. P. Finkelman, *Slavery and the Founders: Race and Liberty in the Age of Jefferson* (New York, 2001), pp. 58–80.
4. A. de Tocqueville, *Democracy in America*, trans., ed., and intro. H.C. Mansfield and D. Winthrop (Chicago, 2002), pp. 331–2. Davis (a man of color) v. Curry, in H. Catterall, *Judicial Cases concerning American Slavery and the Negro*, vol. 1 (Washington, DC, 1926), p. 238. On slavery and southern society see I. Berlin, *Generations of Captivity: A History of African American Slaves* (Cambridge, MA, 2003); on northern free-labor ideology see E. Foner, *Free Soil, Free Labor, Free Men: The Ideology of the Republican Party before the Civil War* (New York, 1995).
5. *Cincinnati Philanthropist*, Dec. 18, 1838; *Lexington Examiner*, Sept. 25, 1847.
6. R.L. Reid (ed.), *Always a River: The Ohio River and the American Experience* (Bloomington, 1991), p. 2; World Resources Institute, Earthtrends: Environmental Information, 'Watersheds of the World: North and Central America—Mississippi Watershed: Ohio Subbasin,' multimedia.wri.org/watersheds_2003/na15.html (accessed Oct. 2012); C. Hodson and B. Rushforth, 'Bridging the Continental Divide: Colonial America's French Quarter,' *OAH Magazine of History*, 25 (2011), 19–24, at 22; C.J. Ekberg, *French Roots in the Illinois Country: The Mississippi Frontier in Colonial Times* (Urbana, 1998), pp. 111–70. On slavery, see Bill of Sale, Sept. 19,

1766, Apr. 3, 1771, Letter Oct. 11, 1777, Lasselle Collection, Indiana State Library, Indianapolis; J.A. James (ed.), *George Rogers Clark Papers, 1781–1784* (Illinois Historical Collections, 19, Springfield, 1926), p. 85. On the French slave trade see, B. Rushforth, 'Slavery, the Fox Wars, and the Limits of Alliance,' *William and Mary Quarterly*, 63 (2006), 53–80; B. Rushforth, 'A Little Flesh We Offer You: Origins of Indian Slavery in New France,' *William and Mary Quarterly*, 60 (2003), 777–809. E. Hinderaker, *Elusive Empires: Constructing Colonialism in the Ohio Valley, 1673–1800* (New York, 1997); D.W. Meinig, *The Shaping of America: A Geographical Perspective on 500 Years of History, Volume 1, Atlantic America, 1492–1800* (New Haven, 1986), pp. 209, 231–5; R. White, *The Middle Ground: Indians, Empires, and Republics in the Great Lakes, 1650–1815* (New York, 1991).

7. John Corlis to Susan Corlis, July 1831, Corlis–Respess Family Papers, Filson Historical Society, Kentucky; D. Feller, *The Jacksonian Promise: America, 1815–1840* (Baltimore, 1995), pp. 22–5; C. Sellers, *The Market Revolution: Jacksonian America 1815–1846* (New York, 1991), pp. 43–4; D.E. Bigham, *Towns and Villages of the Lower Ohio* (Lexington, 1998); M.B. Lucas, *A History of Blacks in Kentucky, vol. 1: From Slavery to Segregation, 1760–1861* (Frankfort, KY, 1992), pp. 96–100; A.R.L. Cayton, *Frontier Indiana* (Bloomington, 1996), p. 275; D. Bigham, 'River of Opportunity: Economic Consequences of the Ohio,' in Reid (ed.), *Always a River*, pp. 130–79.

8. *Vincennes (IN) Western Sun*, June 7, 1828, May 1, 1830; T.C. Buchanan, *Black Life on the Mississippi: Slaves, Free Blacks and the Western Steamboat World* (Chapel Hill, 2004), pp. 53–80; A.R.L. Cayton, 'Artery and Border: The Ambiguous Development of the Ohio Valley in the Early Republic,' *Ohio Valley History*, 1 (2001), 19–26; K.M. Gruenwald, 'Space and Place on the Early American Frontier: The Ohio Valley as Region, 1790–1850,' *Ohio Valley History*, 4 (2004), 31–48.

9. US Census Bureau, '1850 Census of Clark County, Indiana.' With Cincinnati and Louisville there was no need for another major city, and thus one did not develop in southern Indiana. In this particular case, Louisville, a slaveholding city, and Cincinnati, a non-slaveholding city, combined to feed the needs of the larger region. All of this is to suggest that the logic of the market economy in antebellum America did not necessarily heed sectional boundaries. I accessed all of the data analyzed in this chapter using the University of Virginia's online census browser, http://mapserver.lib.virginia.edu (accessed Oct. 2012).

10. Foner, *Free Soil*, pp. ix–xlii; D.B. Davis, *Inhuman Bondage: The Rise and Fall of Slavery in the New World* (New York, 2006), p. 35; W. Johnson, *Soul by Soul: Life inside the Antebellum Slave Market* (Cambridge, MA, 1999); J. Oakes, *Slavery and Freedom: An Interpretation of the Old South* (New York, 1990); S. Smallwood, 'Commodified Freedom: Interrogating the Limits of Anti-Slavery Ideology in the Early Republic,' *Journal of the Early Republic*, 24 (2004), 292; E.E. Baptist, '"Cuffy, Fancy Maids, and One-Eyed Men": Rape, Commodification, and the Domestic Slave Trade in the United States,' *American Historical Review*, 106 (2001), 1619–50; W. Johnson, 'The Pedestal and the Veil: Rethinking the Capitalism/Slavery Question,' *Journal of the Early Republic*, 24 (2004), 299–308.

11. The labor contract is the nexus of the power relations that define the commodification of labor; the contract symbolizes the negotiation between employed and employee. The worker has the freedom to sign the contract, while the employer has the power to assign value to a person's labor. In contrast, on a bill of sale, a person is given a price, effectually flattening everything that person says or does into a single dollar amount. Thus, ownership of labor power is intrinsic in the ownership of a laborer; in a sense, chattel slavery involved the commodification of the laborer *and*

his or her labor power; see Johnson, *Soul by Soul*, pp. 19–44, and A.D. Stanley, *Wage Labor, Marriage, and the Market in the Age of Slave Emancipation* (New York, 1998), pp. 1–59.

12. *Cincinnati Directory for the Year 1842*, comp. C. Cist (Cincinnati, 1842); D.E. Bigham, *On Jordan's Banks: Emancipation and Its Aftermath in the Ohio River Valley* (Lexington, 2006), pp. 5–55; J.W. Trotter, *River Jordan: African American Urban Life in the Ohio Valley* (Lexington, 1998), pp. 3–51; Lucas, *History of Blacks in Kentucky*, pp. 101–17; H.D. Stafford, 'Slavery in a Border City: Louisville, 1790–1860,' PhD dissertation, University of Kentucky, 1982, pp. 112–28; J.D. Martin, *Divided Mastery: Slave Hiring in the American South* (Cambridge, MA, 2004), p. 39; C. Eaton, 'Slave-Hiring in the Upper South: A Step toward Freedom,' *Mississippi Valley Historical Review*, 46 (1960), 663–78; K.C. Barton, 'Good Cooks and Washers: Slave-Hiring, Domestic Labor, and the Market in Bourbon County, Kentucky,' *Journal of American History*, 84 (1997), 436–60.

13. B. Drew, *The Refugee: A North-Side View of Slavery*, in *Four Fugitive Slave Narratives* (Reading, MA, 1969 [1856]), p. 233; M. Tadman, *Speculators and Slaves: Masters, Traders, and Slaves in the Old South* (Madison, 1989), pp. 301–2; J.B. Hudson, *Fugitive Slaves and the Underground Railroad in the Kentucky Borderland* (Jefferson, 2002), p. 14; R.H. Gudmestad, *A Troublesome Commerce: The Transformation of the Interstate Slave Trade* (Baton Rouge, 2003), pp. 62–92; S. Deyle, *Carry Me Back: The Domestic Slave Trade in American Life* (New York, 2005); W.W. Freehling, *The Road to Disunion, vol. 1: Secessionists at Bay, 1776–1854* (New York, 1990), pp. 23–4; Berlin, *Generations of Captivity*, p. 161; Oakes, *Slavery and Freedom*, p. 151.

14. Theorists have suggested that in borderlands, debates over border-crossing target both the actions and the people involved. Acceptable people make acceptable transgressions, but residents restrict the movement of those whom they view as a threat. See R. Castronovo, 'Compromised Narratives along the Border: The Mason–Dixon Line, Resistance, and Hegemony,' in S. Michaelsen and D.E. Johnson (eds), *Border Theory: The Limits of Cultural Politics* (Minneapolis, 1997), pp. 195–220; G. Anzaldua, *Borderlands/La Frontera: The New Mestiza* (San Francisco, 1987); J.S. Migdal, 'Mental Maps and Virtual Checkpoints: Struggles to Construct and Maintain State and Social Boundaries,' in Migdal (ed.), *Boundaries and Belonging: State and Societies in the Struggle to Shape Identities and Local Practices* (New York, 2004), pp. 3–26; M. Simpson, *Trafficking Subjects: The Politics of Mobility in Nineteenth-Century America* (Minneapolis, 2005), pp. 56–91; P. Sahlins, *Boundaries: The Making of France and Spain in the Pyrenees* (Berkeley, 1989).

15. G.W. Geib, 'Jefferson, Harrison, and the West: An Essay on Territorial Slavery,' in D.E. Bigham (ed.), *Indiana Territory, 1800–2000: A Bicentennial Perspective* (Indianapolis, 2001), pp. 99–125; S. Middleton, *The Black Laws: Race and the Legal Process in Early Ohio* (Athens, OH, 2005), pp. 7–17; Bigham, *Towns and Villages*, pp. 11–46; Mangin, 'Freemen in Theory,' 58; A.R.L. Cayton, *The Frontier Republic: Ideology and Politics in the Ohio Country, 1780–1825* (Kent, OH, 1986), pp. 51–80; A.R.L. Cayton, 'Land, Power, and Reputation: The Cultural Dimension of Politics in the Ohio Country,' *William and Mary Quarterly*, 47 (1990), 266–86; E. Eslinger, 'The Evolution of Racial Politics in Early Ohio,' in A.R.L. Cayton and S.D. Hobbs (eds), *Center of a Great Empire: The Ohio Country in the Early American Republic* (Athens, OH, 2005), pp. 81–104; J.C. Hammond, *Slavery, Freedom, and Expansion in the Early American West* (Charlottesville, 2007), p. 78; E.L. Thornbrough *The Negro in Indiana: A Study of a Minority* (Indianapolis, 1957); *Indiana Negro Registers, 1852–1865*, comp. C.D. Robbins (Bowie, 1994), pp. 58–60; Finkelman, *Slavery and the Founders*, pp. 58–80; Cayton, *Frontier Indiana*, p. 179; N. Etcheson, *The Emerging Midwest: Upland*

Southerners and the Political Culture of the Old Northwest, 1787–1861 (Bloomington, 1996); Rankin v. Lydia, Oct. 1820, in Catterall, *Judicial Cases*, p. 294.

16. On the Fugitive Slave Law and the fugitive slave clause of the US Constitution, see Finkelman, *Slavery and the Founders*, pp. 81–104; C. Wilson, *Freedom at Risk: The Kidnapping of Free Blacks in America, 1780–1865* (Lexington, 1994), pp. 40–66; Middleton, *Black Laws*, pp. 42–73, 227–30; E.L. Thornbrough, 'Indiana and Fugitive Slave Legislation,' *Indiana Magazine of History*, 50 (1954), 201–28; Thornbrough, *Negro in Indiana*, 39; M. Holt, *The Fate of Their Country: Politicians, Slavery Extension and the Coming of the Civil War* (New York, 2004), pp. 86–8; P. Finkelman, *An Imperfect Union: Slavery, Federalism, and Comity* (Chapel Hill, 1981), pp. 157–78; S.E. Maizlish, *Triumph of Sectionalism: The Transformation of Ohio Politics, 1844–1856* (Kent, OH, 1983), pp. 121–46. On the relationship between outsiders and identity see A. Norton, *Reflections on Political Identity*, (Baltimore, 1988), pp. 143–84.

17. C. Kettleborough, *Constitution Making in Indiana: A Sourcebook of Constitutional Documents, with Historical Introduction and Critical Notes, vol. 1* (Indianapolis, 1916), pp. 290–4, 361–3; *Indiana Election Returns, 1816–1851*, comp. D. Riker and G. Thornbrough (Indianapolis, 1960), pp. 388–90; H.D. Tallant, *Evil Necessity: Slavery and Political Culture in Antebellum Kentucky* (Lexington, 2003), pp. 141–3.

18. The use of the word 'spaces' references the work of historians of slavery, Stephanie Camp and Steven Hahn. See S.M.H. Camp, *Closer to Freedom: Enslaved Women and Everyday Resistance in the Plantation South* (Chapel Hill, 2004), and S. Hahn, *A Nation under Our Feet: Black Political Struggles in the Rural South, from Slavery to the Great Migration* (Cambridge, MA, 2003). Hahn wrote that enslaved and free African Americans made efforts to 'establish relations and values suitable to a world without enslavement' (*A Nation under Our Feet*, pp. 19, 485).

19. Graham v. Strader, Oct. 1844, in Catterall, *Judicial Cases*, pp. 65–8; J.W. Coleman Jr, *Slavery Times in Kentucky* (Chapel Hill, 1940), p. 100; J. Henson, *The Life of Josiah Henson, Formerly a Slave, Now an Inhabitant of Canada, as Narrated by Himself* (Boston, 1849), pp. 28–9; Lucas, *History of Blacks in Kentucky*, pp. 101–17; Trotter, *River Jordan*, pp. 3–51; Martin, *Divided Mastery*, p. 39; M. Grivno, *Gleanings of Freedom: Free and Slave Labor along the Mason–Dixon Line, 1790–1860* (Urbana, 2011); S. Rockman, *Scraping By: Wage Labor, Slavery and Survival in Early Baltimore* (Baltimore, 2009).

20. H. Bibb, *Narrative of the Life and Adventures of Henry Bibb, an American Slave, Written by Himself* (New York, 1849), pp. 86–7, http://docsouth.unc.edu/neh/bibb/bibb.html. 'Henry' was listed a blacksmith, 'Charles' a carpenter, and 'Billy' a barber and steamboat cook (*Louisville Public Advertiser*, June 1, 1825; *Western Sun*, Dec. 11, 1824, July 18, 1820, Sept. 27, 1827). When Josiah Henson stopped in Cincinnati while traveling from Virginia to Kentucky with his owner's slaves, African Americans in the city tried to convince him and the others to run away: Henson, *Life of Josiah Henson*, pp. 22–5.

21. *Vincennes (IN) Western Sun*, March 11, 1820; *Louisville Public Advertiser*, Jan. 22, 1820; *Indiana Gazette*, May 7, 1821; *Louisville Public Advertiser*, Oct. 10, 1821; Drew, *The Refugee*, pp. 248, 273. A Kentucky judge ruled, 'Experience teaches, that there is no danger to be apprehended from too great an alacrity, or passionate ardour, in apprehending slaves as runaways, without probable cause': Jerrett v. Higbee, in Catterall, *Judicial Cases*, p. 305.

22. Drew, *Refugee*, pp. 189, 106; *Proceedings of the Ohio Anti-Slavery Convention Held at Putnam on the 22, 23, and 24th of April, 1835* (Cincinnati, 1835), pp. 30, 43. Such practices closely follow what Whitman has called 'term slavery'; see T.S. Whitman, *The Price of Freedom: Slavery and Manumission in Baltimore and Early National Maryland*

(Lexington, 1997), pp. 98–101. D.R. Egerton, 'Slaves to the Marketplace: Economic Liberty and Black Rebelliousness in the Atlantic World,' *Journal of the Early Republic*, 26 (2006), 617–39. The idea of 'stealing' versus 'earning' freedom is loosely drawn from E. Baptist, 'Stol and Fetched Here: Enslaved Migration, Ex-slave Narratives, and Vernacular History,' in Baptist and S.M.H. Camp (eds), *New Studies in the History of American Slavery* (Athens, GA, 2006), pp. 243–74.

23. Blassingame, *Slave Testimony*, p. 389; W.W. Brown, *Narrative of William W. Brown, an American Slave, Written by Himself* (London, 1849), p. 30, http://docsouth.unc.edu/ brownw/brown.html (accessed Oct. 2012). On the ways that family shaped slaves' motives for flight, see J.H. Franklin and L. Schweninger, *Runaway Slaves: Rebels on the Plantation* (New York, 1999), pp. 50–3, 66; Hudson, *Fugitive Slaves*, pp. 55–8; Berlin, *Generations of Captivity*, pp. 190–5.

24. Blassingame, *Slave Testimony*, pp. 388–90; Barton, 'Good Cooks,' 447–8; Bigham, *On Jordan's Banks*, p. 16; Hudson, *Fugitive Slaves*, p. 36; Franklin and Schweninger, *Runaway Slaves*, pp. 60–3; Berlin, *Generations of Captivity*, p. 215; W.A. Dunaway, *The African American Family in Slavery and Emancipation* (New York, 2003), pp. 51–114. On the impact of gender on truancy and escape, see Camp, *Closer to Freedom*, pp. 33–47.

25. Blassingame, *Slave Testimony*, pp. 432–3; L.G. Clarke and M. Clarke, *Narratives of the Sufferings of Lewis and Milton Clarke, Sons of a Soldier of the Revolution, during a Captivity of More Than Twenty Years among the Slaveholders of Kentucky, One of the So-Called Christian States of North America* (Boston, 1846), pp. 33–4, http://docsouth. unc.edu/clarkes/clarkes.html (accessed Oct. 2012). Historians agree about the importance of racial solidarity among the African American community in the Ohio River Valley. They argue that fugitive slaves turned to the black community for protection but fail to note how this relative security could reduce slaves' desire to escape north. See J.B. Hudson, 'Crossing the "Dark Line": Fugitive Slaves and the Underground Railroad in Louisville and North-Central Kentucky,' *Filson Club History Quarterly*, 75 (2001), 33–83; Hahn, *Nation under Our Feet*, pp. 35, 42; N.M. Taylor, *Frontiers of Freedom: Cincinnati's Black Community, 1802–1868* (Athens, OH, 2005), pp. 29, 138–60; Berlin, *Generations of Captivity*, pp. 43, 215; K.P. Griffler, *Front Line of Freedom: African Americans and the Forging of the Underground Railroad in the Ohio Valley* (Lexington, 2004), pp. 30–57; Lucas, *History of Blacks in Kentucky*, pp. 92–3.

26. L. Koehler, *Cincinnati's Black Peoples, A Chronology and Bibliography, 1787–1982* ([Cincinnati], 1986), p. 7; *Louisville Public Advertiser*, Nov. 30, 1835; Taylor, *Frontiers of Freedom*, pp. 37–9, 54–5; Bigham, *On Jordan's Banks*, pp. 34–6.

27. Kentucky exported about 22% of its male slaves between the ages of 10 and 19 in the 1850s, whereas Mississippi imported 27% of its male slaves of the same age: Tadman, *Speculators and Slaves*, pp. 301–2; Hudson, *Fugitive Slaves*, p. 14.

28. H. Smith, *Fifty Years of Slavery in the United States of America* (Grand Rapids, 1891), pp. 15–16, http://docsouth.unc.edu/neh/smithhar/smithhar.html; Drew, *Refugee*, pp. 126, 260.

29. In his study of fugitive slaves in Kentucky, the historian J. Blaine Hudson estimated that 44,000 enslaved blacks escaped between 1810 and 1860, or a rate of loss of 0.5% annually; see Hudson, *Fugitive Slaves*, pp. 161–2. Hudson and other historians have suggested that the escape of fugitives was a 'slow bleed' that continuously weakened the institution in the state; see also W. Freehling, *The Reintegration of American History: Slavery and the Civil War* (New York, 1994), pp. 253–74. For a full discussion of the interplay between opportunity and escape see M. Salafia, 'Searching for

Slavery: Fugitive Slaves in the Ohio River Valley Borderland, 1830–1860,' *Ohio Valley History*, 8 (2008), 38–63.

30. Blassingame, *Slave Testimony*, pp. 152–3. On slaves' deceptions of masters, see G. Osofsky, 'Introduction to Puttin' on Ole Massa: The Significance of Slave Narratives,' in Osofsky (ed.), *Puttin' on Ole Massa: The Slave Narratives of Henry Bibb, William Wells Brown, and Solomon Northup* (New York, 1969), pp. 9–44. For broader discussions of slave culture in the Deep South, including slaves' under-standing of and desire for freedom, see M. Gomez, *Exchanging Our Country Marks: The Transformation of African Identities in the Colonial and Antebellum South* (Chapel Hill, 1998); S. Stuckey, *Slave Culture: Nationalist Theory and the Foundations of Black America* (New York, 1987), pp. 3–97; L. Levine, *Black Culture and Black Consciousness: Afro-American Folk Thought from Slavery to Freedom* (New York, 1977), pp. 3–135.

31. Clarke and Clarke, *Narratives of the Sufferings*, pp. 33–4.

32. Bibb, *Narrative of the Life*, p. 83. The historiography well documents the circumscribed lives of free African Americans. See L.P. Curry, *The Free Black in Urban America 1800–1850: The Shadow of the Dream* (Chicago, 1981), pp. 249–51; L.F. Litwack, *North of Slavery: The Negro in the Free States, 1790–1860* (Chicago, 1961), pp. 30–186; I. Berlin, *Slaves without Masters: The Free Negro in the Antebellum South* (New York, 1975), pp. 182–249; Berlin, *Generations of Captivity*, pp. 230–44; J.P. Melish, *Disowning Slavery: Gradual Emancipation and 'Race' in New England, 1780–1860* (Ithaca, NY, 1998), pp. 261–74; E.W. Berwanger, *The Frontier against Slavery: Western Anti-Negro Prejudice and the Slavery Extension Controversy* (Urbana, 1967), pp. 7–59; Middleton, *Black Laws*, pp. 42–73; Thornbrough, *Negro in Indiana*, pp. 92–150. Some recent works that revise the view of a wholly racist North include Griffler, *Front Line of Freedom*; Taylor, *Frontiers of Freedom*; P. Finkelman, 'Ohio's Struggle for Equality before the Civil War,' *Timeline*, 23 (2006), 28–43.

33. S. Rockman, 'The Unfree Origins of American Capitalism,' in C. Matson (ed.), *The Economy of Early America: Historical Perspectives and New Directions* (University Park, PA, 2006), pp. 335–62; Smallwood, 'Commodified Freedom,' 289–98.

34. Recently the historian Stanley Harrold argued that years of conflict over the border between slavery and freedom propelled Americans toward the Civil War. While Harrold argued that constant conflict made accommodation impossible, I argue that constant conflict made accommodation necessary. Residents had to find a way to curb their differences because of the centrality of the Ohio River to the local economy and social stability. See Harrold, *Border War: Fighting Over Slavery before the Civil War* (Chapel Hill, 2010), pp. 183–207.

12
Boundaries of Slavery in Mid-Nineteenth-Century Liberia

Lisa A. Lindsay

On the official seal of Liberia, a sailing ship approaches the coast; a palm tree, plough, and spade stand on the shore; a dove flies overhead carrying an open scroll; and the sun rises over the waters. Above the image is the national motto: 'The love of liberty brought us here.' With the possible exception of the palm tree, all of these symbols reference the founding of Liberia by African Americans, and their hopes for self-sufficiency and peace. Although many of the colony's original white backers had been supporters of slavery, attempting simply to rid the United States of free black people, those African Americans who settled in Liberia from the 1820s saw colonization as their best chance for freedom. On the continent of their ancestors, settlers hoped for the political and economic agency that they had been denied in the United States. In this way, the Americo-Liberians (as settlers and their descendants became known) conceived of Liberia as a place of 'free soil,' not unlike the 'free' states of the American North. They could achieve personal freedom through physical movement to this 'free' territory, where racial slavery was explicitly outlawed.[1]

Yet for the first four decades of its existence, Liberia shared a stretch of West African coastline with rulers and merchants who exported thousands of captive Africans across the Atlantic into slavery, mostly to Cuba.[2] Liberia's leaders and their backers in the United States considered part of the colony's mission to be helping to suppress this external slave trade. During the first half-century after its founding in 1822, a series of treaties with African leaders, backed up by force of arms, expanded the settlement from its initial base at Cape Mesurado more than 300 miles along the coast and some 40 miles up the major rivers in the area. Settlers and their officials justified the expansion of Liberia's borders as part of an effort to dislodge slave traders and suppress the slave trade. They relied heavily on British, and to a lesser extent, American naval patrols to intercept slave ships off the Liberian coast, and they launched their own offensives against nearby slaving operations.

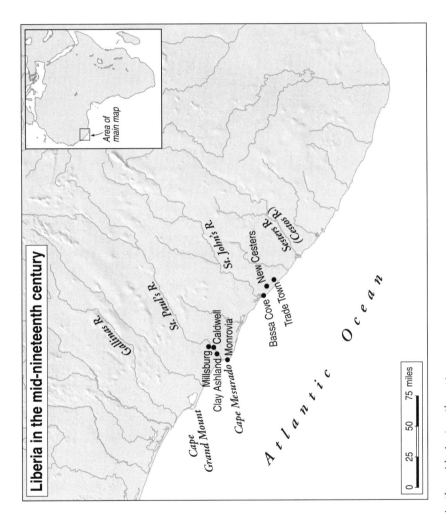

Map 12.1 Liberia in the mid-nineteenth century

Figure 12.1   The official seal of the Republic of Liberia. © Peter Probst/Alamy

During this initial period of expansion, practically all of Liberia entailed borderlands where immigrants came into contact with indigenous peoples. These interactions resulted in considerable cultural mixing; but as in the Americas, they also entailed tension and violence.[3] Although they were vastly outnumbered by indigenous Africans, American settlers expanded their territory and exacted commercial monopolies by force. They also construed their antislavery sentiment very narrowly. Notoriously averse to agricultural or menial work themselves, Liberian settlers engaged indigenous Africans, many of them children, in low-wage or non-wage labor. Criticized for their abuses, they insisted on a semantic distinction: that Liberian labor practices did not entail slavery. By singling out and delineating slave trading as a particular horror, they suggested that other types of unfree labor in the colony—glossed as apprenticeship or pawnship—were acceptable, if only for indigenous Africans.

The interactions along and within Liberia's borders, then, involved slavery in two seemingly inconsistent ways. While African American settlers were largely protected from slaveholders in the United States as well as the predations of local slavers, Africans in and near the borderlands were vulnerable, if not to outright enslavement by Americo-Liberians, then to attack and even conditions resembling slavery. In this way, Liberians were not unlike other nineteenth-century Africans—much as they may have resisted such an assertion. Many (if not most) West African societies contained hereditary servile groups and/or individuals born into slavery. However, the process of *enslavement*—that is, making an individual into a slave—most often involved capture and forced mobility from a place where that individual was an *insider*, enmeshed in kin and community, to a milieu in which she or he was an *outsider*, exploitable in large measure because of her or his social marginality.[4] Since political boundaries more or less defined people as insiders or outsiders, they also determined who was enslavable in the eyes of particular groups, along with who was not.[5] Thus, as this essay details, Liberia's extensive borderlands between the 1820s and 1860s did not so much separate spaces of slavery and freedom as bring different peoples together and mark out who was vulnerable to enslavement, where, and by whom.

Antislavery and Territorial Expansion

Liberia's founding resulted from the confluence of three developments: a base of interest in Africa among some African Americans; the creation of the American Colonization Society (ACS) in 1816 by white advocates of African American emigration; and the US government's 1808 ban on American participation in Atlantic slave trading. The 1819 Act for Suppression of the Slave Trade provided for naval patrols on the African coast to help enforce America's prohibition of the Atlantic slave trade, as well as for the settlement in Africa of captives liberated from intercepted slave ships. The Monroe administration allocated $100,000 to create such a settlement on the African coast. The job was given to the fledgling ACS, which had begun to recruit African American settlers but lacked funds to send them to Africa.

The first ACS ship, the *Elizabeth*, landed 88 African Americans and three white agents at the British colony in Sierra Leone in 1820. Ostensibly, they were to clear land and build houses to accommodate future recaptured Africans. In reality, they were settlers, nearly two-thirds of them children. Vulnerable to tropical diseases and possessing no claim to land on their own, they and another shipload of subsequent colonists remained at the British settlement until 1822. At that point, a US Navy commander and the ACS agents, having selected Cape Mesurado, 250 miles south of Freetown, Sierra Leone, as a suitable site, approached the local Dei political leader, known to history as King Peter, to arrange for a purchase of land. The king refused through hours of haggling, until the Navy commander Robert Stockton held a loaded pistol to King Peter's head, thereby ensuring the beginnings of the American Liberian settlement. The colonists named their base at Cape Mesurado 'Monrovia,' after the president whose accommodation to the ACS had made it possible.[6]

Relations with nearby Africans proceeded along the lines set by the initial transaction with King Peter. Though ACS agents and settlers understood their mission to be creating a free society and bringing civilization to Africans in need of it, from the point of view of Gola, Bassa, Dei, and other leaders of the 16 African political groups in the area, these Americans were unwelcome usurpers of land and interlopers in trade. Moreover, the colonists' intolerance of the slave trade, reiterated in virtually every treaty negotiation, rankled with African elites and their foreign-born trading partners. Although regional and longer-distance trade included a range of products, African political leaders profited from the sale of war captives, victims of raids, and criminals to foreign slave dealers, some of them long-term residents in the area. In the early 1820s, perhaps as many as 2000 captives were exported per year from the less than 100-mile stretch of coastline between Cape Grand Mount in the north and Monrovia in the south.[7] Don Pedro Blanco, a mixed-race slaver from Brazil, and Theodore Canot, a French-Sardinian trader, 'gave a touch of glamour to the whole business,' according to an early twentieth-century retrospective account. 'While maintaining an outward aspect of courtliness and intelligence, they were yet the most efficient slave merchants on the coast,' maintaining their own 'factories,' or fortified trading establishments, with nearby barracoons for captives.[8]

These and various other foreign slave traders shared with their African trading partners a sense of the threat posed to them by the American settlement, and they supported African operations against the colonists. King Peter, still stinging from the initial treaty negotiations, threatened to sell the American settlers into slavery unless he received tribute from them. Just in time, a US brig arrived with a reinforcement of 18 recaptive Africans (that is, people rescued from slave ships) and 37 immigrants, led by a Methodist minister named Jehudi Ashmun, who took over the military security of the settlement. Tensions came to a head in November 1822, after the colonists had raided a slave factory on the coast, liberating the captives awaiting shipment and burning the holding pens to the ground. King Peter organized a coalition of local African groups to attempt to destroy the colony. The king and his allies attacked the settlement twice, aided by Cuban slave traders who had settled in the adjoining Gallinas country. Eight hundred Africans swept into Monrovia, which contained 135 settlers, only 35 capable of bearing arms. Still, the attackers were cut down by the settlers' cannon fire. King Peter launched a second unsuccessful attempt several weeks later, and the decades that followed continued to see military conflict between local Africans and the colonists, although the Americans maintained the balance of firepower.[9]

After a rough truce with King Peter, Jehudi Ashmun as de facto and then official leader set about organizing the settlement, expanding its territory, trading for African produce, and attacking the slave trade. During 1825 and 1826, Ashmun endeavored to lease, annex, or buy African lands along 150 miles of the coast and on major rivers leading inland, and he demanded treaties with nearby African leaders for exclusive rights to their trade. In 1825 Ashmun organized a settler attack against a nearby Spanish-owned slave factory, freeing 99 Africans. The next year, with the aid of a Colombian privateer named Captain John Chase, he led a second assault on slave traders. At Trade Town, where an African leader called King West was holding some thirty recaptives hostage, the combined forces attacked the town and destroyed slave factories. But Ashmun's efforts to attack slaving operations went well beyond the US government's mandate to intercept American vessels engaged in slave trading, and they were not continued after his death in 1826. Ten years later, ACS agent A.D. Williams reported that the slave trade continued and was 'fearfully increasing in our vicinity.'[10] Moreover, it was rumored that even colonists were participating in the slave trade, secretly selling slaves to Spanish dealers at the Gallinas River and Cape Mount. In 1830 the ACS established stiff penalties for any settler convicted of slave trafficking.[11]

The hostility of Liberia's leaders to the slave trade continued to fuel conflicts with local Africans and their foreign trading partners. In 1832, tensions between settlers and Dei leaders who had lost lands to the Liberians came to a head when some captives awaiting sale by the Dei king ran away to the colony. ACS administrator Joseph Mechlin refused to return them, prompting a coordinated attack by Dei forces and their Gola allies. With recaptive Africans as the first line of fire, a force of settler volunteers marched on their enemies' settlements, overwhelmed their defenses, and burned their towns. In the peace that followed, Mechlin exacted a treaty that deprived Dei of their previous access to inland trade routes.[12] Two years later, the recently built settlement of Port Cresson was attacked by Kings Joe and Peter Harris, rivals to the Bassa chief who had provided the land. Survivors attributed the attack to the instigation of the slaver Theodore Canot, who intended to reestablish the slave trade at the mouth of the St John's River after driving the Port Cresson people away.[13] In 1838, I.F.C. Finley, governor of the American colony at Bassa Cove and son of a founder of the ACS, was murdered. Liberian officials blamed slave traders and suspected the influence of Canot. In response to the murder and to continued slave trading in the surrounding area, Governor Thomas Buchanan organized a party of Liberians to attack the slave factory at Trade Town.[14] According to Peyton Skipwith, a Liberian settler from Virginia who participated in the attack,

> we went down and broke up the factory and brought away all the effects say in goods and destroyed about fifty puncheons [of] Rum which was turn loose on the ground say the effect in goods &c to the amt of ten thousand Dollars. After we had taken the goods or a part we had to contend with the natives which fought us two days very hard but we got the victory and form a treaty before we left with one of the chiefs but not with the other and only got four slaves so we cannot say that we concluded a final peace ...[15]

Just as the ACS had financed its first voyages to Africa by linking them to the American government's allocations for suppressing the slave trade, opposition to slaving became the justification for expanding settler jurisdiction in West Africa.[16] ACS and Liberian officials insisted that settler control over a continuous stretch of coastline was the only sure means of suppressing the slave trade. As the initial base at Cape Mesurado expanded in size to some nine hundred settlers by 1840, the ACS also began to send new emigrants to what became a string of towns stretching up the nearby St Paul's River. Competing state and local colonization societies in the United States also recruited their own emigrants for settlements that ultimately merged with the ACS colony.[17] In 1849, Liberia's first president, Joseph Jenkins Roberts, appealed for funds to purchase the Gallinas territory by emphasizing that Liberia's previous acquisition of territory had helped to suppress the slave trade. Adding Gallinas, he argued, would enable the Liberians to keep the whole coast 'free from the demoralizing and wilting influence of the Slave trade.'[18]

Antislavery on the High Seas

For settlers and their officials, Liberian antislavery required not only territorial sovereignty, but also naval patrols. Several months before the attack on Trade Town, the same Peyton Skipwith who had fought there had written to his former master,

> I also wish to remind you that I see daily the Star Spangled Banner unfurled on the coast of Africa as a protection for the slaver to keep the British man of wars from taking them[,] which we think as a hand full of people to that of the United States a disgrace to her Banner. We if we had vessels could defy them to take our cross & stripes [the Liberian flag] and Hoist them to her mast head for the protection of the slave trade.[19]

Skipwith was right: because American diplomats had consistently refused to grant British naval officers the right to search vessels flying the American flag, slavers used the Stars and Stripes as cover.[20]

Since its own ban on slave trading was passed in 1807, the British government had used its diplomatic and naval power to convince other Europeans to stop trading in slaves as well. As the Napoleonic Wars came to an end, British diplomats secured treaties by which France and other major continental powers agreed to abolish their slave trades, with Portugal agreeing to end slave trading north of the equator. Moreover, between the 1810s and 1830s, Britain successfully negotiated treaties involving the reciprocal right of search in pursuit of slavers with France, Spain, Portugal, Brazil, the Netherlands, Denmark, the Hanseatic cities, and several smaller powers. This left ships flying the American flag as almost the only ones that the Royal Navy was not permitted to stop in the Atlantic and search, or seize if found to be slavers. To intercept such vessels themselves, a small American naval force sailed for Africa in 1820. It had multiple missions, however, only one of which was seizing American-flagged slavers (and landing those rescued at Liberia); and American navy ships cruised the African coast only occasionally.[21]

In spite of European and American bans on slave trading, the trans-Atlantic commerce continued, including from the coasts of Liberia and nearby Sierra Leone, where the British had their own colony of liberated Africans and other immigrants. Between 1820 and 1840, some thirteen thousand slaves were exported from the vicinity of the Liberia colony.[22] According to a report presented to the British parliament, there were 11 American ships in the slave trade in 1837, and 19 in 1838. One ship, the *Valador*, had formerly carried emigrants to Liberia; renamed the *Scorpion* and then the *Viper*, it was converted into an American-flagged slave ship. In 1840 it made the first of several runs between New Cess, near Bassa Cove, and Cuba, in one instance carrying 680 Africans into slavery. In 1844, it successfully eluded British and American cruisers, carrying 300 more people into slavery.[23]

In 1842, the American government agreed to regularize its previously haphazard antislavery squadron with the ratification of the Treaty of Washington (also known as the Webster–Ashburton Treaty) with the UK. In lieu of granting British naval officers the right to search American vessels, the treaty required the US to maintain its own naval squadron on the African coast, primarily to suppress the slave trade. This hardly entailed an American commitment to antislavery, however, particularly as sectional tensions over slavery increased at home. As the squadron prepared to sail for Africa, Secretary of the Navy Abel P. Upshur made its instructions clear: 'The rights of our citizens engaged in lawful commerce are under the protection of our flag; and it is the chief purpose as well as the chief duty of our naval power to see that these rights are not improperly abridged or invaded.' In other words, suppressing the slave trade was secondary to protecting American maritime trade. 'It is to be borne in mind,' the Secretary continued, 'that while the United States sincerely desire the suppression of the slave trade, and design to exert their power, in good faith, for the accomplishment of that object, they do not regard the success of their efforts as their paramount interest, nor as their paramount duty.' Indeed, slave trade suppression was not worth risking American interests or 'the exposure of [Americans] to injurious and vexatious interruptions in the prosecution of their lawful pursuits.' Over its entire 18-year life span, the American antislavery squadron never consisted of more than six vessels, and it captured only 36. In contrast, the British averaged 19 ships on patrol at any one time and between 1843 and 1861 captured 595 slave ships.[24]

Its concern for American commercial interests is reflected in the operations of the American naval squadron during its first two years in Africa. Although it intercepted only one slaver during this time, Commodore Matthew Perry forcibly intervened to protect American trade and Liberian interests. On separate occasions in 1842 and 1843, African assailants had attacked American merchant vessels off the Liberian coast. Perry first proceeded to Sinoe, south along the coast from Monrovia, where one of the attacks had taken place. There, a massive show of naval force brought about a treaty between the inhabitants and the Liberian government. Next, near Cape Palmas, Perry brought ashore some two hundred men to burn to the ground four villages whose inhabitants were deemed responsible for one of the attacks on American ships, effectively reinforcing the power of the settlers over their African neighbors. As one settler recounted several months later,

> Peace and harmony exists among us, with our Savage natives. The U.S. Fleet has done great good on the Coast of Africa. They [have] in a measure dispersed the slave trade, & also subdued the Natives & brought them to Know their place, more so than they did before the arrival of said fleet.[25]

Later that year, however, the US government, determined not to extend its influence in Africa, clarified that Liberia was not its colony and American armed forces would not be used to interfere in Liberian commerce. Meanwhile, the British government had refused to acknowledge the sovereignty of the ACS, thereby supporting European traders from Sierra Leone who ignored Liberian land claims and refused to pay customs fees. Without backing from the US government, Liberian leaders claimed sovereignty for their own republic, declaring independence from the American Colonization Society in 1847. They crafted a constitution based on that of the United States but also outlawing slavery and restricting citizenship and property-holding to 'persons of color.'[26] As an independence gift, and likely in recognition of Liberia's insecurity, Britain's Queen Victoria gave the Liberian government a steamer, the *Lark*, which became the sole vessel in the Liberian Navy.[27]

Independence and Antislavery

One of the first acts of the new nation-state was to purchase a tract of territory south along the coast about fifty miles from Monrovia called New Cesters (sometimes called New Cess). This area had been locally notorious as a market for slaves for export, and Liberian officials described the purchase as an attempt to suppress the slave trade. Immediately after the transaction, the Liberian government gave notice to specific slave traders that they should stop their business and export no more slaves. Yet in the absence of serious enforcement measures, slave trading in the area continued.[28] The next year, the settler newspaper the *Liberia Herald* reported that slaves were often transported in the vicinity of New Cesters and nearby Trade Town, and that 'the Spaniard at New Cesters ... is as deeply engaged in [the slave trade] now as he ever was.'[29] This remained a thorn in the side of the new republic, although perhaps less as a direct threat to safety than to national pride. As the *Liberia Herald* editorialized, 'If we are able to break up that establishment [the Spanish slave fort] and yet suffer it to remain … we will most certainly be accused of winking at the slave trade. Yet to say we are not able to remove a few slavers is humiliating.'[30]

Less than a year later, the Liberians were ready to shore up their control over the newly acquired territory and the recalcitrant chiefs and Spanish slave traders within it. Baptist missionary John Day wrote that tensions had been mounting at New Cesters, with attacks on Liberians and their property and chiefs refusing to deliver those whom the Liberian authorities accused of the crimes. As one of the Liberian volunteers for the operation wrote to a correspondent in Virginia, after President Roberts returned from a trip to France, 'we were all commanded to get ourselves in order for the war, as we had to contend against an African tribe, called the New Cesters' tribe.'[31]

'No insult or wrong will as soon fill their ranks with volunteers, as the suppression of slavery,' the missionary Day said of Liberian settlers.[32] Some 350 to 400 volunteers were mustered at Monrovia in March 1849 for an expedition accompanied by President Joseph J. Roberts, commanded by General J.N. Lewis, comprised of two regiments, and transported by the French steamer *Espado*.[33] At New Cesters, opposing African troops were no match for the Liberian onslaught and cannon fire from the French steamer. A Spanish slave trader known as Don Francisco fled his establishment, leaving it empty. The Liberians discovered behind its thick mud walls an enclosure of about two acres, with a two-storey frame house and fine furniture, along with two large thatched houses for slaves. Additional structures on both sides of the enclosure had accommodated keepers of the slaves and attendants of the Spanish traders. Over the following week, the Liberian regiments proceeded to settlements known as Joe West's Town and Trade Town, where another Spanish trader surrendered and promised to deliver some two thousand slaves to the Liberians at a later date.[34]

The victorious Liberian forces returned to Monrovia with a renewed sense of their power and mission. Solomon Page, one of the volunteers, wrote that 'We were successful during the war, something which I did not expect or anticipate before we went.' Another boasted,

> a great many told me when I was in America that we could not take the Spanyards. We have got them in our town waiting for tryal. It proved as in all of the wars that God is on our side & if he be for us who can be against us. We have been oppressed long enough. We mean to stand our ground & contend for our rights until we die.

The missionary Day pointed out in a letter to the Southern Baptist Convention's Foreign Mission Board that this victory over slave traders was also a victory for Liberian territorial expansion. Eleven new headmen had put themselves under the 'protection' of the Liberian republic, he reported, and many more were talking of doing so. His conclusion: 'Our civil jurisdiction will now extend far and wide.'[35] The editor of the *Liberia Herald* noted the enormous cost of the expedition, but asserted that its achievements were worth the outlay. 'We cannot expect peace and quietness while the slave trade is going on near us,' he wrote. 'Nor can we hope to exert our full influence upon the surrounding tribes until the accursed traffic is wholly destroyed. When that most desirable object is accomplished, we shall then breathe freely.'[36]

Slavery in Liberia

In the same letter as his description of the attack on slavers, the Rev. Day also observed the state of slavery and the slave trade within Liberia itself: 'There are slave holders,' he wrote, 'but the slavery around us amounts to but little when slaves cannot be sold, and the slave trade must forever cease the length and breadth of Liberia.'[37] Three weeks later, he added in another letter to his mission superiors that the 'intelligent part' of the settler community opposed slavery and would take up arms against it; but others had been tried for selling slaves themselves.[38] Day's comments hint that Liberian antislavery rhetoric overlaid a more complex situation within the colony. Slavery was explicitly prohibited in the ACS charter and later in the Liberian constitution. Yet from the time of the colony's founding, a two-tiered system developed, with American settlers, especially the earlier arrivals, as a privileged class and indigenous Africans as a subordinate labor force. Some of those Africans were those 'liberated' from nearby slavers during settler attacks. When in 1825, for instance, Ashmun led an attack which freed Dei captives, he distributed those individuals to settlers, who put them to work in their own enterprises.[39] But the majority came from two sources: indigenous communities under the 'protection' of the colony, and captive Africans liberated from slave ships and landed at Liberia by the American Navy.

The larger of the two groups comprised 'servants' from indigenous communities. In a system regulated by law starting in 1838, settler families and entrepreneurs incorporated impoverished Africans, typically with the promise of educating and 'civilizing' them.[40] In exchange, these mostly children and teenagers worked in settler households and enterprises. Throughout West Africa there existed a similar system, in which human beings served as collateral for loans on which their labor represented the interest.[41] But in Liberia, pawnship— or apprenticeship, or wardship, as the settlers preferred to call it—represented a significant portion of domestic service as well as agricultural labor, since settlers strongly preferred trade to farming. As the settler Payton Skipwith wrote in 1834,

> There is Some that have come to this place that have got rich and a number that are Suffering. Those that are well off do have the natives as Slaves and

poor people that come from America have no chance to make a living for the natives do all the work.[42]

Observers at the time and in retrospect differed about whether 'apprentice-ship' was a benign institution of assimilation, or whether, in the words of an observer from the 1840s, 'These pawns are as much slaves as their sable proto-types in the parent states of America.'[43]

One of the most scathing descriptions of the labor system came from William Nesbit, a Pennsylvania African American who published an account of his four months' travel to Liberia in 1853. Although his trip had been spon-sored by the ACS, Nesbit became a bitter critic of colonization. According to him, forced labor within Liberia's borders was widespread. 'Every colonist keeps native slaves (or as they term them servants) about him, varying in number from one to fifteen, according to the circumstances of the master,' he charged. 'These poor souls they beat unmercifully, and more than half starve them, and all labor that is done at all, is done by these poor wretches.'[44] Nesbit's account devoted a full chapter to slavery. Liberian slaves faced the same kinds of toil as those in America did, he wrote, the major difference being that Liberian slaves were fed less because they were more easily replaced. A large supply was available for cheap purchase from their parents, presumably because their own opportunities for earning a living were being eroded by the expansion of Liberian settlement. Nesbit's characterization of Liberian slaveholders was stark: 'They are mostly manumitted slaves themselves, and have felt the blight-ing effects of slavery here, only to go there to become masters.' This, he scoffed, was in spite of Liberian claims about civilizing Africans. 'They [settlers] profess to have broken up the foreign slave trade, which is far from the truth; but sup-pose they had done so, is that *even* a blessing, under the circumstances?,' he asked rhetorically.[45]

Other observers acknowledged abuses in the apprenticeship system, but insisted that this was not slavery. The Rev. Samuel Williams, who arrived in Liberia as a missionary shortly before Nesbit's trip, refuted his fellow Pennsylvanian in his own publication, *Four Years in Liberia*. He took strong issue with Nesbit's assertions about slavery in Liberia, reminding readers that slavery was against the law there and that strict rules governed apprenticeship.

> Nearly all [settlers] have natives as helps in their families, and this is as it should be; but I confess that black people are no better than white people, as many, when they have power, abuse it, and so it is with some in Liberia; wicked persons there do abuse the native youths.

Still, he insisted that the entire settler population should not be blamed for the transgressions of a few.[46]

The other major source of captive labor for Liberian settlers was the Africans liberated from slave ships by American naval patrols, who came to be known in Liberia as 'Congos' because many of them originated near the Congo River in Central Africa. For the first several decades of Liberian settlement, and in contrast to the recaptives landed by the British at Sierra Leone, their numbers were extremely small.[47] Between 1820 and 1843, the US Navy sent to Liberia only 287 recaptured Africans, a number that represented 6 percent of the immigrants for the period. But from 1846 and especially with several landings in 1860, the recaptured African population jumped markedly. During those years, a total of 5457 recaptives landed at Liberia (although all but 756 came in 1860).[48] The recaptives were placed as 'apprentices' with settlers, who received an annual allocation from the US government for their support and education. Yet few of those 'apprenticed' were actually educated, and credible critics alleged that apprenticeship was simply a cover for involuntary, unpaid labor.[49]

Recaptive labor became especially significant along the St Paul's River, which flowed inland from just north of Monrovia. African American settlers had begun settling and farming there in the 1820s. By the early 1850s, three towns— Caldwell, Millsburg, and Clay Ashland—anchored a series of trade routes and agricultural estates along the river. Although very little farming occurred in Monrovia, along the St Paul's, settlers recreated the kind of rural societies many of them had known in America.[50] There, large estates produced sugar and later coffee; a few prosperous settlers built large brick houses; and largely unfree laborers toiled in the fields. For Liberia's boosters, the region offered the promise of refined prosperity, with 'neat houses and well cultivated farms, standing on the borders of the deep green wilderness, overlooking the majestic stream as it glides silently but powerfully along, bearing on its bosom the rich productions of its banks.'[51] According to the Rev. Samuel Williams, 'The St Paul's farmers are in general, industrious and prosperous. Many very fine plantations are to be seen ... who have as good sugar plantations as I ever saw in the neighborhood of New Orleans.' Amazingly, the comparison between plantations in Liberia and in antebellum Louisiana raised barely a comment about slavery. Another admirer, perhaps Edward Blyden, reported on an 1852 visit to several large estates, singling out 'Iconium,' owned by Allen B. Hooper, as 'one of the handsomest places we have ever seen,' with flower beds, fields of coffee and sugar cane, and plots of garden vegetables. On a walk around the farm, the observer noted favorably the 'quiet manner [in which] the laborers perform their seemingly agreeable task.'[52]

Although African recaptives had been 'apprenticed' to planters in the St Paul's settlements from the 1840s, their numbers grew dramatically around 1860. Between 1840 and 1856, the US African squadron had captured only 18 slave ships, just over one per year. But in the four years between 1857 and 1861, the squadron captured 20 ships, landing their human cargoes in Liberia. In 1860 alone, nearly 5000 recaptive Africans arrived in the colony, which at the time contained barely that number of Americo-Liberians. 'What shall we, what *can* we do with such an appalling amount of heathenism, superstition, and barbarity all at once?,' asked the settler press.[53] Drawing on an allocation from the US Treasury, the ACS agent John Seys paid upriver farmers 50 cents and Monrovians 25 cents for each recaptive taken in 'apprenticeship.' Commercial firms also joined in, requesting and receiving recaptives to work in sawmills and other enterprises. Missionary stations received recaptives. And groups of liberated Africans formed their own settlements outside of apprenticeship, prompting vague Liberian complaints about their 'depredations.'[54] Still, the Liberian legislature ultimately rejected measures to control closely the movement of the 'Congos' from one county to another. The existing statues on apprenticeship were sufficient regulation, lawmakers argued, and additional measures would unconstitutionally apply to one group of Liberians and not to others. Moreover, restricting the movement of recaptives 'would not only compel the *natives* to *serve us*, but also our own sons and daughters, if otherwise, then it would be slavery to all intents and purposes.'[55]

Aftermaths

No other recaptives were landed at Liberia after the influx of 1860. During the American Civil War, the US government recalled the African squadron to US waters in order to strengthen its blockade of the Confederate states. At this point, the long-running standoff between the US and UK over the right to search suspected slavers was finally resolved. In the absence of an American naval presence off the African coast, and in order to maintain efforts to suppress the slave trade at its source, Secretary of State Seward suggested a treaty authorizing the British Navy to search and seize American-flagged slavers. The treaty became effective in 1862.[56]

A new, official US commitment to antislavery, combined with British patrols and, most crucially, the end of Cuban slave importation in 1866 finally ended the external slave trade from the African coast.[57] Liberians continued to face the hostility of local Africans determined to protect their lands and trade routes from colonial encroachment, and this took the form of armed conflict through the end of the nineteenth century. But settlers no longer shared a coastline with trans-Atlantic slavers, and they no longer could extend their territorial claims in the name of fighting the slave trade. The question of slavery itself, however, was hardly settled.

In 1919, the League of Nations launched an official inquiry into forced labor in Liberia and the forcible recruitment of unfree Liberian labor for shipment to plantations on the nearby Spanish-owned island of Fernando Po. As Ibrahim Sundiata has detailed, the involvement of the US government in the inquiry—which it in fact instigated—was intended to mute criticism of the US-based

Firestone Corporation, whose Liberian rubber plantations relied on coerced and unpaid African labor.[58] But the Liberian government was also to blame, having developed an arrangement with Spain to supply workers to cocoa plantations on Fernando Po in exchange for $45 per recruit, most of which was pocketed by state officials. In addition, labor exactions for road-building and other state projects were widespread, with government officials and a rapacious Liberian Frontier Force demanding that African chiefs produce specified numbers of workers ostensibly for limited and paid contracts. In reality, in the words of one historian, the system came close to 'the chain gang of the American South, except that Liberian forced laborers were required to supply their own tools and food.'[59]

Conclusion

The paradoxes of Liberian history are striking. Founded and sustained largely through the efforts of American white supremacists, Liberia was envisioned as a place where African Americans could achieve freedom outside the slave system of the United States. Surrounded by slave traders, it was supported and protected by an American navy only half-heartedly committed to antislavery. Excluded from American citizenship, the emigrants nonetheless endeavored to transplant American democratic institutions—but only in their bounded enclave, among their own people. Deeply opposed to bondage themselves, settlers inflicted slave-like conditions, carefully described in words other than slavery, on Africans in their midst.

Borderlands have typically been spaces of considerable fluidity, where cultural or political norms adapted or were challenged in the context of multicultural interactions. As the paradoxes in its history suggest, Liberia's extensive borderlands witnessed inconsistent approaches to slavery. As in other parts of Africa (and elsewhere in the world), slavery and freedom were determined by social and political as well as physical geography. Insiders and outsiders, constructed by political borders and cultural attributes, enjoyed different opportunities and faced differing impositions. It was not until 1904, in fact, that the majority of people living in Liberia were considered to be citizens.[60] The stark divisions between Americo-Liberians and indigenous Africans, whether marking 'freedom' or other attributes of citizenship, would persist until the late twentieth century, when they erupted in decades of violence from which the country has only recently emerged.

Notes

1. This is, of course, an overstatement. By the 1850s the Dred Scott decision and the Fugitive Slave Law eroded protections for African Americans even in the 'free' states.
2. See http://slavevoyages.org (accessed Apr. 30, 2013).

3. On cultural mixing in early Liberia, see W.E. Allen, 'Liberia and the Atlantic World in the Nineteenth Century: Convergence and Effects,' *History in Africa*, 27 (2010), 7–49; for the Americas, see the Introduction to this volume.

4. The classic statement of all this is I. Kopytoff and S. Miers, 'African Slavery as an Institution of Marginality,' in Miers and Kopytoff (eds), *Slavery in Africa: Historical and Anthropological Perspectives* (Madison, 1977), and more generally, O. Patterson, *Slavery and Social Death: A Comparative Study* (Cambridge, MA, 1985).

5. Precolonial African borders were much more fluid and subject to change than they became in the colonial and postcolonial eras. Even so, African political authorities and others considered 'outsiders' and 'insiders' distinctly. For the classic articulation of this argument with regard to ethnicity (generally a gloss on political entities), see F. Barth, *Ethnic Groups and Boundaries: The Social Organization of Culture Difference* (Boston, 1969).

6. Of the many accounts of the founding of Liberia, the most engaging is chapter 1 of James T. Campbell's *Middle Passages: African American Journeys to Africa, 1787–2005* (New York, 2006). Two classic accounts are T.W. Shick, *Behold the Promised Land: A History of Afro-American Settler Society in Nineteenth-Century Liberia* (Baltimore, 1977), and P.J. Staudenraus, *The African Colonization Movement, 1816–1865* (New York, 1980).

7. C.A. Clegg III, *The Price of Liberty: African Americans and the Making of Liberia* (Chapel Hill, 2004), p. 101. According to the *Transatlantic Slave Trade Database*, the estimated numbers of slaves exported from the Windward Coast were 2369 in 1820; 678 in 1821; 956 in 1822; zero in 1823; 2023 in 1824; and 1234 in 1825 ('Windward Coast' estimates by year, www.slavevoyages.org [accessed Sept. 29, 2011]). Clegg's estimates are higher.

8. C.S. Johnson, *Bitter Canaan: The Story of the Negro Republic* (New Brunswick, 1987), p. 62. Johnson, a Fisk University sociologist, was the American representative to the 1929 International Commission of Inquiry into the Existence of Slavery and Forced Labor in the Republic of Liberia.

9. L. Saneh, *Abolitionists Abroad: American Blacks and the Making of Modern West Africa* (Cambridge, MA, 2000), pp. 206–8; Campbell, *Middle Passages*, p. 54; Staudenraus, *African Colonization Movement*, p. 67.

10. E.S. van Sickle, 'Reluctant Imperialists: The US Navy and Liberia, 1819–1845,' *Journal of the Early Republic*, 31 (2011), 107–34, at 117; also see Staudenraus, *African Colonization Movement*, ch. 13.

11. Clegg, *Price of Liberty*; Johnson, *Bitter Canaan*, p. 72.

12. Clegg, *Price of Liberty*, pp. 105–11; Johnson, *Bitter Canaan*, pp. 65–7.

13. Shick, *Behold the Promised Land*, p. 61.

14. Johnson, *Bitter Canaan*, p. 71.

15. Peyton Skipwith to John H. Cocke, Nov. 11, 1839, in B.I. Wiley, *Slaves No More: Letters from Liberia, 1833–1869* (Lexington, 1980), pp. 51–2.

16. Shick, *Behold the Promised Land*, p. 61.

17. For a list of settlements with their populations in 1843, see Shick, *Behold the Promised Land*, p. 34.

18. Joseph Jenkins Roberts, 'Appeal to the Government and People of the United States,' in *American Colonization Society Papers*, Library of Congress, May 19, 1849, at http://www.loc.gov/exhibits/african/afam003.html (accessed Sept. 26, 2011).

19. Peyton Skipwith to John H. Cocke, May 20, 1839, in Wiley, *Slaves No More*, p. 49.

20. Van Sickle, 'Reluctant Imperialists'; also see Johnson, *Bitter Canaan*, p. 71.

21. D.L. Canney, *Africa Squadron: The US Navy and the Slave Trade, 1842–1861* (Washington, DC, 2006), chs 1–2.

22. *Transatlantic Slave Trade Database*, 'Windward Coast' estimates, www.slavevoyages.org (accessed Sept. 29, 2011).

23. P. Duignan and L.H. Gann, *The United States and Africa: A History* (Cambridge, 1987), p. 30; Clegg, *Price of Liberty*, p. 159.

24. Quoted in Canney, *Africa Squadron*, pp. 56–7. On the 1842 treaty and the question of British search of American vessels, also see van Sickle, 'Reluctant Imperialists.' For total numbers of captured slavers, see Canney, *Africa Squadron*, p. 222, and C. Lloyd, *The Navy and the Slave Trade: The Suppression of the African Slave Trade in the Nineteenth Century* (reprint, London, 1968), pp. 275–81.

25. Peyton Skipwith to John H. Cocke, Sept. 29, 1844, in Wiley, *Slaves No More*, p. 60.

26. *Liberian Constitution of 1847*, at 'The Liberian Constitutions,' http://www.onliberia. org/con_1847.htm (accessed May 1, 2013).

27. 'The Liberian Cutter "Lark",' *Liberia Herald*, June 25, 1849.

28. Roberts, 'Appeal to the Government and People of the United States.'

29. 'Slave Canoes Captured,' *Liberia Herald*, Sept. 30, 1848.

30. *Liberia Herald*, June 30, 1848.

31. John Day to Rev. James B. Taylor, Apr. 3, 1849, AR 551–2, Southern Baptist Convention, International Mission Board (IMB) archives, Richmond, VA; Solomon S. Page to Charles W. Andrews, Apr. 22, 1849, in Wiley, *Slaves No More*, pp. 106–7.

32. Day to Taylor, Apr. 23, 1849, AR 551–2, IMB archives.

33. James C. Minor, adjutant of the First Regiment, wrote a detailed account of the operation which was published in three installments in the *Liberia Herald* of June 25, July 27, and Sept. 28, 1849. Also see Wiley, *Slaves No More*, pp. 322–3.

34. 'An extract of the account of the war expedition marched against the Kings of New Cesters and Trade Town, and the Spaniards residing in these territories, for the purpose of buying slaves, as given by J. C. Minor, Adjutant of the first Regiment,' *Liberia Herald*, July 27, 1849.

35. Solomon S. Page to Charles W. Andrews, n.d. [1849], in Wiley, *Slaves No More*, pp. 107–8; Sion Harris to William McLain, May 20, 1849, in Wiley, *Slaves No More*, p. 227; Day to Taylor, Apr. 3, 1849, IMB archives.

36. *Liberia Herald*, May 18, 1849.

37. Day to Taylor, Apr. 3, 1849, IMB archives.

38. Day to Taylor, Apr. 23, 1849, IMB archives.

39. Staudenraus, *African Colonization Movement*, p. 151.

40. See Shick, *Behold the Promised Land*, pp. 65–6, for a benign portrait of this institution.

41. See P.E. Lovejoy and D. Richardson, 'Trust, Pawnship, and Atlantic History: The Institutional Foundations of the Old Calabar Slave Trade,' *American Historical Review*, 104 (1999), 332–55, and Lovejoy and Richardson, 'The Business of Slaving: Pawnship in Western Africa,' *Journal of African History*, 41 (2001), 67–89.

42. Peyton Skipwith to John H. Cocke, Feb. 10, 1834, in Wiley, *Slaves No More*, p. 36.

43. Quoted in Johnson, *Bitter Canaan*, p. 73.

44. W. Nesbit, *Four Months in Liberia, or, African Colonization Exposed* (orig. 1855), repr. in W.J. Moses, *Liberian Dreams: Back-to-Africa Narratives from the 1850s* (University Park, PA, 1998), p. 90.

45. Nesbit, *Four Months in Liberia*, pp. 102, 104.

46. S. Williams, *Four Years in Liberia: A Sketch of the Life of Rev. Samuel Williams, with Remarks on the Missions, Manners, and Customs of the Natives of Western Africa; Together with an Answer to Nesbit's Book* (orig. 1857), reprinted in Moses, *Liberian Dreams*, p. 172.

47. By 1814 there were 10,000 recaptives in Sierra Leone, mostly in Freetown. Over the next half century, about 3000 new recaptives were landed at Sierra Leone per year,

eventually numbering more than 150,000. See J. Peterson, *Province of Freedom; A History of Sierra Leone, 1787–1870* (London, 1969).

48. Shick, *Behold the Promised Land*, p. 66.
49. Johnson, *Bitter Canaan*, p. 81.
50. See Shick, *Behold the Promised Land*, ch. 5.
51. *Liberia Herald*, 18 Aug. 1852.
52. Williams, *Four Years in Liberia*, p. 146; *Liberia Herald*, March 3, 1852.
53. *African Repository*, 37 (Jan. 1861), quoted in Shick, *Behold the Promised Land*, p. 68. On numbers of ships captured, also see Duignan and Gann, *United States and Africa*, p. 36.
54. Stephen A. Benson to R.R. Gurley, Oct. 31, 1860, in *African Repository*, 37 (Jan. 1861), quoted in Shick, *Behold the Promised Land*, p. 70.
55. Minutes of the Liberian House of Representatives, Dec. 30, 1864, quoted in Shick, *Behold the Promised Land*, p. 72.
56. Duignan and Gann, *United States and Africa*, p. 39.
57. For the declining numbers of human exports from Africa, see the Transatlantic Slave Trade Database, www.slavevoyages.org.
58. I. Sundiata, *Brothers and Strangers: Black Zion, Black Slavery, 1914–1940* (Durham, NC, 2004). A brisk and useful summary of the scandal is in Campbell, *Middle Passages*, ch. 6.
59. Campbell, *Middle Passages*, p. 244.
60. J.G. Liebenow, *Liberia: The Evolution of Privilege* (Ithaca, NY, and London, 1969), p. 25.

Part VI
Reading Borders:
Individuals and Their Borderlands

13
Unofficial Frontiers: Welsh–English Borderlands in the Victorian Period

Roland Quinault

Life on a shifting frontier was a defining experience for generations of Americans and one that received academic recognition by Frederick Jackson Turner in his seminal 1893 essay.[1] In Britain, by contrast, the sea has always been the most important and immutable border—separating the island from the European mainland. But Britain also has its internal borderlands, with their own history and peculiar character. For centuries the borderland between Scotland and England was as notorious for its lawlessness as the Wild West in its heyday. The border between England and Wales was also originally a place of contestation and conflict.

In the eighth century, King Offa built a dyke from the Dee estuary in the north to the Wye estuary in the south, to demark his Anglo-Saxon kingdom of Mercia from the lands of the Welsh princes to the west. Three centuries later, King William I, after conquering England, protected his new realm from the Welsh by creating powerful marcher lordships along the frontier. Those lordships were abolished during the reign of Henry VIII, when the Acts of Union of 1536 and 1542–43 gave Wales largely the same administrative and legal structure as England.[2] The abolition of the Court of Great Sessions in 1830 further undermined the separate identity of Wales.[3] Nevertheless, most of the people of Wales still regarded themselves as Welsh, rather than English, and the majority of them continued to speak Welsh as their first language.[4] Consequently there were linguistic borderlands between Welsh-speaking and English-speaking areas.

Throughout the Victorian era, the Welsh administrative border with England only partially conformed to the linguistic border between England and Wales, and there were also zones of bilingualism where both languages were spoken.[5] In that respect the Anglo–Welsh borderland differed markedly from that between England and Scotland, being more similar to linguistic borderlands found elsewhere in Europe, for example in the contemporary Austro-Hungarian Empire.[6]

The Welsh counties that were contiguous with the English border had very different linguistic loyalties. In North Wales, the eastern part of Flintshire was

279

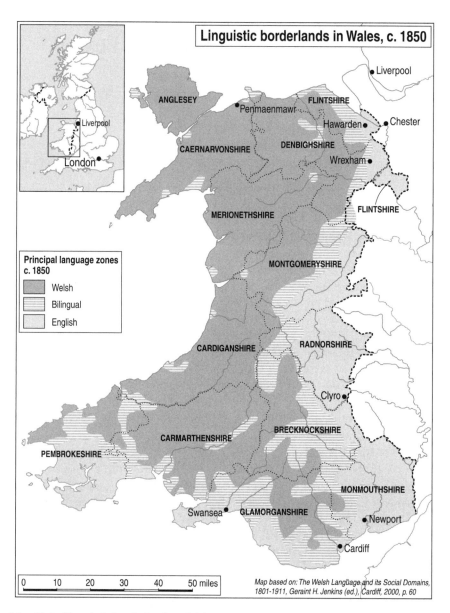

Map 13.1 Linguistic borderlands in Wales, c. 1850

English-speaking and the western part Welsh-speaking. In 1888 the ratepayers of English Maelor, a detached part of Flintshire that was both English-speaking and surrounded by England, voted overwhelmingly in favor of the transfer of their district to the English county of Shropshire.[7] But in Montgomeryshire, to the south of Flintshire, the predominant language was Welsh. Still further south, by contrast, Radnorshire was almost wholly English-speaking and in Breconshire English was dominant in the eastern part of the county and the Wye valley. In Glamorgan, the most populous and industrial county in Wales, Welsh predominated until the last quarter of the nineteenth century. Then Welsh lost ground in the coastal and southeastern areas of the county, partly because of immigration from England. By 1901 just over half the population of Glamorgan spoke only English and nearly all the rest were bilingual. Monmouthshire, to the east of Glamorgan, was technically an English county but the Welsh language was predominant in its western and more industrialized half.[8]

There were also linguistic frontiers far to the west of the land border between England and Wales. On the coast of south Wales, the Gower peninsula was English-speaking as was the south of Pembrokeshire, which was known as 'Little England beyond Wales' (a moniker already current in the sixteenth century—as recorded in George Owen's *Description of Penbrokshire*, published in 1603).[9] English-speaking south Pembrokeshire was separated from Welsh-speaking north Pembrokeshire by a linguistic border known as the Landsker. In 1888 *The Times* commented on that divide: 'in this narrow compass, the two most obstinate languages in the world are placed in juxtaposition to one another and ... neither has in the course of several centuries gained any appreciable advantage.' The article noted that marriage across the Landsker had been extremely rare until recently and was still far from common.[10] The linguistic divide in Pembrokeshire remained remarkably stable until the later twentieth century—in marked contrast to the situation in the Welsh Marches, where Welsh rapidly retreated westward in many areas.[11]

Historians of the United Kingdom in the Victorian period have paid little attention to the Anglo–Welsh border—either the fixed administrative frontier or the shifting linguistic one. That neglect is unsurprising because the border received little specific attention in contemporary English language sources. Nevertheless the border impinged, often in a subtle and highly localized way, on the lives of many people, English as well as Welsh. That was evident in the experiences of three English newcomers to Wales in the Victorian period. They were: George Borrow, a writer and traveler; William Gladstone, the foremost statesman of the age; and Francis Kilvert, a young clergyman. Their varied experiences of the Anglo–Welsh borderlands are revealed in their diaries and also, in Gladstone's case, in his reading and speeches. These sources provide neglected evidence about the Anglo–Welsh border and its cultural significance, which is examined in the rest of this chapter.

The Experience of George Borrow

George Borrow (1803–81) was an idiosyncratic writer who published several books about his travels in Europe. They included *The Bible in Spain* (1843), which proved popular, *Lavengro* (1851), which was part autobiography and part fantasy, and *The Romany Rye* (1857), which recounted his time with gypsies. After his marriage, Borrow confined his wanderings to Britain. He had a natural aptitude for languages and he had some proficiency in Welsh. As a young man in the city of Norwich—far from Wales—he had translated the Welsh poetry of Ab Gwilym and learnt how to pronounce the language from a Welsh groom.[12]

In the summer and autumn of 1854, Borrow made a tour of Wales, partly on foot, during the course of which he conversed with both English and Welsh-speakers. He later remarked that 'it was my knowledge of Welsh, such as it was, that made me desirous that we should go to Wales.'[13] He recorded his experiences in penciled notebooks, from which he compiled a book about his journey, eventually published in 1862 as *Wild Wales, Its People, Language and Scenery*.[14] The book attracted little interest on its publication but it gradually became popular as an informal travelogue and was reprinted, in various library editions, in the twentieth century.

Previous English travelers in Wales, such as Daniel Defoe, who made an extensive tour in the early eighteenth century, had made virtually no reference to the Welsh language in their accounts.[15] Borrow, by contrast, provided considerable information about the extent of Welsh-speaking in the mid-nineteenth century and the outlook of its practitioners. Sir Lewis Morris, whose grandfather and namesake was an eighteenth-century Anglesey bard much admired by Borrow, wrote to him in 1862:

> I have just finished reading your work on *Wild Wales* and cannot refrain from writing to thank you for the very lifelike picture of the Welsh people, North and South, which, unlike other Englishmen, you have managed to give us. To ordinary Englishmen the language is of course an insurmountable bar to any real knowledge of the people, and the result is that within six hours of Paddington or Euston Square is a country nibbled at superficially by droves of holiday-makers, but not really better known than Asia Minor.[16]

Borrow's practice of recording accounts of his conversations with the people that he met on his tour—both English- and Welsh-speakers—was criticized as commonplace by *The Cornhill* but commended by *The Spectator*.[17] It enabled him to map the everyday borderlines between Welsh and English culture in a way that was not attempted by other contemporaries.[18] His interest in Welsh literature was subsequently praised by John Southall, a late-Victorian expert on the distribution of the Welsh language:

> His work is marred by the introduction of a good deal of public-house chat, but it betrays an acquaintance with Welsh literature far more extensive than is to be found in the works of half-informed English tourists of an earlier date, whose works are looked up to as standards, and in vain we search Pennant and Nicholson, or such county histories as Fenton's and Coxe's for the kind of information we get here.[19]

Nevertheless the evidence that Borrow collected and published in *Wild Wales* has been largely ignored or dismissed by modern scholars of Welsh culture. Kenneth Morgan, for example, claimed that Borrow depicted Wales as a semi-civilized, picturesque survival from a druidic past.[20] Hywel Morgan has recently referred to Borrow's 'clichés of the picturesque,' which were allegedly 'crafted to suit a high Victorian literary taste.'[21] Borrow, however, was a highly idiosyncratic writer, who was by no means an exemplar of high Victorian literary taste. He did not patronize and rarely romanticized the Welsh but he was impressed by their love of their own classic literature, which, he felt, was not matched by working-class Englishmen.[22] Borrow has also been criticized for failing to comment on the decline of the Welsh language, but his approach was to provide a quick snapshot of life in Wales, not a long-term perspective.[23] Although he embellished some of his diary entries relating to the itinerant Irish that he met, he did not fabricate his encounters with Welsh-speakers.[24]

In 1854, Borrow encountered the Welsh before he reached Wales—at Chester, which had been the capital of the English marcher lordship along the Welsh border. He was informed that there were many Welsh-speaking people in Chester—both adults and children—and many chapels with services in Welsh.[25] Yet when Borrow crossed the border and visited Wrexham, the largest town in northeast Wales, he found few Welsh-speakers or chapels. The only Welsh that he heard was 'cwrw da'—good ale. But when he walked through the industrial district beyond Rhiwabon he spoke in English to a woman who replied: 'dim saesneg'—no English. 'This is how it should be,' Borrow said to himself. 'I now feel I am in Wales.'[26] He soon discovered, however, that linguistic diversity characterized the Welsh borderland. Walking on the hills above Wrexham, he met two women who spoke no Welsh but also a family who spoke only Welsh. 'Truly' he then said to himself, 'I am on the borders. What a mixture of races and languages!'[27] He noted that the Welsh-speaking carters in the borderland swore at their horses in English.[28] The linguistic diversity was particularly evident in tourist centers such as Llangollen, where the visitors attended English church services, while there was an evening service in Welsh for the poor. But the poorest people Borrow met on the road were Irish tinkers, who spoke English but no Welsh.

One avenue of Anglicization was intermarriage between Welsh and English. Borrow met a number of Anglo–Welsh couples, even in remote rural districts. One Welsh girl was married to a clog-maker from Bolton who made a living by sending wood squares for clogs to his home town. They spoke only English when together.[29] The tentacles of English business reached far into the interior of North Wales. Borrow met many people who had lived in Liverpool or represented commercial firms in that city.

The attempts by Borrow to converse in Welsh or listen to the locals speaking it were not always well received by the locals. He concluded,

> The Welsh are afraid lest an Englishman should understand their language and, by hearing their conversation, become acquainted with their private affairs ... All conquered peoples are suspicious of their conquerors. The English have forgot that they ever conquered the Welsh, but some ages will elapse before the Welsh forget that the English have conquered them.[30]

That was a striking conclusion given that the Edwardian conquest of North Wales had taken place over five hundred years previously.

On the west coast of Wales, Borrow met an English couple that farmed near Machynlleth but had virtually no contact with their Welsh neighbors. The farmer's wife explained, 'Oh, sir, the English cannot make friends amongst the Welsh. The Welsh won't neighbour with them, or have anything to do with them, except now and then in the way of business.' She informed him that the Welsh particularly disliked strangers who spoke or understood their language. When her husband went to the local inn, all the Welsh customers left and the landlord complained that the Englishman was ruining his trade.[31] The same attitude was evident in the Welsh-speaking part of South Wales. In a pub in rural Glamorgan, a customer said to Borrow, 'I tell you plainly that we don't like to have strangers among us who understand our discourse, more especially if they be gentlefolks.' Borrow replied, 'That's strange; a Welshman or foreigner, gentle or simple, may go into a public house in England and nobody cares a straw whether he understands the discourse of the company or not.'[32] The clannishness of the Welsh was regional as well as linguistic. In North Wales Borrow encountered a strong prejudice against people from South Wales. One person complained that 'people from Wrexham speak neither English nor Welsh, nor even South Welsh.'[33]

When Borrow reached the suburbs of Swansea, which was the world capital of copper smelting and tinplate making in the Victorian era, he crossed another linguistic divide:

> As I passed under what appeared to be a railroad bridge, I inquired in Welsh of an ancient-looking man in coaly habiliments, if it was one. He answered in the same language that it was, then instantly added in

English:—'You have taken your last farewell of Wales, sir; it's no use speaking Welsh farther on.' I passed some immense edifices, probably manufactories, and was soon convinced that, whether I was in Wales, or not, I was no longer among the Welsh. The people whom I met did not look like Welsh. They were taller and bulkier than the Cambrians and were speaking a dissonant English jargon. The women had much the appearance of Dutch fisherwomen; some of them were carrying huge loads on their heads. I spoke in Welsh to two or three whom I overtook.

'No Welsh, sir!'

'Why don't you speak Welsh?' said I.

'Because we never learnt it. We are not Welsh.'

'Who are you then?'

'English, some call us Flamings.'[34]

The description of the English-speaking folk of the Swansea district as 'Flamings' had some historical pedigree. In the medieval period, people from Flanders, encouraged by English monarchs, had settled along the South Wales coast. They had then been joined by immigrants from southwest England, particularly north Devon. The emergence of Swansea as an industrial metropolis in the late eighteenth and early nineteenth centuries occasioned further immigration from England to the district.

Borrow completed his tour of Wales by traveling through Monmouthshire, which was nominally an English county but whose people were mostly Welsh-speakers. There, as in northeast Wales, he found that linguistic diversity characterized the borderland. One woman informed him that all the people for at least eight miles to the east of Newport could speak Welsh. Another, however, claimed that the English-speakers in the locality outnumbered the Welsh by ten to one. Borrow concluded that there was a rough parity between the two local communities because about half of those he had met and addressed in Welsh had answered him in that tongue.[35]

In *Wild Wales* Borrow provided a picture of the principality in which language not only divided the Welsh from the English but also, to some extent, from themselves. He encountered linguistic borderlands not only close to the frontier between England and Wales but also in the western parts of the principality far from England. Although Borrow made no attempt to generalize or theorize from his own personal experiences, he clearly demonstrated that the Anglo–Welsh borderlands were socially and linguistically diverse. It is noteworthy that he did not link the linguistic divide between Welsh- and English-speakers to the religious divide between Anglicanism and Nonconformity, which was such a pronounced feature of Welsh life in the Victorian era. That may have been because Borrow, as an Anglican (albeit one of decidedly Protestant leanings), had no direct experience of Welsh Nonconformist chapels. Also his tour of Wales took place before the emergence of Welsh Nonconformity as a major political and cultural force in the 1860s and subsequent decades. Even so, the absence of religious references in Borrow's account is a reminder that the Welsh language had both a heritage and a contemporary usage that was not exclusively linked to Nonconformity. In that respect he provided a portrait of mid-Victorian Wales significantly different from the usual narrative of the period.

The Experience of William Gladstone

George Borrow's direct experience of the Welsh–English borderlands was short-lived and he had no obvious impact on how the Welsh and English regarded each other. By contrast the statesman William Ewart Gladstone (1809–98) lived on those borderlands for over half a century and his changing perception of them helped to refashion national attitudes. Yet Gladstone was Scots, not Welsh, by ancestry and English by upbringing. He was born and partly brought up in Liverpool, which at that time was only a short journey away from North Wales by road and coastal ferry. There were, moreover, close commercial connections between Liverpool and North Wales, which Borrow noted in *Wild Wales*. As a teenager, in 1826, Gladstone had taken a holiday along the north Welsh coast, which he described in a travel journal. He admired the scenery and the castles but made no specific reference to the Welsh people or language.[36] First at Eton and then at Oxford, Gladstone was a friend of Stephen Glynne, whose family owned a large ancestral estate in Flintshire, in the northeast corner of Wales. In 1839 Gladstone married Stephen's sister, Catherine, and from the 1850s on the couple spent up to half the year at Stephen's home, Hawarden Castle. Hawarden is in Wales but it is only a few miles from the border with the English county of Cheshire. The district around Hawarden was English-speaking both before and during Gladstone's residence there.[37] Indeed Hawarden Castle had been a center of English authority in North Wales in the early medieval period. In 1282 the castle had been attacked by Dafydd ap Gruffudd, the brother of the last independent Welsh prince. That attack led King Edward I to launch the military campaign that finally suppressed Welsh independence.

Gladstone's connection with Wales was initially confined to the eastern, largely English-speaking, part of Flintshire. But in the later 1850s and 1860s he and his family regularly spent a month, in the late summer, on the North Wales coast at Penmaenmawr, to the west of Conwy, where the locals were mostly Welsh-speaking. One Sunday, in 1855, Gladstone wrote in his diary, 'Parish church mg & evg. The evening service was in Welsh ... I saw some irreverence of kinds wh wd not occur in England. The language does not make a favourable impression.'[38] It is clear from this entry that Gladstone had no familiarity with spoken Welsh or any interest in it. Nor, a few years later, did a brief perusal of Charlotte Guest's translation of the ancient Welsh tales known as *The Mabinogion* arouse in him an interest in ancient Welsh literature.[39]

In the 1860s, however, Gladstone's attitude to Welsh culture began to change—mainly for political reasons. In 1866 he became the leader of the Liberal Party and his commitment to the disestablishment of the Irish Church gained him the support of Welsh Nonconformists, who constituted a majority of the population. One of their leaders, Henry Richard, sent his *Letters on the Social and Political Condition of the Principality of Wales* to Gladstone, who read them in 1867.[40] The letters alerted him not just to the grievances of Welsh Nonconformists but also to the strength of Welsh-speaking culture as displayed in literary and musical competitions, known as eisteddfods. These events had been held in Wales since the early medieval period but had later fallen into abeyance until their revival in the late eighteenth century. In 1858 a National Eisteddfod was held at Llangollen, which was regularly followed by others, at different locations, in subsequent years. The role of the National Eisteddfod movement in promoting Welsh language and culture has recently been questioned.[41] Nevertheless the institution provided a forum that enabled Gladstone to directly address the issue of Welsh culture. In 1873, when he was Prime Minister, Gladstone spoke to the Welsh National Eisteddfod at Mold, the county town of Flintshire, which was close to Hawarden. In his speech, which attracted national attention, he said that his 'connexion with Wales' was 'very dear' to him and that, as Prime Minister, he had found the relations between England and Wales 'full of interest.' He admitted that he had previously shared the prejudices of some Englishmen about the Welsh language but then declared, 'I have come here to tell you how and why I have changed my opinion.' He pointed out that the Welsh language was both 'a venerable relic of the past' and still spoken by 800,000 people. Yet he also advised the Welsh that it was in their economic interest to study English in school and not to discourage its use at home.[42] He favored free trade in languages as in commodities.

Gladstone's 1873 speech to the eisteddfod at Mold had a stimulating effect on Welsh national feeling.[43] Later that year, the Cymmrodorian Society of London, which promoted Welsh arts and sciences, was revived after a lapse of 30 years.[44] Gladstone chaired a meeting of the society, in 1877, when he declared that 'the nationality of Wales' had not yet 'obtained a perfect, just, and due recognition.'[45] His support for Welsh nationality was concurrent with his championship of the Balkan peoples, such as the Bulgarians, in their struggle for independence from the Ottoman Empire. Gladstone's support for the principle of national self-determination climaxed with his support for Irish home rule in 1886. Thereafter Ireland commanded most of his attention but he also supported the aspirations of the Welsh for greater autonomy.

In 1888 Gladstone made another speech to an eisteddfod—at Wrexham, the largest town in northeast Wales. Afterwards, he wrote in his diary, 'Taken all together a day of much interest.'[46] In his speech, he professed a keen interest in Welsh history and admiration for 'the musical talent and feeling in Wales.' He hailed the eisteddfod movement as 'a Welsh national institution,' which not only commemorated the past but also had economic value in the present.[47] Gladstone also referred to his Welsh wife and to his hope that their children 'will not forget that they are in part Welsh people.' None of the family spoke Welsh, however, and his wife, Catherine, though born in Wales, had an English mother. In his speech, moreover, Gladstone admitted that because Hawarden was close to the border with England it had not 'come so absolutely under Welsh influence.' Indeed he had previously denied that Hawarden was located 'upon a border really Welsh.'[48] Even so, he noted in his Wrexham speech, Hawarden still retained much evidence of its original Welshness.[49] The local church was dedicated to St Deiniol, a saint unknown in England, and many local names were of Welsh origin.[50]

After speaking at Wrexham, Gladstone read a number of works relating to Wales, including an English translation of Daniel Owen's 1885 novel, *Rhys Lewis*.[51] Owen was the first major novelist to write in Welsh and is still regarded as one of the greatest. He was born and lived, for most of his life, at Mold, which was on the English–Welsh linguistic border. In his copy of *Rhys Lewis*, Gladstone marked up passages that referred to the sectarian and linguistic divide in Wales.[52] The novel includes a brief account of events at Mold, in 1869, when a dispute between an English colliery manager and his Welsh miners led to arrests, convictions, and a riot in which four people were killed by the military.[53] Gladstone noted Owen's observation, in *Rhys Lewis*, that none of those in authority, who dealt with the arrested miners, understood Welsh.[54]

In 1896 the octogenarian Gladstone received an honorary doctorate from the newly instituted University of Wales, at Aberystwyth. By then he was a hero to most of the people of Wales—partly because he recognized their national identity.[55] He observed that the Welsh had awakened to self-consciousness and made successful efforts to create institutions in harmony with their national history, traditions, and feeling.[56]

Gladstone's gradual emergence as a champion of distinctive Welsh culture owed much to wider political developments but it also reflected his personal experiences from living in Wales. Initially, however, as a resident of an English-speaking village close to the border with England, his contact with Welsh-speaking Wales was minimal at best. Even his later engagement with distinctively Welsh culture was brief and relatively uninformed. Nevertheless, Gladstone's residence on the Welsh side of the border did affect both his attitude to Wales and his sense of family identity. If Gladstone had lived on the English side of the border—where Conservatism and anti-Welsh sentiment were strong—it is less likely that he would have proved so sympathetic to Welsh and more generally Celtic nationalism. In that context, his residence at Hawarden had a political importance that should not be underestimated.

The Experience of Francis Kilvert

Another English resident on the Welsh–English border in the later Victorian period was the Rev. Francis Kilvert (1840–79). He was an Anglican clergyman who recorded his life as a country curate in his diary, which was posthumously published in the 1930s. It was hailed as a minor classic and it still attracts a wide readership.[57] The diary provides an intimate picture of a young clergyman and country life in the mid-Victorian period. From 1865 to 1872 Kilvert was curate of Clyro, a village in Radnorshire, just across the Welsh border from the English county of Herefordshire. He wrote of his parishioners,

> The people of Clyro are still sufficiently Welsh to be suspicious of strangers, and an Englishman would probably not be thoroughly liked and trusted till he had lived for some years in the country. But there is not in Radnorshire the same hostility and bitterness of feeling that is still shown towards the Saxon in many parts of Wales. In fact the people, as a whole, are singularly civil, courteous and obliging ...[58]

The locals' lack of hostility towards the English was partly linguistic in nature because although they were predominantly Welsh in origin, nearly all of them were English-speakers. When and why Welsh largely disappeared from Radnorshire is not entirely clear but it was already apparent before the Victorian period.[59]

One day in 1870, Kilvert wrote in his diary, 'Drove past St Harmon's the only Welsh-speaking parish in Radnorshire.'[60] Six years later, the living of St Harmon's was presented to Kilvert, although he was not a Welsh-speaker. The local Archdeacon visited the parish and concluded that a Welsh service was not needed in a local church.[61] Kilvert's incumbency of St Harmon's was short-lived and his diary for that period is largely missing so it is not possible to ascertain how he dealt with his Welsh-speaking congregation. His diary does, however, record that a clergyman's inability to speak Welsh sometimes led to misunderstandings. Kilvert was told about an assistant curate who had been sent to a Welsh-speaking parish, in Breconshire, although he had very little command of the language. On one occasion, when he published wedding banns, he meant to say 'why these two persons may not lawfully be joined together in holy matrimony.' But what he said in Welsh was, 'why these two backsides may not be lawfully ...'[62]

Kilvert was also informed that the last woman who had spoken the Radnorshire dialect of Welsh had died only four years earlier. In her old age, no one else in the parish could speak Welsh to her except the vicar. Kilvert himself rarely heard the Welsh language and, while on a visit to North Wales, he noted that 'it was strange to hear the little children chattering

Welsh.'[63] Even the Breconshire harriers became Anglophone. They were used to commands in Welsh but when Sir Francis Ford took charge of them he had no Welsh and only one or two of the hounds responded to English. But Sir Francis made those dogs understand him and the rest of the hounds followed their lead.[64] Nevertheless, the Welsh heritage of the Radnorshire borderlands remained evident in the local place names and customs. Kilvert referred to innumerable Welsh place names and gave the Welsh name for Halloween: 'nosCalan Guaf.'[65]

There are a number of references to the English–Welsh border in Kilvert's diaries. On one occasion he visited the Pant—a house with a Welsh name—where the borderline went along a notch in the chimney. He recalled that an old woman had told him about the birth of a child in the cottage and the care that had been taken that the child should be born in England. 'Stand here Betsey, in this corner,' said the midwife. And the girl was delivered of the child *standing*.'[66] Such concern to be born in England was unnecessary from an official point of view because there was no distinction between Welsh and English citizenship. There was, however, a cultural difference that was evident on both sides of the border. Kilvert wrote in 1872: 'After midnight I ... crossed the border from England into Wales. The English inn was still ablaze with light and noisy with the song of revellers, but the Welsh inn was dark and still.'[67] In Wales there was a strong temperance movement particularly among the Nonconformist majority. That pressure led Gladstone's government to pass, in 1881, an Act closing Welsh public houses on Sundays. That was the first time, in the modern era, that Parliament had passed legislation specifically relating to Wales. But Kilvert's diary of his time in Clyro presents no evidence of serious tensions on either side of the Welsh–English border. That probably reflected the absence of a linguistic divide.

Conclusion

The diverse experiences of Borrow, Gladstone, and Kilvert illustrated the varied character of the English–Welsh borderlands in the Victorian era. In some areas the border was both an administrative and linguistic frontier—as was largely the case between Welsh Montgomeryshire and English Shropshire. In some other areas it was purely linguistic—as between Welsh-speaking north Pembrokeshire and the English-speaking south. In Radnorshire, by contrast, English was spoken on both sides of the border. There were also growing areas both of bilingualism and multilingualism to the west of the border with England. Yet in all those varied borderlands there was a distinct sense of separate identity on either side of the divide. The borders between Welsh and English areas remained significant frontiers, which illustrated the disunited nature of the United Kingdom at the height of its world power.

Notes

1. Frederick Jackson Turner, *The Frontier in American History* (Huntington, NY, 1976 [1920]).
2. For the earlier history of Wales see J. Davies, *A History of Wales* (London, 1993).
3. M. Cragoe, *Culture, Politics and National Identity in Wales, 1832–1886* (Oxford, 2004), p. 23.
4. For the Welsh language in the Victorian period see G.H. Jenkins (ed.), *The Welsh Language and Its Social Domains 1801–1911* (Cardiff, 2000).
5. I.G. Jones, *Mid-Victorian Wales: The Observers and the Observed* (Cardiff, 1992), p. 57.
6. For the linguistic borderlands in the Austrian Empire see P. Judson, *Guardians of the Nation: Activists on the Language Frontiers of Imperial Austria* (London, 2007).
7. *Times*, Feb. 3, 1888, p. 6.
8. *Encyclopaedia Britannica*, vol. 28 (11th ed., New York, 1911), p. 268: entry on Wales. See also Jenkins, *Welsh Language*.
9. B. John, *Pembrokeshire* (Newton Abbot, 1976), p. 68.
10. *Times*, Aug. 20, 1888, p. 13: 'Letter from Wales, xvi, Pembrokeshire.'
11. B.S. John, 'The Linguistic Significance of the Pembrokeshire Landsker,' *Pembrokeshire Historian*, 4 (Haverfordwest, 1972), 27.
12. M. Collie, *George Borrow, Eccentric* (Cambridge, 1982), p. 23.
13. G. Borrow, *Wild Wales*, with an introduction by Brian Rhys (London, n.d. [c.1935]), p. 21.
14. C.K. Shorter, *George Borrow and His Circle* (London, 1913), pp. 364–6.
15. D. Defoe, *A Tour Thro' the Whole Island of Great Britain*, vol. II, letter III, ed. J. McVeagh (London, 2001), pp. 172–85.
16. Sir Lewis Morris to Borrow, 29 Dec. 1862: Shorter, *Borrow and His Circle*, pp. 371–2.
17. W.I. Knapp, *Life, Writings and Correspondence of George Borrow* (2 vols, London 1899), II, pp. 214–15.
18. R.A.J. Walling, *George Borrow, the Man and His Work* (London, 1908), p. 321.
19. J.E. Southall, *Wales and Her Language* (London, 1893), p. 163.
20. K.O. Morgan, *Wales 1880–1980* (Oxford, 1982), p. 3.
21. *Times Literary Supplement*, Oct. 14, 2011, p. 30.
22. Borrow, *Wild Wales*, p. 116.
23. D.A. Halsell and F.A. Montgomery, 'Refocusing on George Borrow (1803–81): An Eccentric View of Wales?,' *Geography*, 88 (2003), 118.
24. A.M. Fraser, 'George Borrow's Wild Wales: Fact and Fabrication,' *Transactions of the Honourable Society of Cymmrodorian* (1980), 163–74.
25. Borrow, *Wild Wales*, p. 39.
26. Borrow, *Wild Wales*, pp. 42–3.
27. Borrow, *Wild Wales*, p. 350.
28. Borrow, *Wild Wales*, p. 382.
29. Borrow, *Wild Wales*, p. 126.
30. Borrow, *Wild Wales*, p. 294.
31. Borrow, *Wild Wales*, pp. 417–18.
32. Borrow, *Wild Wales*, p. 545.
33. Borrow, *Wild Wales*, p. 377–9.
34. Borrow, *Wild Wales*, p. 558.
35. Borrow, *Wild Wales*, p. 585.
36. Gladstone's journal of his tour in North Wales, Aug. 31 to Sept. 2, 1826 (British Library, London: Gladstone Papers, Add MS 44718, ff. 68–71).

37. Southall, *Wales and Her Language*, p. 353.
38. H.C.G. Matthew (ed.), *The Gladstone Diaries, vol. 5: 1855–1860* (Oxford 1978), p. 76 (Sept. 23, 1855).
39. Matthew, *Gladstone Diaries, vol. 5*, p. 435 (1 Nov. 1859).
40. H.C.G. Matthew (ed.), *Gladstone Diaries, vol. 6: 1861–1868* (Oxford, 1978), p. 553 (Oct. 21–2, 1867); Cragoe, *Culture, Politics and National Identity*, pp. 42–3.
41. H.T. Edwards, 'The Welsh Language in the Eisteddfod,' in Jenkins, *Welsh Language*, pp. 293–316.
42. *Times*, Aug. 20, 1873, p. 5.
43. K.O. Morgan, *Wales in British Politics, 1889–1922* (Cardiff, 1970), p. 41.
44. *A Sketch of the History of the Cymmrodorian* (London, 1877), p. xix.
45. *Times*, May 14, 1877, p. 10.
46. H.C.G. Matthew *Gladstone Diaries, vol. 12: 1887–1891* (Oxford, 1994), pp. 144–5 (Sept. 1–5, 1888).
47. *Times*, Sept. 5, 1888, p. 6.
48. Gladstone to Sir T.D. Acland, Jan. 13, 1882 (Gladstone Papers, Add MS 44545, f. 88).
49. *Times*, Sept. 5, 1888, p. 6.
50. Late in his life, Gladstone established a library at Hawarden for the education of the clergy and people of Wales, which he named Saint Deiniol's after the local saint. The library, recently renamed Gladstone's Library, remains today a monument to Gladstone's interest in the Welsh heritage of his adopted homeland.
51. D.O. Lewis, *Rhys Lewis, Minister of Bethel, an Autobiography*, trans. from Welsh by James Harris (London, 1888); Matthew, *Gladstone Diaries, vol. 12*, p. 145 (Sept. 5–8, 1888).
52. Gladstone's annotated copy of *Rhys Lewis* is in Gladstone's Library at Hawarden.
53. For the Mold riot see *Times*, June 4 and 5, 1869.
54. Owen, *Rhys Lewis*, pp. 122–6.
55. Cragoe, *Culture, Politics and National Identity in Wales*, p. 36.
56. *Times*, June 27, 1896, p. 14.
57. W. Plomer (ed.), *Kilvert's Diary 1870–1879* (3 vols, London, 1977).
58. Quoted in F. Grice, *Francis Kilvert and His World* (London, n.d.), p. 41.
59. Southall, *Wales and Her Language*, pp. 338–40.
60. *Kilvert's Diary*, I, p. 104 (Apr. 20, 1870).
61. *Kilvert's Diary*, III, p. 288 (May 4, 1876).
62. *Kilvert's Diary*, I, pp. 346–7 (May 22, 1871).
63. *Kilvert's Diary*, I, p. 353 (June 12, 1871).
64. *Kilvert's Diary*, II, p. 335 (March 11, 1873).
65. *Kilvert's Diary*, II, p. 77 (Oct. 30, 1871).
66. *Kilvert's Diary*, I, p. 262 (Nov. 18, 1870).
67. *Kilvert's Diary*, II, p. 191 (May 1, 1872).

14

'Home on the Range': Rootedness and Identity in the Borderlands of the Nineteenth-Century American West

Nina Vollenbröker

Borderlands are often dynamic and heterogeneous. They are areas in-between—contact zones—and as such speak to issues of cross-cultural interaction, transitions, and change. Home, on the other hand, is a concept and a space usually associated with stability, alignment, and continuity. As a central anchor of human life, home is commonly understood to speak about personal rootedness and an ongoing sense of self. And yet thousands of individuals make—and have historically made—their home amidst the fluidity and hybridity of borderlands. This chapter looks in detail at the daily realities of the men and women who inhabited a particular borderland: the trans-Mississippi West of the mid- to late nineteenth century. Unlike other studies, it does not concentrate on the region's borders with Canada or Mexico—distinct geographical and political boundaries which created obvious borderlands. Rather, the essay approaches the West itself as a space of ongoing contestation, interaction, and exchange: an area which remained an experienced borderland for pioneer settlers interacting with its vast terrain, shifting meanings, and indigenous population while the area's official outlines continued to be redrawn.[1] In four sections the essay analyzes the original, unedited diaries kept, the houses constructed, and the artifacts used by homesteaders in present-day Nebraska and by travelers on the Overland Trails to California, Oregon, and Utah to reveal how nineteenth-century men and women found their place in the West. In its conclusion, the chapter uses the pioneers' experience and practice to question the supposedly unsettling nature of borderlands and the assumed stability of home and self.

The Departed Home

In the trans-Mississippi West of the nineteenth century, *home* was an important concept. Virtually all adult Euro-Americans who inhabited the region's vast prairies and plains, its deserts, mountains, and coastal areas, had recently migrated there. Settled homesteaders were staking their claim while other emigrants traveling on the complex system of Overland Trails to the Pacific or Utah were still in the process of migrating. All had covered considerable distances; they had come from New England, the southern United States or Mexico, from Germany, Britain, Scandinavia, and other parts of Europe. Many of these newly arrived Westerners' personal diaries survive as manuscripts in archives across the United States today, and unsurprisingly most contain manifold references to a previous place of residence recently departed.[2] Among their new, unfamiliar surroundings, the men's and women's thoughts regularly returned to the familial house and to the loved ones they had left behind—to the 'childhood home where first I learned to romp and play'[3] as Pennsylvania-to-Oregon emigrant Agnes Stewart Warner noted in 1853. These memories typically brought with them painful feelings of separation and loss. While loading possessions ('eight trunks, one valise, three carpet bags … one washtub of little trees, utensils for cooking and two provision boxes') into her covered wagon bound for the Kansas frontier, Miriam Davis Colt wrote in her diary, 'I find that my sympathies and the warm feelings that live in my heart for my friends and neighbors are not as easily gathered up and boxed as are our goods.'[4] Two months into her journey Overland Trail traveler Esther Belle Hanna recorded,

> Sabbath 9 [May], This is a beautiful morn, I think of home and the dear ones there; each day I am getting further from them. I feel a sadness steal over me at times when I think I shall see them no more on earth, but it is all for the best …[5]

Passages like Agnes's, Miriam's, and Esther's articulate nostalgia, remembrance, and sorrow. They express a longing for a now temporally and geographically distant homeplace. Most Westerners had left their past place of residence for good, yet a complete severance from the old familiar home rarely occurred. The majority of migrants retained delicate yet powerful bonds to these sites of former rootedness and their inhabitants, connections that are particularly apparent in the artifacts they chose to take West. Today, in an Oregon archive, we can study a bundle of mismatched fabric scraps wrapped in blue cloth and brought across the continent by Overlander Rachel Bond. Her biography recounts the collection of swatches being reclaimed from the waters of the Platte River after the wagon in which Rachel was traveling capsized. Negotiating rivers was feared among emigrants (virtually every diary notes a drowned traveler, sometimes almost casually: 'Thursday, June 14th … Nothing of importance happened except that there were four men drowned while attempting to swim their stocks [across the Platte River]').[6] The rescue mission of Rachel's fabrics underlines the importance of what, to the untrained eye, looks like an assortment of frayed rags.

Rachel's collection of fabrics is, in fact, a 'piece bag.' In the nineteenth century these items were not uncommon and of some material value, containing
off-cuts for use in future patchwork. But they were of even greater sentimental
value: in most families, piece bags were handed down through generations of
women and held cherished fabric scraps saved decades previously. It is likely that
Rachel's bundle contained pieces from her mother's and grandmother's sewing
projects. In addition to these gradually accumulated swatches, the bag probably
also included scraps given to Rachel by friends—pieces of cloth that had been
purposefully cut from a dress or blouse to express bonds of friendship. This kind
of gifting practice was common in the nineteenth-century United States and
became particularly popular during the years of the Westward migration. As torn
or disjointed fragments severed from their context—from a mother's bonnet,
from a friend's skirt, from the larger home where friendship and family were
once enacted—these pieces of fabric taken on the Western Trails inevitably bear
the memory of rupture, irreconcilability and loss associated with leave-taking.
However, Rachel Bond's colorful, multipatterned fabric collection can also be
seen as *healing* broken connections. Upon her arrival in Oregon, Rachel made
a quilt using some of the very patches originally placed into her piece bag on
the other side of the continent.[7] Her sewing practice thus transformed jagged
fabric edges into tight seams, traces of hasty severance into symbols of together-
ness, solitary fragments into a cohesive text speaking about who she is, where
she lives, and what matters to her. The bonds she created between the quilt's
many pieces of fabric mirror the bonds she preserved with those who gifted the
swatches to her and who were then two thousand miles away.

Like Rachel Bond, numerous Euro-American women inhabiting the Western
borderlands chose needlework to fashion meaningful connections between
an old and a new home. Emigrants exchanged letters containing fabric scraps
with loved ones who had stayed behind. One frontierswoman wrote to her
family: 'I have been looking for something to send you, but I could not find
anything that I could send in a letter [but] a piece of my new dress.' Over the
next ten years, she dispatched a number of cloth mementoes from Nebraska
to New York—'some pieces of my new dresses for patch work,' 'a piece of my
gingham,' 'a piece of my bonnet, trimmed with green plaid ribbon.'[8] Another
nineteenth-century needleworker, Lucy Webster, made a panel for a friendship
quilt (a quilt whose individual components are designed, made, and inscribed
by a group of women as a physical expression of their comradeship) intended
as a gift to a friend about to migrate West. The inscription she placed on her
piece asks the prospective traveler to 'remember me' when she is 'far away
from this dear spot, In a distant Land.'[9] In a similar vein the manuscript diary
of Overlander Barsina Rogers French, now at the Henry E. Huntington Library,
contains an unfinished embroidery sampler spelling out the sentence 'Think of
Me'[10]—the stitched, tactile words bridging the distance between an unfamiliar,

transitory landscape and a previous, stable home, between a *mobile* and a *static* way of being in place.

The New Home

In addition to these needlework projects, the use of other artifacts and of particular spatial practices also underlines that the recently departed stable sites of family, kinship, traditions, and childhood remained important to the men and women in the borderlands of the American West.[11] In fact, most emigrants worked hard to (re-)create *home* amidst the unfamiliar expanse of the Western landscape on a number of levels. They were, for example, quick to map the trusted vocabulary of domesticity onto the less conventional spaces that contained their lives. Overlanders commonly voiced feelings of being 'quite at home'[12] in their wagons or 'comfortable as in a house'[13] in their tents. Esther Belle Hanna's diary shows that she regarded her covered wagon's internal space as the private 'bedroom' she would have had in a permanent East-coast home:

> 11.May … our carriage is very comfortable and we have a real nice little bedroom of it at night, shut it all up close let down the backs of the seats, spread our mattress, hang up our clothes on the hooks which are out in all round, I have my looking-glass, towel etc. hung up & everything is in order. Got some beautiful wild flowers today, they grow in profusion on the prairie.[14]

Esther's description lists a number of objects which are potent carriers of homeliness: a bed, a closet for personal belongings, a decorative bouquet of cut flowers. Other writers similarly emphasize the presence of 'beds, as white as snow' and of 'Books, glass, china, and other furniture in polite usage' in settlers' cabins.[15] The importance of such artifacts amidst the Western wilderness is confirmed by the magnificent visual archive of Solomon D. Butcher. Butcher was a Virginia emigrant and a self-taught yet technically highly skilled photographer who recorded everyday life and living quarters around the late nineteenth-century Nebraska borderland.[16] His more than three thousand glass plate negatives show the facades of countless 'sod houses'; utilitarian structures made from chunks of soil laid in horizontal strips like courses of bricks. The majority of Nebraska pioneers lived in this type of dwelling while 'proving up' on their claim, and moved into timber-frame structures once they had obtained the papers to the land.[17]

Most homesteads depicted by Butcher bear witness to their owners' daily hard work: the compact houses are commonly surrounded by uneven ground and recently dug wells, by livestock, carts, and wagons, windmills and heavy farming equipment. But they also reveal the settlers' touching attempts to furnish an otherwise unrefined and functional environment with the markers of a 'happy home.' A considerable number of houses, for example, displayed a rather unexpected adornment: birdcages.[18] As acoustic ornaments these brought the song of pet canaries to the often overbearingly silent landscape; as visual ornaments birdcages heeded the housekeeping advice offered to Western homemakers by local magazines including the *Nebraska Farmer*, which, in an 1880 article entitled 'Beautify Your Homes,' argued for the importance of domestic decoration in and around pioneer cabins:

> How little mankind enjoys this life, compared with what they might enjoy, by simply expending a little labor and care, around and within their homes; they might become paradises as far as it is possible for anything on earth to become a paradise. ... [P]lant evergreens, shrubs and flowers profusely by every walk, at every door ... have a cage of sweet little canaries to claim a place on the veranda and a share of your love; decorate the walls of every room with portraits and landscapes ...[19]

Butcher's archive demonstrates that homesteaders did generally pay attention to the appearance of their borderland living quarters. Most settlers chose architecturally pleasing forms, proportions and symmetry when constructing their cabins. They made curtains and flower boxes for windows, hung wreaths on front doors, planted decorative trees and flower beds. What is more, the frontier photographs confirm the importance of a different set of key markers of home, also emphasized by Esther Hanna's above-cited depiction of her Overland wagon. They highlight the significance of what sociologists Graham Allan and Graham Crow, in their academic study of home, define as 'the power to exclude others.'[20] Esther's text underlines a distinct sense of interiority, of privacy, security, and control. She expresses the wagon as a contained space of which she felt in control, 'shut ... up close' as it was against the unpredictable outside realm which, her diary insists, was a treeless desert populated by Indians, rattlesnakes and other 'wild beast[s].'[21] Butcher chose to compose his pioneer portraits in a manner which also underlines the perceived importance of domestic interiority and of control over who accesses one's private sphere. In virtually all his images, settlers are depicted standing (or sitting) by the front door of their sod house. In this location they appear to be guarding their private space from the beholder: while we are allowed to inspect the homesteaders' land and a selection of private possessions they present outside their houses—items as diverse as baby cradles and high chairs, framed family portraits, chests of drawers, washing machines, and even a parlor organ—the inside of their home explicitly and deliberately remains off-limits. Butcher, who possessed both the equipment and the technical expertise to take pictures indoors, never photographed the Nebraskan pioneers inside their homes and the spectator is at no time allowed to cross the threshold.[22]

Solomon Butcher's portraits of everyday borderland living quarters respect the homesteaders' private sphere and thus confirm secluded interiority as a potent carrier of homeliness. Like a multitude of Overland diary entries, his images indicate that in the West being at home was frequently and strongly associated with being inside, separated from the external world. But they also challenge this generally accepted understanding of home. As mentioned above, most photographs depict settlers outside their cabins surrounded by personal belongings which are typically found within the home. Some of these objects had clearly been brought outside for the occasion; many Nebraska emigrants intended to send their portraits to family and acquaintances who had stayed behind and were keen to present themselves as surrounded by the same 'markers of genteel comfort'[23] as their New England contemporaries. Other items, however, appear to be installed on the open prairie on a permanent basis: domestic artifacts such as dining tables, rugs, wicker rocking chairs, or sewing machines seem to have been displaced by everyday life and to have naturally found their place outside the sod houses. These objects' position on the open prairie implies their owners' daily actions—they indicate that the borderland inhabitants regularly ate, socialized, and sewed not *inside* but *outside* their cabins. A number of photographs even show fully functioning and somewhat permanent kitchens, including storing and washing-up facilities as well as stoves, installed along external sod house walls or elsewhere in the yard around the main homestead.

The Nebraskan settlers' habit of bringing their domestic items across the threshold of their cabins indicates their understanding of home less as defined, bounded location—a building—and more as performance. This change in focus is beautifully confirmed by pioneer Miriam Davis Colt. Preparing her family's daily meals on an open fire, Miriam realizes that domestic space is defined not by the walls that enclose it but by a particular domestic practice:

> May 20th.—Have been busy all day in my kitchen, whose dimensions are by no means confining. It is roofed by the blue dome of heaven, the partition wall on the south is the timber that fringes the Neosho [River]; on the north, east, and west, the smooth green prairies, gently swelling, declining, then swelling higher again, until in the distance it is joined with the roofing of blue.[24]

So far, this chapter has shown that everyday life in the borderlands of the nineteenth-century American West was defined by change and that its inhabitants' focus on home—as a distinct place left behind, as an emotional and spatial concept, as a physical structure created using trusted architectural language—allowed them to bring stability and predictability to the shifting and unfamiliar environment that surrounded them. The way the pioneers on the frontiers and on the Overland Trails remembered, (re)constructed and enacted their homes enabled them to anchor themselves by using the tools and under-standings they had brought with them from a more predictable life and a more predictable place.

Home thus brought stability to an environment of continual change. However, the shifting, challenging nature of the Western borderlands and the daily lives of those inhabiting them in turn also brought change to the perceived stability and permanence of home. In the first instance, nineteenth-century Westerners challenged the perceived link between home and a distinct location. In their day-to-day lives, Overland travelers and Nebraska homestead-ers performed home not as a fixed point in space, not as a singular, unchang-ing site of original belonging to which an individual has a unique, eternal tie. Instead, their practice reveals home as something adaptable or movable, as an entity capable of migrating across great distances to a geographically entirely different place; from New York or Connecticut to Nebraska or California.

What is more, the documents left behind by the inhabitants of the trans-Mississippi West question not only the intrinsic connection commonly thought to exist between home and geographical location but also the per-ceived link between home and architectural enclosure. The nineteenth-century photographs and diaries show home itself as also capable of transforming or evolving—in fact, change seems to be at the core of the Western home. Butcher's photographs and the diaries from homesteaders and Overland Trail travelers suggest that home was, in fact, home-*making*: home was acted out—as cooking, as eating, as child-minding, as socializing—in locations deemed most appropriate at the time. As I will now go on to show, this unconventional understanding of home as practiced, as flexible and as independent of strict, containing boundary walls had a considerable impact on how Westerners thought of themselves as individuals.

The Recognizable, Stable Self

In 1856, Kansas homesteader Miriam Davis Colt recorded in her diary,

> I have cooked so much out in the hot sun and smoke, that I hardly know who I am, and when I look in the little looking glass I ask: 'Can this be me?' Put a blanket over my head and I would pass well for an Osage squaw. My hands are the color of smoked ham.[25]

A few years earlier and on the Overland Trail to California, Henry Rice Mann anticipated her sentiments: 'if I could have been seen by any of my friends I should not have been recognized, I was so completely covered with dirt.'[26] The questions 'Who am I?,' 'Can I be recognized as myself in this strange place?,' and 'Am I still different from those who are not like me?' surface in many

nineteenth-century Overland diaries. These existential deliberations are frequently offset—almost displaced—by a profusely efficient portrait of normality
and predictability; of life carrying on as it always had done. They emphasize
that mothers continued to tell bedtime stories and fathers read newspapers,
that young children passed the time 'toddling along behind the wagons' or
doing homework, and that older children quarreled or fell in love (although
some teenagers expressed their discontent with their admirers, grumbling that
'There is a great many young men loves me ... They don't suit my taste').[27]

The seemingly trivial nature of these concerns might initially appear strange
in the face of the Westerners' overall situation. Their lives were enfolded in an
exceptional context; Overland travelers and pioneer settlers lived with adverse
weather, with attacks from wild animals, with the daily threat of severe illness,
with lack of food and water. Yet in the midst of this environment of extremes
and unfamiliarity, their diaries frequently record comparatively unremarkable
routine occurrences. This is not coincidental. Women's studies scholar Margo
Culley observes that the disturbance of spatial continuity associated with moving away from a familiar home is commonly seen to disrupt identity; that it
'creat[es] a sense of a discontinuity of self—I was that, now I am this; I was
there, now I am here.'[28] If the Westerners feared losing their identity in the
fluid, unfamiliar and expansive borderlands they inhabited, then using their
diaries to report commonplace events helped create an environment in which
they could recognize themselves.

Overlanders and homesteaders also used extratextual tools to maintain a
familiar sense of self. For women, physical appearance was important. In the
1860s Sarah Josepha Hale, editor of the influential Philadelphia-published
Godey's Lady's Book, had proposed that a woman's 'power is in her beauty,'[29] and
Western popular magazines delivered that same message to households on the
other side of the Mississippi River. Iowa's *Prairie Farmer* claimed that 'Women,
like flowers, have at least a part of their mission to do in beautifying the earth'
and the *Wichita Eagle*, published in Kansas, added that 'to be sunshine in a shady
place should be the aim of every true woman.'[30] In keeping with these sentiments, Solomon Butcher's archive confirms that their appearance did indeed
remain important to Western females and that Nebraska women commonly
used their attire to underline their sophistication and femininity amidst their
'uncivilized' surroundings. In 1886, he photographed four sisters (Fig. 14.1—
from left to right, Harriet, Elizabeth, Lucie, and Ruth Chrisman) who had come
from Virginia with their parents and three brothers to each file a homestead
claim. It is clear that the young women had dressed up for the portrait; all have
accessorized their clothes using belts, lace collars, velvet trims, and brooches.
Ruth Chrisman also wears a contoured, hip-length basque jacket which is
adorned with elaborate embroidery and features the high collar and fitted
sleeves then fashionable.[31] All four girls are neatly groomed and Elizabeth and

Lucie Chrisman hold a hat, indicating that the sisters were keen to protect their complexion from the fierce prairie sun. Indeed, Overland women commonly attempted to maintain a ladylike pallor. Ada White Royer writes about covering her arms with stockings to shield her skin[32] and Henry Rice Mann observes travelers wearing face masks on their way to the Pacific: 'I saw the two ladies as they passed us at one of our resting places and the poor souls had on false faces to preserve their delicate features, the dear creatures.'[33] Like their clothes, their comparatively pale faces and limbs signified the women's Euro-American selves. Their white complexion was associated with middle-class cleanliness and immediately set female migrants apart from the colored appearance and perceived poor hygiene of the native tribespeople roaming the trans-Mississippi West.[34]

It should be noted that, over time, many emigrants inhabiting the nineteenth-century West integrated elements of Native American culture into their everyday lives. Overlanders commonly relied on the indigenous population to supplement their limited Trail diet. They also accepted Native American weavings (Miriam Davis Colt's pioneer diary records her family sleeping 'rolled in Indian blankets like silk worms in cocoons'),[35] moccasins, and bison robes. Some learned to use animal hides to make furniture or work clothes suitable

Figure 14.1 The Chrisman sisters on a claim in Goheen settlement on Lieban Creek, Custer County (photograph by Solomon D. Butcher, 1886). Nebraska State Historical Society, Lincoln, Nebraska

for the harsh prairie environment. A few white Westerners are known to have established lasting friendships with their native counterparts. However, the emigrants' actions and writings more often confirmed the official national narrative of white superiority and Native American Otherness. Butcher's archive for one clearly underlines that Nebraska settlers were keen to set themselves apart from the 'wilderness' and 'savages'[36] by portraying themselves as maintaining the values at the heart of Euro-American society, including civility, sophistication, and cultural refinement. A number of his photographs depict homesteaders playing music and reading books. One image shows a sod house whose front garden, vaguely enclosed by a row of fledgling trees, is dotted with white metal hoops. Close to the building, wearing starched white aprons and collars and leaning on croquet rackets, two women position themselves like comfortably leisured East-coast ladies. Art historian Melissa Wolfe has identified Butcher's composition as mimicking 'images of croquet players, such as "Summer in the Country" by Winslow Homer that were widely reproduced in national periodicals and were read as images of middle-class leisure.'[37] Diarists similarly recorded attempts to cling to genteel pastimes, emphasizing that they studied Shakespeare and 'the life and poems of Schiler [sic]'[38] while traveling to the Pacific, that they displayed social grace by formally visiting from wagon to wagon, and that they commonly engaged in polite ballgames and musical performances.[39]

The materials considered here also indicate that Westerners clung not only to accepted cultural and social norms but also to their gendered roles in order to maintain a familiar sense of self. The diary of Overlander Charlotte Pengra, for example, is a catalogue of everyday domestic thoughts and practices: she ponders the best color for pillow cases (anything dark or patterned, nothing white), contemplates the technicalities of washing clothes, and writes about designs for sunbonnets and looking after her sick child. Interestingly, Charlotte usually returns to these types of thoughts at the end of the day, predictably finishing many diary entries with a record of her domestic achievements:

> Saturday 4th [June] … I have unpacked aired and packed all the clotheing done a large washing; baked a tub full of bread, stewed apples, washed out the waggon and cooked two meals … by a fire made of willow bushes about as large round as a mans thumb.[40]

Diary writing was highly encouraged in the nineteenth century and educators were quick to point out that a dedicated 'evening review,' a final writing session appraising one's 'conduct of the day,' was particularly effective in strengthening the character and thus of upmost importance.[41] Charlotte Pengra can be seen to have appraised her day and her identity in this manner. And when she paused to look at her self though her diary's text, she was able to recount a set of daily actions—baking, washing, airing luggage—that projected a familiar person: a maternal woman, presiding over a clean and orderly domestic sphere as advocated by the advice literature of the time.[42]

Every day, Charlotte thus successfully created her self as a recognizable person through her actions in space and through her recording of them. Many Trail women used their domestic practice to align themselves with the accepted behavioral norms and ideals of nineteenth-century femininity and male Westerners acted in a similar manner. If, in the nineteenth-century ideology of socially constructed separate spheres, home was cast as the arena of women, the same forces allocated the outside world to men. Male Trail diaries, accordingly, tend to focus on tasks outside the domestic sphere. Most men record standing guard along the boundary of their party's encampment at night. They write about venturing away from the wagon and into the unknown wilderness, sometimes for extended periods of time, to deal with Native Americans, find lost cattle, search for safe river-crossings and firewood or replenish life-saving supplies.[43] Missouri to California emigrant Edward Willis recorded in 1849, 'Thursday, June 7th … At 3 o'clock cam[e] on herd of Buffalo. Had hard chase—wounded several. Killed one about sundown. Made our supper of the meat and crackers. Storm coming up.'[44] In keeping with this, Butcher's pioneer portraits show men sporting hunting rifles as well as featuring another noticeable element: the majority of his images depict piles of antelope horn prominently displayed in front of the sod houses or on their roofs—visual markers of the head of the family's hunting successes. As historian John M. MacKenzie points out, in nineteenth-century society pursuing and killing animals symbolized 'all the most virile attributes of the imperial male; courage, endurance, individualism, sportsmanship … resourcefulness, a mastery of environmental signs and a knowledge of natural history,'[45] and Western men clearly bore this Eastern perception in mind when they presented themselves to the photographer.

The Fluid Self

If they perceived themselves as being in danger of losing their identity amidst the shifting, expansive trans-Mississippi borderlands, then clinging to accepted social, cultural, and gender norms helped Westerners in their quest to remain recognizable to themselves. But a more detailed look at the portrait of the Chrisman sisters (Fig. 14.1) indicates that while emigrants worked hard to preserve parts of an approved and established self in their unpredictable context, they also used their borderland position to continually question and actively redefine who they perceived themselves to be.

The four young women are clearly dressed for Butcher's camera, but underneath the fashionable accessories their core attire is comparatively plain. Harriet and Lucie (left and third from left) both wear what was known as a 'Mother Hubbard Wrapper,' a long dress with ruching trims on the bust, a bodice without corseting and a full, flowing skirt. The other two sisters both wear a 'sack skirt'—an unadorned and inelegantly cut garment seen frequently on the homesteading frontier as it was utilitarian and simple to sew. All four girls' clothes are patterned: Ruth's outfit is known to have been a brown and white percale, Harriet's appears to be of the same fabric, Elizabeth and Lucie wear gingham.[46] A fellow homesteader's writings observe that 'gingham is very good and serviceable'[47] on the dusty frontier while emigrant Charlotte Pengra asserts that 'white is not suitable'[48] on the Overland Trail.

The Chrisman portrait highlights that Western women typically had to forego stylish dress patterns as well as what costume historian Sally Helvenston poignantly terms a 'symbol of middle-class gentility'[49]—white fabric. What they might have gained in return for these losses is implied by the two horses present in Butcher's image. The photograph has been identified as depicting the sod house in which Elizabeth and Lucie lived and it seems that Harriet and Ruth traveled there on horseback.[50] Interestingly, both young women's horses wear a standard saddle (as opposed to the sidesaddle which female riders generally used at the time), meaning they must have ridden astride.[51] The photograph thus recalls the diary of a Kansas homesteader:

> I mounted my pony on a man's saddle, and we started off; but … I could make but very little headway … having on my Bloomer with calico pants, I just put a foot in each stirrup … it was surprising to me, to see with what ease, safety and speed I could now ride horseback.[52]

Newly arrived emigrants commonly smirked at the sight of Western females wearing unfitted dresses (like the Chrisman's sack skirts) or even 'Bloomer Costumes,' shorter than usual skirts worn over loose trousers tapered at the ankle. Bloomers were widely ridiculed in polite New England society, which generally supported longer and tighter outfits for ladies.[53] Many Overlanders and homesteaders, however, soon deemed these garments 'very appropriate'[54] for Western life as they permitted women to negotiate daily life more successfully. Miriam Davis Colt confirms, 'Am wearing the bloomer dresses now; find they are well suited to a wild life like mine. Can bound over prairies like an antelope, and am not in so much danger of setting my clothes on fire while cooking.'[55] While Eastern fashions continued to create woman as a passive 'ornament for her household,'[56] Western borderlands supported an attire which, while assisting women in carrying out approved activities in the domestic sphere (such as cooking), also enabled them to extend their position actively and to alter their daily actions. Freed from tight-fitting skirts and constricting corsets, female emigrants record embracing both spaces (the open prairies away from the domestic sphere) and practices (riding a horse at great speed) that were usually reserved for their male contemporaries.

If their dress allowed Western women like Miriam to transgress the restrictive boundaries created by the internalizing narratives of nineteenth-century spatial womanhood and to position themselves differently, then the above-described understanding of home as relying less on rooms, walls, and definite thresholds and more on everyday practice opened up similar possibilities. Taking meals away from the formal dinner table, for example, presented the opportunity to change social relationships. Sitting down to eat is a highly symbolic and loaded rite, as food historian Amy Bentley explains, 'Mealtime rituals and patterns, such as the "alpha male" sitting at the head of the table … often designate a person's place in the larger social hierarchy and function to maintain and mediate the social hierarchy amongst different groups of people.'[57] Away from the spatial rigidities of the dining room, the gender roles constructed by food could become less defined in the nineteenth-century West. On the Trails, family members commonly ate their meals sitting in random locations on the desert floor or with their feet in a river; on the Nebraska frontier, meals were frequently taken outside the small cabin or in the fields.[58] As a consequence, identities typically confirmed by standard mealtime setups—'men as presidents and presenters, women as coordinators and servers'[59]—became open to contestation and re-articulation. And as the importance usually given to the distinct spatial thresholds enclosing the home crumbled in the Western borderlands, socio-spatial relations between genders were negotiated in other ways too. Illinois-to-California emigrant Mary Burrell, for instance, reports patrolling the containing perimeter of her party's encampment: 'May 7 … Although it is Sunday still we are traveling … / Put [Mary's cousin Putnam Robson] & I stood guard till 1 o'clock.'[60] Numerous borderland women recorded that they stood guard alongside male Overlanders, that they left the security of the homestead or camp to shoot buffalo, catch fish, and trade with Native Americans, and that, in the words of nineteenth-century novelist Eliza Farnham, they 'yoked and unyoked the oxen, gathered fuel, cooked … drove the team, hunted wood and water.'[61] In the West, these unconventional and somewhat liberated actions were often tolerated as something that ultimately benefitted the female domain of home and family. The *Kansas Union Sentinel* editor wrote in 1864,

> We read (and see a little of it) of instances in which women drive into city and town … Under ordinary instances this would be improper and unbecoming, but … [i]n almost all these situations women's actions were praiseworthy … because they helped preserve the home.[62]

However, rather than seeing their new extra-domestic position as the sacrifice required to obediently 'preserve' the home and the submissive homemaker, many women actively used their skills to break away from the isolated domestic sphere allocated to them and into the public arena. They produced butter, sewed clothes and gloves ('had many more orders than I could fill'),[63] or plaited straw bonnets and took them into settlements for sale. Most were aware of the increased authority this gave them—frontierswoman Mary Ellen Todd admits she experienced 'a secret joy in being able to have a power that set things going'[64] when she cracked the whip and drove a team of oxen—and a number of women used their Western migration to separate themselves almost completely from normative visions of feminine identity, feminine place, and feminine practice. A poignant example of this is the homesteading venture of Mary and Agnes Price. At a time when unmarried women had only recently gained the right to apply for land in their own name (under the Homestead Act of 1862) and when many females still filed their claim to extend a husband's or a father's holdings, the two unmarried sisters used the space of home to overcome common gender expectations and hierarchies. Having emigrated from Ohio to Nebraska—they could 'not see why a girl could not do anything a boy could do'[65]—Mary and Agnes chose two adjoining plots of land next to one another and each woman filed her own claim for 160 acres. In order to eventually request the title, each woman had to build a home on her own plot and live in it for five years—and they did so. The sisters, as Mary confirms, constructed one single 'house across the line [the boundary dividing their adjoining claims] ... so that we could each eat on our own land.'[66] Practicing home in the borderlands of the American West allowed Mary and Agnes Price to express themselves independently of patriarchal structures. Their spatial text does not speak about dependence or subordination; instead, it speaks about balance, cooperation, and ongoing interaction with an equal.[67]

'Home on the Range': Conclusion

This essay has approached the mid- to late nineteenth-century trans-Mississippi West as a borderland: an area of complex cultural exchange, a region of continually shifting boundaries and meanings, a space of contestation and identity formation.[68] Considering the borderland West through the lens of 'home' and 'self' has presented a number of remarkable insights into the social and spatial experience of its inhabitants. In an environment defined by ambiguity, fluidity, and ongoing change, the lives of those who made their home amidst the vast and fluid trans-Mississippi area were shaped not solely by unsettlement and discomfort but just as much by continuity and stability. The diaries surviving from the Overland Trails and Solomon D. Butcher's visual archive highlight that emigrants' interactions with their environment were guided by an intense aspiration to preserve and strengthen the elements that anchored their lives and provided a sense of groundedness. Through the physical interventions they created in the Western landscape, through their daily interaction with their surroundings and through the way they presented themselves, the nineteenth-century men and women were able to foreground a strong sense of rootedness in their borderland existence—a process that involved clinging to some knowledge and habits but also entailed revising or even erasing other norms and practices.

While querying the complete fluidity and uncertainty of borderlands, this essay has also reflected on 'home' and 'self,' raising in turn questions about the presumed stability and permanence of both dwelling and identity. The emigrants' way of making their daily lives amidst the less defined spatial boundaries of the borderland West reminds us that home is not a single, spatially delineated location to which a person is somehow naturally and eternally connected, but a hybrid which continually evolves as it is produced and reproduced by daily actions.[69] And as the borderland West's lack of distinct boundaries challenged the perceived purity and stability of home, the identities of those who were 'at home' in the trans-Mississippi area were revised in parallel. As they continued to cross spatial thresholds (the boundaries of the domestic home and the boundaries of the nation), the Western men and women also crossed social, cultural, and gender thresholds.[70] Their fluid practice in a *hybrid space* enabled a corresponding navigation of an equally hybrid self and the borderland inhabitants emerge as individuals with allegiances to—and the ability to speak from—a multitude of positions.[71]

Notes

1. Publications which address national boundaries and relating borderlands in the western United States include J.F. Brooks, *Captives and Cousins: Slavery, Kinship, and Community in the Southwest Borderlands* (Chapel Hill, 2002); B.H. Johnson and A. Graybill (eds), *Bridging National Borders in North America: Transnational and Comparative Histories* (Durham, NC, 2010); R. White, *The Middle Ground: Indians, Empires, and Republics in the Great Lakes Region, 1650–1815* (Cambridge, 1991).

2. For this chapter I consulted manuscript and transcript diaries at the Henry E. Huntington Library, San Marino, California (abbreviated as HEHL in references for individual diaries), the Beinecke Rare Book and Manuscript Library, New Haven, Connecticut (BRBML), and Utah State University, Logan, Utah (USU). To balance this, I also worked with 'Trails of Hope,' an excellent digital archive in which Brigham Young University, University of Utah (BYU), and Utah State University make selected diaries of Mormon travelers available to researchers in the form of high-resolution, color-corrected PDF scans. This collection can be found at http://overlandtrails.lib.byu.edu/about.html.

3. A.S. Warner, *Diary of a Journey from Pennsylvania to Oregon* (manuscript diary, 1853: HEHL).

4. M.D. Colt, *Went to Kansas; Being a Thrilling Account of an Ill-fated Expedition to That Fairy Land and Its Sad Results* (Watertown, 1862), p. 25.

5. E.B. Hanna, *Diary of Overland Journey from Pittsburgh, PA, to OR* (manuscript diary, 1852: HEHL).

6. C.L. Long, *Overland* (manuscript diary, 1849: BRBML).

7. Walter McIntosh, Rachel Bond's son-in-law, wrote a book about her journey entitled *Allen and Rachel: An Overland Honeymoon in 1853*. He makes reference to a quilt with

a tulip pattern constructed in Bond's Oregon log cabin. The quilt remains in the private collection of Sue O'Neal. The bundle of mismatched fabrics taken on the trail remains at Lane County Historical Museum, Eugene, Oregon ; an image of it can be seen in M.B. Cross, *Treasures in the Trunk: Quilts of the Oregon Trail* (Nashville, 1993).

8. In E. Hedges, 'The Nineteenth-Century Diarist and Her Quilts,' *Feminist Studies*, 8 (1982), 296–7.

9. See E. Hedges et al., *Hearts and Hands: The Influence of Women and Quilts on American Society* (San Francisco, 1987), p. 52.

10. B.R. French, *Diary* (manuscript diary, 1867: HEHL).

11. I borrow the term 'spatial practice' from Henri Lefebvre, who defines it as actions producing, altering, and reproducing spatial meanings and spatial relationships. *The Production of Space* (Oxford, 1991), p. 73. This term seems to best reflect the migrants' ongoing questioning of relationships between rootedness and mobility, East and West, home and public sphere—a complex, conflictual process acted out in space.

12. M.S. Bailey, *A Journal of Mary Stuart Bailey, Wife of Dr Fred Bailey, from Ohio to Cal. April–Oct 1852* (manuscript diary, 1852: HEHL).

13. A.M. Crane, *Journal of a Trip from Lafayette, Ind., to Volcano, Calif., via Fort Laramie, Salt Lake City and the Humboldt River. 1852, Mar. 26–Aug 28* (manuscript diary, 1852: HEHL).

14. Hanna, *Diary*.

15. M.A. Holley, *Texas: Observations, Historical, Geographical and Descriptive, Written during a Visit to Austin's Colony, with a View to a Permanent Settlement in That Country, in the Autumn of 1831* (Baltimore, 1833), pp. 39–40. Mary Austin Holley traveled West and published an account of what she saw; her book was the first history of the Republic of Texas in English.

16. Butcher's images are held at Nebraska State Historical Society and can be studied via the Library of Congress website at http://memory.loc.gov/ammem/psquery.html.

17. Upon their arrival in the area, a great number of homesteaders started in a 'dugout,' a basic shelter carved out of the ground. From their dugout, most pioneers 'upgraded' to a freestanding sod house. Finally, a third move might bring homesteaders into the more permanent home they had been striving for, usually a timber-framed structure with horizontal weather boarding.

18. The frequent occurrence of bird cages is also observed by M. Melissa Wolfe in her excellent '"Proving Up" on a Claim in Custer County, Nebraska: Identity, Power, and History in the Solomon D. Butcher Photographic Archive (1886–1892),' PhD dissertation, Ohio State University, 2005.

19. Quoted in Wolfe, '"Proving Up",' p. 87.

20. G. Allen and G. Crow, 'Introduction,' in Allen and Crow (eds), *Home and Family: Creating the Domestic Sphere* (Basingstoke, 1989), p. 4.

21. Hanna, *Diary*.

22. Butcher did take photographs inside other, non-domestic buildings, including shops and hotels.

23. Annette Kolodny uses this term to refer to items in frontier cabins: A. Kolodny, *The Land Before Her: Fantasy and Experience of the American Frontiers, 1630–1860* (Chapel Hill, 1984), p. 100. One account of an Overlander pondering the standard of her domestic sphere can be found in the diary of Charlotte Pengra who writes: 'Thursday 16th [April] pitched our tent in the subberbs [suburbs] of Tipton ... had several calls from the ladies in town soon after we got into our house felt quite happy in showing them the conveniences we have on the road': C.E.S. Pengra, *Diary of an Overland Journey from Illinois to Oregon via South Pass* (manuscript diary, 1853: HEHL).

24. Colt, *Kansas*, p. 52.

25. Colt, *Kansas*, p. 72.

26. H.R. Mann, *The Diary of Henry Rice Mann from June 22 to September 18 1849* (manuscript diary, 1849: BRBML).

27. J.R. Jeffrey, *Frontier Women: 'Civilizing' the West 1840–1880* (New York, 1998), p. 84. Falling in love: M. Burrell (1854), *Mary Burrell's Book* (manuscript diary, 1854: BRBML). Bedtime stories: M.L.R. Powers, *The Journal of a California Emigrant* (manuscript diary, 1859: BRBML). Newspapers: S.D. Butcher, 'North Custer County, Nebraska. 1888' (photograph held at Nebraska State Historical Society, Lincoln, Nebraska). Children toddling behind wagons: Long, *Overland*. Homework: J.A. Blood, *Journey to California across the Plains* (manuscript diary, 1850: BRBML).

28. M. Culley (ed.), *A Day at a Time: The Diary Literature of American Women from 1764 to the Present* (Old Westbury, 1985), p. 9.

29. Quoted in M. Ryan, *The Empire of the Mother: American Writing about Domesticity, 1830–1860* (New York, 1982), p. 34.

30. Quoted in S.I. Helvenston, 'Ornament or Instrument? Proper Roles for Women on the Kansas Frontier,' *Kansas Quarterly*, 18 (1986), 35.

31. For information on American clothing in the nineteenth century see J.L. Severa, *Dressed for the Photographer: Ordinary Americans and Fashion, 1840–1900* (Kent, OH, 1995).

32. See Helvenston, 'Ornament,' 39.

33. Mann, *Diary*.

34. A number of diarists depict the indigenous population as unclean. See, for example, W. Ajax, *Journals, 1861–1863* (manuscript diary, 1861–3: BYU), and C.E. Hines, *Overland Journey from Hastings, New York, to Portland, Oregon* (manuscript diary, 1853: BRBML).

35. Colt, *Kansas*, p. 39. For positive encounters with Native Americans see, for example, Burrell, *Book*; M.M.B. Moore, *Journal of a Trip to California, 1860* (manuscript diary, 1860: BRBML), and C. L'Hommedieu Long, *Overland, 1849* (manuscript diary, 1860: BRBML).

36. Many Westerners as well as popular nineteenth-century writers referred to the indigenous population as savages. Examples include Hanna, *Diary*; W. Irving, 'Traits of Indian Character,' in *Sketch Book of Geoffrey Crayon* (New York, 1819–20); C. Beecher and H. Beecher Stowe, *The American Woman's Home: Or, Principles of Domestic Science; Being a Guide to the Formation and Maintenance of Economical Healthful, Beautiful, and Christian Homes* (New York, 1869).

37. Wolfe, '"Proving Up",' pp. 84–5.

38. E. Hillyer, *Overland Journey from Ohio to California* (manuscript diary, 1849: BRBML). A reference to reading Shakespeare can be found in C.C. Cox, *Overland Trip from Texas to California* (manuscript diary, 1849: HEHL).

39. Ball games: S. Newcomb, *Journal of an Overland Trip from Darien, Wis., to California and Oregon* (manuscript diary, 1850–51: HEHL); visiting wagons: Hines, *Overland*; concert: Burrell, *Book*.

40. Pengra, *Diary*.

41. W.A. Alcott, *The Young Man's Guide* (Boston, 1834), pp.135, 161.

42. Popular domestic advice literature aimed at women included L.M. Child, *The American Frugal Housewife* (New York, 1832), C. Sedgwick, *Home* (New York, 1835), C. Beecher, *Treatise on Domestic Economy* (New York, 1841), C. Beecher and H. Beecher Stowe, *The American Woman's Home* (New York, 1869), as well as the magazine *Godey's Lady's Book* (1830–78).

43. See, for example, Cox, *Overland* (reports searching for river-crossings); W. Montgomery, *A Journey to California* (manuscript diary, 1850: HEHL—reports men searching for firewood); S.B. Eakin, *Journal of a Trip across the Plains* (manuscript diary, 1866: HEHL—reports going after lost cattle); J.A. Stuart, *Notes on a Trip to California, and Life in the Mines* (manuscript diary, 1849–53: HEHL—reports a buffalo hunt); Hillyer, *Overland* (reports fight with Native Americans).

44. Willis, *Diary*.

45. J.M. MacKenzie, 'The Imperial Pioneer and Hunter and the British Masculine Stereotype in Late Victorian and Edwardian Times,' in J.A. Mangan and J. Walvin (eds), *Manliness and Morality: Middle-class Masculinity in Britain and America, 1800–1940* (Manchester, 1987), p. 179. On manliness in a British context, see J. Tosh, *A Man's Place: Masculinity and the Middle-Class Home in Victorian England* (New Haven and London, 1999).

46. Estell Chrisman Laughlin identifies Ruth's dress as a brown-and-white percale in a letter to Nebraska State Historical Society librarian Myrtle Berry. See: http://memory.loc.gov.

47. Quoted in Severa, *Dressed*, p. 299.

48. Pengra, *Diary*.

49. Helvenston, 'Ornament,' 39.

50. Nebraska State Historical Society states the image was taken in front of Elizabeth's sod house. The Autry Centre states that Elizabeth and Lucie 'took turns living with each other so they could fulfill the residence requirements without living alone.' See http://theautry.org/explore/exhibits/sod/history.html (accessed Feb. 10, 2012).

51. The fact that the horses wear standard saddles is observed in Wolfe, '"Proving Up",' p. 77.

52. Colt, *Kansas*.

53. The Bloomer Costume was invented in New York and worn there by a number of women activists campaigning for dress reform, suffrage, and equal rights. However, it was ridiculed in the popular media and failed to become commonly accepted.

54. Quote from Mrs Francis H Sawyer's diary, cited in M.M. Miyamoto, 'No Home for Domesticity? Gender and Society on the Overland Trails,' PhD dissertation, Arizona State University, 2006, pp. 124–5.

55. Colt, *Kansas*, p. 65.

56. This is how costume historian Sally Helvenston describes the well-attired woman in nineteenth-century America; Helvenston, 'Ornament,' 35.

57. A. Bentley, *Eating for Victory: Food Rationing and the Politics of Domesticity* (Urbana, 1998), p. 62.

58. For accounts of Overlanders eating see C.B. Call, *Diary from Salt Lake City, Utah, to Los Angeles California in 1886* (manuscript diary, 1886: HEHL). Butcher's archive contains manifold images of homesteaders eating outside their cabins; one photograph also depicts a 'packed lunch,' indicating farmer-pioneers ate in their fields.

59. Bentley, *Eating*, p. 60.

60. Burrell, *Book*.

61. Quoted in Kolodny, *Land*, p. 97.

62. Quoted in Helvenston, 'Ornament,' 36.

63. Texas pioneer Ella Bird-Dumont as quoted in S. L. Myres (ed.), *Ho For California! Women's Overland Diaries from the Huntington Library* (San Marino, CA, 1980), p. 150.

64. Mary Ellen Todd's words are quoted in L. Schlissel, 'Mothers and Daughters on the Western Frontier,' *Frontiers: A Journal of Women Studies*, 3 (1978), 32.

65. Quoted in S.A. Hallgarth, 'Women Settlers on the Frontier: Unwed, Unreluctant, Unrepentant,' *Women's Studies Quarterly*, 17 (1989), 25.

66. M.P. Jeffords, 'The Price Girls Go Pioneering,' in E.R. Purcell (ed.), *Pioneer Stories of Custer County, Nebraska* (Broken Bow, NE, 1936), pp. 74–6.

67. Although it must be acknowledged that the Price sisters' actions did follow a patriarchal script in some ways as it was an official political decision to grant unmarried women land in their own name (Homestead Act 1862). It should also be noted that most women used their right to file a claim not as a means to secure independence but instead used their 160 acres to extend a husband or father's holdings; my interpretation contrasts with some scholars' conclusions that the breakdown of traditional roles did not empower Western women. See, for example, J.M. Faragher and C. Stansell, 'Women and Their Families on the Overland Trail to California and Oregon, 1842 to 1867,' *Feminist Studies*, 2 (1975), 150–66, and L. Schlissel, *Women's Diaries of the Westward Journey* (New York, 1982).

68. For a definition of borderlands as complex and shifting ecumenes, see the Introduction to this volume.

69. Key works about human actions as producers of space include I. Borden, *Skateboarding, Space and the City: Architecture and the Body* (Oxford, 2001); E.S. Casey, *The Fate of Place* (Berkeley, 1997); M. de Certeau, *The Practice of Everyday Life* (Berkeley, 1988 [1984]), esp. ch. 3; M. Douglas, 'The Idea of Home: A Kind of Space,' *Social Research*, 58, no. 1 (1991), 288–307; D. Massey, *For Space* (London, 2005); Lefebvre, *Production of Space*; B. Penner, *Newlyweds on Tour: Honeymooning in Nineteenth-Century America* (Durham, NH, 2009).

70. Due to its restricted length, this chapter has mainly focused on how female Westerners challenged gender boundaries. Men also embraced activities and physical spaces which were not typically associated with them. Male Overlanders record doing laundry, cleaning wagons, and cooking and are known to have helped out with childcare.

71. It should be noted that the Westerners' spatial practice never seriously threatened imperial expansion. While they lived with fluid boundaries and were often open to difference, this did not necessarily lead to liberal or inclusive politics in successive generations. Euro-American Western settlers excluded racial and ethnic others, as is only hinted at in their attitudes toward Native Americans in this chapter.

Concluding Reflections: Borderlands History and the Categories of Historical Analysis

Lloyd Kramer

The history of borderland regions, peoples, and cultural exchanges has become one of the most innovative areas of contemporary historical scholarship, as the wide-ranging, perceptive chapters in this collection clearly demonstrate.[1] Although the preceding essays focus on the two centuries between 1720 and 1920 and deal mainly with European and American societies, they examine themes that are also relevant for other historical eras and for borderlands on every continent (as the chapters on Australia, Sumatra, and Africa confirm). The authors often note that borderlands are geographical, political, and social spaces where the lines between cultures become blurred, and this blurring of boundaries extends also to the influence of borderlands history on the familiar categories of historical analysis. This book shows specifically how borderlands become contact zones where cross-cultural exchanges are constantly evolving, but (taken as a whole) it also shows how the historical study of liminal places or peoples contributes to evolving 'boundary-crossings' between the subjects of historical research and the methodological subdisciplines of modern historiography. Borderland cultures facilitate creative, transnational interactions, so this historical reality may help to explain why historical studies of the borderlands encourage scholarly travel beyond the frontiers of traditional historical analysis.

I want to discuss these scholarly border-crossings by noting how the essays in this volume challenge us to rethink some well-known '-isms' that historians describe and draw upon in their accounts of modern history. Histories of borderland cultures often use (or bring empirical specificity to) categories and concepts such as nationalism, imperialism, racism, capitalism, structuralism, and poststructuralism. Borderland studies also raise questions about these '-isms' because the overlapping cultures in borderland societies show the dangers of reifying analytical categories. The social, cultural, and material complexities of borderland peoples complicate or defy the abstract '-isms' that are used to describe them. The new borderlands history thus encourages historians to question rigid analytical dichotomies. In contrast to well-demarcated conceptual distinctions, this kind of history stresses the fluidity of identities, the multicultural and multilayered dimensions of social interactions, and the limits of state power. Each part of this volume therefore suggests that studying the fluidity of borderland social life produces new scholarly streams between the ideological '-isms' that are used to channel or contain the flow of human history within clearly delineated historical categories.

Borderlands history, however, does not simply challenge the categorical boundaries of famous '-isms'; it also takes readers across the methodological and thematic boundaries that separate the subdisciplinary fields of political history, cultural history, social history, environmental history, economic history, microhistory, and gender history. These different components of historical knowledge represent distinctive approaches to historical scholarship, and they all appear in this collection of insightful essays. Yet the research on borderland societies shows why historical studies should never privilege one methodology over all others. Historical analysis of the borderlands reconfirms the essential historical claim that all human experiences and social changes are shaped by multiple, overlapping historical forces. The borders that ostensibly differentiate human societies actually reveal the diversity, ambiguity, and multicausality that exist everywhere—even in those places that historians have portrayed as coherent, unified, or clearly bounded by the categories of historical knowledge. Historians of the borderlands therefore reject monocausal explanations for human conflicts and social changes, and their research provides fresh perspectives on the 'centers' of past societies as well as the 'peripheries.' The history of borderlands, in short, guides us into new intellectual territories, but it also takes us on a critical journey across the national and methodological terrains in which historians have been traveling for a long time.

Borderlands and the '-isms' of Historical Thought

Historians think and write about human societies by using the categories and theoretical assumptions of the famous '-isms.' These conceptual and ideological constellations encompass extremely diverse ideas and actions that range from conservatism to communism, romanticism to realism, Darwinism to evangelicalism, or determinism to existentialism—to name only a few notable examples. There is an '-ism' for almost every belief, social movement, or identity, and one can well imagine borderland studies that might use virtually any influential '-ism' to develop analytical themes or comparative perspectives. The essays in this volume, however, refer mostly to ideas that are associated with nationalism, imperialism, and racism; and they develop empirical perspectives that are especially relevant for our understanding of capitalism, structuralism, and poststructuralism. I cannot possibly examine all the references to these '-isms' in each preceding chapter, but I would like to note more generally how such categories are present throughout this book and how borderlands history challenges us to rethink the meaning of the ideas and actions to which they refer.

Nationalism

The meaning of nations depends on a belief in clearly defined borders. These geographical boundaries also mark cultural and political boundaries for the people who live within those borders and who are expected to constitute a coherent cultural, linguistic, historically connected population that is protected by a sovereign state. The idea that national populations should be both culturally coherent and politically united under a strong government has been a key theme of modern nationalisms and a powerful ideological force in almost all modern societies. Without clearly delineated borders, however, the nationalist assumptions about collective identities, national territories, state power, cultural coherence, and other nations begin to break down. Nations require borders, and nationalists rely on maps, passports, languages, racial categories, religions, and government institutions to construct the bounded meaning of what Benedict Anderson famously described as 'imagined communities.'[2] Modern historical studies have often adhered to nationalist ideological assumptions by stressing the distinctive traits of national political systems, or describing the characteristics of national cultures, or accepting national boundaries as the organizing framework for historical scholarship.

Nationalism thus emphasizes the historical reality and validity of national borders, but the new borderlands history questions this whole framework of nationalist thought and scholarship. People in the borderlands have always moved across the rivers, mountains, and valleys that modern nationalists have imagined as the 'natural borders' of sovereign national states. Historians who use borderlands as their main category of analysis thus challenge popular nationalist ideologies and the nationalist categories of most historical narratives. Borderlands historians rightly argue that people often have more than one cultural or linguistic identity, that borders are never simply national (even when they follow a 'natural' barrier), and that all borderlands and hence all nations consist of multicultural communities. To be sure, the governments that impose well-marked borders, immigration controls, and legal systems on their borderlands often have great influence on borderland peoples. Borders are never stable cultural zones, however, and human migrations are never easy to police. The much-desired order of national institutions therefore requires ongoing negotiations with border communities, most of which retain more autonomy than nationalist historiographies can ever explain. Viewed from the transnational borderlands, nationalism becomes a less pervasive, coherent, and powerful force in the political-cultural history of the modern world.

Imperialism/Colonialism

Empires are often described as multicultural systems that differ from the imagined, unified political cultures of nation-states. Empires rely on local 'go-betweens' or social elites to sustain their political power, and imperial borders are often vaguely delineated frontiers rather than clearly defined cultural-political boundaries. Yet historians of empires also resemble the historians of nations and nationalisms insofar as they assume that power is concentrated in an imperial capital.[3] Imperial authorities (in this view) therefore try to bring local people into a consolidated, transnational system, which became the enduring but unachievable aspiration of Napoleon Bonaparte, the tsar of Russia, the British Empire and every other imperial regime in the modern world. Moving the study of empires from the imperial center to the borderlands of imperial systems demonstrates that even the most powerful governments cannot really control the peripheries of large colonized territories. The borders always threaten imperial aspirations. People on the margins of empires ignore the imperial laws, move into the 'highlands' or wander beyond the reach of customs agents, travel across rivers that are supposed to block passage, and evade imperial taxation. Borderland peoples are rarely the hapless pawns of a centralizing imperial system; they protect local traditions, ignore boundaries that appear on imperial maps, and speak their own languages.

Imperial and colonial agents therefore embark on endless 'civilizing missions' to bring the distant borderland people into the empire's unifying languages, legal systems, religions, and schools, but the new borderlands history shows that this imperial project often fails. Borderland peoples express a remarkably similar historical defiance in every part of the world. From Mexico to Scotland, from Australia to Liberia, from the Napoleon-controlled Rhineland to the Napoleon-controlled mountains of Italy, from the borders of Canada and the United States to the borders of Poland and Russia—and in almost every other colony or imperial system you could study—borderland peoples resist the imposition of a centralizing imperial order. Borderland history confirms, of course, that empires (like nations) want to control their borders, yet borderland administrators must be pragmatic. They have to recognize regional traditions or languages and negotiate with local leaders. Brightly colored imperial borders may look coherent on the maps in government offices, school classrooms, and history books, but no empire finally manages to impose its unmediated will on borderland peoples. Historians who describe the metropole from the social and cultural position of the borderlands become more skeptical about the alleged, far-flung power of empires. It would be wrong, however, to conclude that the limitations of imperial systems completely undermine their influence. In fact, as the history of borderlands also demonstrates, modern empires have promoted racial and ethnic categorizations that reshaped border-land cultural identities and partially controlled borderland peoples, even those who steadfastly resisted their ascribed racial positions and exclusions.

Racism

Borderlands create particular problems and opportunities for subaltern groups that are defined using racial categories. The cultural construction of race becomes another kind of boundary because the ascribed racial identities open and close social, political, economic, and cultural opportunities to both individuals and whole groups of people. The 'color line' creates well-policed racial borders, which become barriers that national and imperial racists defend like the checkpoints at a political border. Historians of modern societies have also used race as a social and explanatory category, so that racism looms like nationalism and imperialism as one of the border-defining '-isms' in modern cultures and scholarship.[4] The historical research on borderlands, however, reshapes the meaning of race and racism in ways that resemble the borderland challenges to nationalism and imperialism. Borderland historians describe the malleability of race and (without denying the power of racism) explain how subaltern groups have used borders to defy, escape, or sometimes enhance their racially defined social positions. Indians in the Americas, for example, moved across national borders to protect themselves from the soldiers or government officials of European-American governments; and enslaved persons in the United States crossed borders to flee from slavery in border states or from the brutal slave labor system on plantations in the deep south. Other borderlands in Liberia, Russia, Mexico, and Australia were also places where racial/ethnic categories became contingent or more easily challenged through migration or escape. Like the Indians who protected their interests by crossing Mexican and Canadian borders or the enslaved black people who gained their freedom by reaching Canada, Russian serfs escaped oppressive obligations by moving into Poland, Chinese workers found more autonomy by crossing state borders in Australia, and lower-class English lovers were able to enter 'irregular marriages' by eloping across the border with Scotland. Borders have offered freedoms for all kinds of people who were mired in social and race-defined constraints.

The categories of slave and free labor—partly defined by race—also became unstable in borderlands, where one's status as an enslaved or free worker could be separated from rigid racial categories. Racism may well be as pervasive in the borderlands as in the capital cities of nations or empires, but borderlands history suggests that racism loses some of its power along the borders where people flee from a race-defined labor system, migrate away from strict social hierarchies, or marry into different 'racial' communities. Borderlands history thus shows how racist ideologies and institutions (while always present) were perhaps more frequently destabilized, more readily challenged, and more often escaped than histories of racism usually recognize. If borderlands have provided social spaces for challenging racism, however, they have never really offered a social route for escapes from capitalism.

Capitalism

The racial dimensions of free labor/slave labor in borderland economies can be taken as an example of how the history of borderlands contributes new perspectives on the development of capitalism during the industrializing era between 1750 and 1920. As many chapters in this volume explain, border regions have often served as somewhat shadowy commercial zones for economic activities that flourish outside or beyond state controls and the institutional controls of large capitalist enterprises. Although modern governments have always attempted to restrict or tax cross-border trade by imposing customs duties, tariffs, and other import/export fees, borderlands have remained a haven for smugglers and illicit traders.[5] Valuable commodities and labor flow across national and imperial borders—whiskey, food, drugs, luxury goods, clothing, enslaved workers—so borderlands may be one of the best places to study how 'unofficial' economies have always influenced the economic development of modern states and transnational commercial networks. Equally important, borderlands become a site of economic competition for essential resources, including water, fertile agricultural soils, rare products, valuable minerals, and food. Controlling borderlands trade has been one of the main aspirations of would-be governing powers in every part of the world, from eighteenth-century Sumatra and Mexico to precolonial Africa, colonial Australia, and nineteenth-century Europe.

Historians often emphasize the importance of mercantilist economic policies and colonial trading systems in the rise of modern state power, but the history of borderland trade, smuggling, and transnational commercial exchanges moves our historical attention from state-supported enterprises and large-scale industrial capitalism toward less visible market systems that have never disappeared from modern national economies. The simple dichotomies of precapitalist and capitalist trade or the distinctions between domestic and international commerce do not hold up well on borderlands. Indeed, almost every simple binary opposition seems to break down on the very borders that are supposed to define the clearest possible dividing lines for modern societies; and this breakdown of binary polarities in historical categories such as capitalism and nationalism suggests how the borderlands also challenge or expand the more theoretical abstractions of structuralism and poststructuralism.

Structuralism

The study of borderlands relies on numerous themes that evolved out of intellectual innovations in structural anthropology and linguistics, including the influential books of Claude Levi-Strauss.[6] Structuralist themes can be found, for example, in historical accounts of the long-enduring borderland cultural traditions, the social rituals in borderland communities, the contrasting terminologies in borderland languages, the collective identities that evolve through borderland descriptions of 'self' and 'other,' and the borderland material objects that also convey symbolic meanings. Historians of borderlands are especially attuned to the cultural significance of clothing, food, diseases, family and marriage customs as well as the symbolic complexities of both the natural and built environments. The material components and symbolic layers of daily life are present, of course, in all cultural contexts, from a capital city to a borderland village, but the analytical focus on borderland communities helps historians better recognize the non-written cultural exchanges that shape personal experiences and social behaviors.

Structuralist themes also help borderland historians recognize the constant interactions between the material world and symbols. The trade of material objects among people on both sides of a border deepens personal awareness of one's own cultural practices and identities, as borderland historians emphasize in their descriptions of 'things' that carry both monetary and symbolic value. This emphasis on cross-border exchanges, however, also leads borderland studies beyond structuralism because the binary oppositions that are posited in structuralist theory (for example, nature/culture, raw/cooked, outsider/insider) almost never exist so clearly or distinctly in borderland communities. Although historians need to make analytical distinctions as they separate people into linguistic, cultural, political, and ethnic groups at the 'border' (for example, Polish/Russian, free/slave, Swahili/Mijikenda), borderland cultures and peoples show how all such categorical boundaries become blurred or break down with the careful study of historical realities.

Poststructuralism

The challenge to binary oppositions leads borderland historians from structuralism toward poststructuralist perspectives that rarely attract explicit support from professional historians. Indeed, the claim that borderland history draws on the themes of poststructuralism might well be rejected by many of the contributors to this volume, none of whom refers specifically to poststructuralist literary studies or invokes the ideas of theorists such as Jacques Derrida. Yet the recurring emphasis on the 'margins' of modern nations and the frequent historical critiques of simple binary oppositions can both be seen as historical outgrowths of poststructuralist theories that reshaped literary and cultural studies during the last decades of the twentieth century.[7] Although the authors in this volume do not use the dense jargon or endless 'play of language' that has often characterized the best-known poststructuralist texts, most of the preceding chapters show how the study of borderlands offers a historically grounded context for analyzing themes that others have examined mainly in literature and philosophy.

These themes develop from several theoretical assumptions. Poststructuralists constantly argue, for example, that all identities evolve through interactions and entanglements with various 'others' that are always part of the 'self.' A sense of selfhood thus emerges in the endless social and cultural encounters with difference; and neither identities nor ideas can have a pure or unmediated essence because of this inescapable connection with otherness (the meaning of 'male' is linked to the meaning of 'female,' for example, and the meaning of 'white' is linked to the meaning of 'black'). Social groups and individuals regularly try to define or defend coherent, unified identities by placing the 'other' completely outside the self (usually attributing negative traits to outsiders), but this other always remains somehow present within an imagined (pure) selfhood. Poststructuralists therefore assume that dominant ideas, identities, and anxieties can be best understood by exploring the suppressed and marginal aspects of cultures, texts, or social groups—the 'others' who help to constitute the apparently coherent identities of cultures and imagined communities. The ideological 'center' of a culture becomes more visible when one examines the outsiders, subalterns, and absent others who are always already present, even when they are passed over in silence or condemned through cultural scapegoating. The road toward analytical comprehensions of nations, empires, social hierarchies, gender identities, and racial ideologies thus winds through the cultural margins, or what we might simply call the 'history of the borderlands.'

In this way, borderlands history extends historical research into the diverse, multicultural realms of nationhood and collective identities that have also been widely analyzed in fields such as cultural studies, postcolonial theory, critical race theory, gender and sexuality studies, and the psychology of hybrid, postmodern identities.[8] The scholarly work in these fields is quite diverse, and it typically focuses on present cultures rather than the past. Most such cultural studies thus lack the kind of empirical historical research that appears in the chapters of this volume. Amid these diverse, poststructuralist-inflected fields of research and cultural criticism, however, there is a recurring emphasis on the distorting errors of binary oppositions, the multiple identities that overlap within social groups or individuals who strive for (but never attain) a coherent, unified 'selfhood,' and the valuable knowledge that emerges from 'liminal' places, communities, and texts. Poststructuralism has provided theoretical support for literary critics, anthropologists, and all kinds of other cultural analysts who study marginalized groups or ideas. The social positions of such people and the cultural assumptions about them (as the poststructuralists have argued) are always important, in part because they reveal the ideological foundations of the 'cultural center' as well as the cultural effects of political and social power.

The chapters in this book show how similar themes also appear within the work of borderlands historians who give historical and material specificity to theories that have spread from late twentieth-century poststructuralism into the contemporary humanities and social sciences. Borderlands history thus gives cultural studies a stronger empirical and social foundation, even if social historians rarely refer to poststructuralism and cultural theorists rarely pursue archival historical research.

The history of modern borderlands also generates new perspectives on other conceptual '-isms' that appear regularly in historical narratives, but even this brief survey suggests why borderlands history contributes to a critical rethinking of the analytical categories and theoretical frameworks that are used to explain how human societies change across time. Although borderland historians draw on the binary oppositions of well-known '-isms' and (implicitly) on the theories of structural anthropology or poststructuralism, the authors in this volume focus on specific places and historical cultures to blur analytical dichotomies and to give diachronic, empirical complexity to theories that are often applied only to literary texts and contemporary cultures. Borderlands history has thus become a site for creative, interdisciplinary reinterpretations of the '-isms,' but borderland studies are also helping to reshape the methodologies that historians employ to describe the diversity of historical changes, continuities, and conflicts.

Borderlands and the Subdisciplines of Historical Scholarship

Borders have always been important for diplomatic, military, and political historians, and the evolving history of national or imperial borders has been one of the traditional subjects of historical writing. Historical discussions of early modern dynastic wars, Great Power diplomatic conferences, and national independence movements have long referred to complex, contested demarcations of political borders and borderland populations. This volume shows, however, that the new borderlands history differs from the classic diplomatic history of border negotiations. Borderlands history, as I have suggested by linking some of its themes to poststructuralism, focuses on the ambiguities rather than the clarity of border lines. It emphasizes the ways in which the flowing social and cultural life of human beings crosses all of the borders that diplomats construct. The new borderlands history therefore challenges and redefines the meaning of diplomatic history, but I would like to summarize briefly how the borderlands may also be used to expand or revise other subdisciplines of modern historical studies.

Political History

The history of borderlands shows both the institutional processes that help states expand their power into previously uncontrolled lands and the challenges that centralizing governments confront when they enter borderland territories. Political historians can thus turn to borderlands history to examine the wide range of political opportunities and obstacles that emerge whenever centralizing governments seek to impose their priorities on national populations. Studying borderland policies and conflicts pushes historians beyond the traditional political history of governing elites and institutions because borderland peoples do not simply accept the dictates that are sent out from capital cities. Borderlands are usually contested social spaces where political identities are actively negotiated and redefined through immigration controls, language policies, exile migrations, mapmaking, and daily police actions. Careful studies of borderlands therefore move political history away from 'top-down,' state-centered narratives and into new research on the daily lives, social practices, and cultural values of marginalized peoples. Borderlands history contributes to the kind of political history that describes the 'personal as political,' the political system as a 'political culture,' and major political events as struggles between contending social groups rather than a story of kings, constitutions, and legislative assemblies; and it extends the study of state power into the many unofficial social and cultural components of political life.

Cultural History

Historians have revitalized their study of cultures with the methods of the 'new cultural history.' These methods draw on literary and anthropological theories to examine how languages, symbols, and discursive systems construct cultural meanings for every human society and collective identity. The history of borderlands uses this analytical approach to develop distinctive cross-cultural comparisons of languages, symbols, historical memories, ballads, poems, and overlapping identities in borderland communities. Above all, borderland history shows that people often express and defend multiple cultural identities at the same time. Every borderland is multicultural because the people in such places speak several languages, practice different religions, circulate through diverse educational systems, and develop contrasting memories of famous historical events. These experiences shape complex, interacting cultural identities in each individual and community. Borderland studies thus examine the cultural contradictions and fusions in hybrid societies, thereby adding essential multicultural themes to the cultural history of identities, symbols, and national discourses.

Social and Economic History

Social life in the borderlands helps historians see how people protect their interests through social hierarchies, family networks, and imaginative survival strategies such as immigration, crime, and transnational commerce. The social history of these networks and activities gives historians a notable interest in the material objects of daily life and encourages detailed analysis of how people use things to give meaning to their social relationships, their migrations, and their losses. Borderlands are social contact zones where diseases are exchanged, trade moves between nations, and people can more readily choose the social communities they want to join. Government campaigns to impose tariffs and customs fees, control agriculture and commerce, and restrict the migration of workers all provide important borderland examples of how modern states try to manage social relations and economic activities. All such campaigns enter directly into the lives of borderland peoples; and the history of borderland responses to state-directed economic controls adds to the social history of subaltern populations. Borders are liminal places in which social historians can see how oppressed and enslaved persons have crossed social boundaries, searched for work, claimed new freedoms, or settled into the anonymity of exile lives. Historians who want to understand the social and economic experiences of oppressed populations—persons entrapped by racism, isolated by debilitating poverty, or punished by expulsions from their homelands—will find that the people whom they are investigating have often settled in the borderlands.

Environmental History

The study of borderlands fosters a deepening historical awareness of human interactions with the natural world. Borderlines often follow the meandering course of rivers, high mountains, and barren deserts, but the people who live in such environments use natural resources that were available long before the political boundaries ever existed. Human beings thus exploit natural environments in ways that ignore the lines that diplomats draw on maps, and borderland historians are able to show why environmental history must also be transnational history. The evolving human uses and interpretations of ecological systems can never be contained within national political boundaries. Like other natural environments, however, borderland landscapes are transformed by human interventions that raise complex questions about culture and geography. What kinds of buildings or monuments mark a border space? How do the overlapping cultures and commercial systems of a borderland compete for environmental resources? How does the depletion of resources affect people on both sides of a border? Such questions (and the research to answer them) make the borderlands a key site for some of the most innovative research in environmental history.

Microhistory

Detailed studies of small communities or local social conflicts can give historians new insights into the broader, identity-shaping social processes in national, urban, family, and religious cultures. Microhistory thus helps to support broad claims about social relationships or the struggle for power in social institutions, but it also complicates the generalizations and abstract categories that enable historians to describe large groups of people. Microhistories of borderland communities suggest that most people in multicultural border zones are pulled toward contradictory ideologies, cultural traditions, and personal aspirations that create multicultural, hybrid personal identities. But the claims for this apparent cultural tendency can only be confirmed through careful analysis of the cross-cultural conflicts and collaborations in border towns, where people argue about the architectural style for their buildings, or where travelers encounter children speaking both Welsh and English, or where lonely settlers write about cross-cultural experiences in their diaries. Borderland exchanges offer especially rich empirical materials for the microhistorical analysis of shared social spaces, cultural conflicts, and personal identities.

Gender History

Most of the essays in this volume give relatively little attention to the evolving gender identities in borderland societies, and yet there is evidence to show how borderland experiences destabilize the traditional borders between manhood and womanhood. Gender identities resemble national, racial, and class identities because they are constantly constructed and reaffirmed through cultural interactions with others and because (as the poststructuralists remind us) they are never completely coherent. The manly or womanly 'self' must always navigate among competing cultural expectations for proper gender behaviors, but the women and men in frontier territories discover that traditional boundaries and expectations do not really fit their unsettled social experiences. Attempts to reaffirm or challenge gender identities may therefore become particularly notable in borderland contexts such as the nineteenth-century American plains, and historians who set out to explain the contingency and cultural construction of gender will gather new examples in the work, rituals, and social relationships of borderland people. The history of gender—like the history of politics, cultures, social hierarchies, and environmental changes—thus becomes another field in which borderlands can expand contemporary research methodologies.

Crossing the Borders of Historical Thought

Readers who have traveled with this volume's authors through various borderlands will find more historiographical implications than I have noted in these concluding reflections. There is definitely more to say about how borderland studies might transform the historical analysis of famous '-isms' and the methodologies of different historical subdisciplines. This book can in fact be read as an invitation for historians to revise or rethink all of their familiar concepts and methodologies as they develop new, detailed accounts of the borderlands that can be found in every historical era and human society. Meanwhile, this book contributes to a lengthening historical journey across territories that historians have too often ignored as they immersed themselves in the capital cities of centralizing nations and empires.

The preceding essays provide an outstanding tour of the developing border-lands scholarship, but they will not be our last sight of the territory. We need more descriptions of the terrain and more research on the people who live there. Borderlands history opens a new vista on both the traditional subjects of historical research and the emerging fields of twenty-first century scholar-ship. Innovative historical analysis, like other creative work and critical think-ing, requires the exploration and transgression of inherited boundaries. These imaginative, well-researched essays therefore push readers in different direc-tions, but they all contribute to a shared creative project on the margins of modern national historiographies. They also strongly—and rightly—encourage us to expand our study of history beyond the conceptual borders within which we construct our historical knowledge and write our historical narratives.

Notes

1. I served as a 'commentator/discussant' at the UNC/King's College London confer-ence where the essays in this volume were first discussed, so the following comments summarize some of my responses to discussions that took place during this lively intellectual 'workshop.' Although my concluding reflections do not refer to the specific arguments or detailed research in the various chapters, I would like to thank the following conference participants for the valuable perspectives that they pro-vided throughout their essays and conversations: Timothy Barnard, Jim Bjork, Frank Bongiorno, Chad Bryant, Benjamin Johnson, Lisa Lindsay, Oksana Mykhed, Roland Quinault, Cynthia Radding, Daren Ray, Paul Readman, Michael Rowe, Matthew Salafia, Nina Vollenbröker, and Jason Yaremko.
2. See the discussion of this concept in B. Anderson, *Imagined Communities: Reflections on the Origin and Spread of Nationalism* (2nd rev. ed., London and New York, 2006), pp. 1–46. Helpful surveys of the many different interpretations of nationalism and nationalist thought are available in P. Lawrence, *Nationalism: History and Theory* (Harlow, 2005), and J. Leerssen, *National Thought in Europe: A Cultural History* (Amsterdam, 2006). I have also discussed the multilayered history of nationalism and its interpreters in L. Kramer, *Nationalism in Europe and America: Politics, Cultures, and Identities since 1775* (Chapel Hill, 2011), a work that would have been enriched by the borderland perspectives that emerge from the essays in this volume.
3. The history of empires shows that they are often governed with quite different institu-tions and policies, but the ideologies and management of imperial systems typically stress the importance and grandeur of imperial capitals. For interesting studies of both the similarities and differences in modern empires, see J. Darwin, *After Tamerlane: The Global History of Empire since 1405* (New York, 2008), and J. Burbank and F. Cooper, *Empires in World History: Power and the Politics of Difference* (Princeton, 2010).
4. For examples of how historians have analyzed the ideological themes and political power of racism in both European and global history, see the helpful analysis in N. McMaster, *Racism in Europe: 1870–2000* (New York, 2001), and the wide-ranging essays in M. Berg and S. Wendt (eds), *Racism in the Modern World: Historical Perspectives on Cultural Transfer and Adaptation* (New York, 2011).

5. The various influences of governments, mercantilism, and transnational trade on the development of early modern capitalism have received much historical attention. See, for example, the classic work of I. Wallerstein, *Mercantilism and the Consolidation of the European World Economy, 1600–1750* (Berkley and Los Angeles, 2011 [1980]), and the account of transnational economic exchanges in J. Smith, *Europe and the Americas: State Formation, Capitalism, and Civilizations in Atlantic Modernity* (Leiden and Boston, 2006), esp. pp. 140–92. The history of borderland commercial practices is connected to the more recent development of transnational trade and smuggling in helpful, concise books by G. Popescu, *Bordering and Ordering the Twenty-First Century: Understanding Borders* (Lanham, MD, 2012), and A.L. Karras, *Smuggling: Contraband and Corruption in World History* (Lanham, MD, 2010).

6. The evolving themes of structuralism appeared in several influential books by Claude Levi-Strauss, including *Tristes Tropiques*, trans. J. and D. Weightman, intro. and notes by P. Wilcken (New York, 2012 [1955]); *Structural Anthropology*, trans. C. Jacobson and B. Grundfest Schoepf (Garden City, NY, 1967); and *The Raw and the Cooked*, trans. J. and D. Weightman (New York, 1969). The later literary and cultural uses of structuralist theories are discussed in J. Culler, *Structuralist Poetics: Structuralism, Linguistics, and the Study of Literature* (Ithaca, NY, 1975), and in E. Kurzweil, *The Age of Structuralism: From Levi-Strauss to Foucault* (2nd ed., New Brunswick, NJ, 1996).

7. The most influential theoretical contributions to poststructuralism emerged from the work of Jacques Derrida, *Of Grammatology*, trans. G.C. Spivak (Baltimore, 1976). Poststructuralist themes and methods always attracted more interest among literary critics and cultural theorists than among historians, but I summarized my own early views of how historians might draw on poststructuralist perspectives in L.S. Kramer, 'Literature, Criticism, and Historical Imagination: The Literary Challenge of Hayden White and Dominick LaCapra,' in L. Hunt (ed.), *The New Cultural History* (Berkeley and Los Angeles, 1989), pp. 97–128. LaCapra's provocative books were notable examples of the late twentieth-century interest in poststructuralist theory, as one can see, for example, in D. LaCapra, *Rethinking Intellectual History: Texts, Contexts, Language* (Ithaca, NY, 1983). More recent discussions of poststructuralist influences on historians can be found in M. Poster, *Cultural History and Postmodernity: Disciplinary Readings and Challenges* (New York, 1997), and in the insightful work by S. Gunn, *History and Cultural Theory* (Harlow, 2006). The interdisciplinary borderlands were often an intellectual place where poststructuralists liked to settle.

8. The vast literature in these diverse, cultural subdisciplines cannot be cited here, but borderland historians could draw especially on postcolonial studies to expand their theoretical analysis of the 'cultural margins' in modern societies. Helpful introductions to the main postcolonial themes are available in B. Ashcroft, G. Griffiths, and H. Tiffin (eds), *The Empire Writes Back: Theory and Practice in Post-Colonial Literature* (London and New York, 1989); R.J.C. Young, *Postcolonialism: An Historical Introduction* (Oxford, 2001); and D. Chakrabarty, *Provincializing Europe: Postcolonial Thought and Historical Difference* (Princeton, 2000). Other early examples of research that used similar perspectives to rethink the history of nationalism can be found in H. Bhabha (ed.), *Nation and Narration* (London, 1990).

Select Bibliography

Adelman, J. and Aron, S., 'From Borderlands to Borders: Empires, Nation-States, and the Peoples in between in North American History,' *American Historical Review*, 104 (1999), 814–41.

Alexander, F., *Moving Frontiers: An American Theme and Its Application to Australian History* (Carlton, 1947).

Allen, H.C., *Bush and Backwoods: A Comparison of the Frontier in Australia and the United States* (East Lansing, 1959).

Allina-Pisano, E., 'Borderlands, Boundaries, and the Contours of Colonial Rule: African Labor in Manica District, Mozambique, c.1904–1908,' *International Journal of African Historical Studies*, 36 (2003), 59–82.

Andaya, B.W., *To Live as Brothers: Southeast Sumatra in the Seventeenth and Eighteenth Centuries* (Honolulu, 1993).

Andreas, P., *Smuggler Nation: How Illicit Trade Made America* (New York, 2014).

Anzaldúa, G., *Borderlands/La Frontera* (San Francisco, 1987).

Armour, I.D., *A History of Eastern Europe 1740–1918* (London, 2012).

Aron, S., *American Confluence: The Missouri Frontier from Borderland to Border State* (Bloomington, 2006).

Asiwaju, A.I., *Western Yorubaland under European Rule, 1889–1945* (London, 1976).

Asiwaju, A.I. (ed.), *Partitioned Africans: Ethnic Relations across Africa's International Boundaries, 1884–1984* (London, 1985).

Asiwaju, A.I. and Adeniyi, P.O., *Borderlands in Africa* (Lagos, 1989).

Banta, R.E., *The Ohio* (New York, 1949).

Baptist, E. and Camp, S.M.H. (eds), *New Studies in the History of American Slavery* (Athens, GA, 2006).

Barkey, K. and von Hagen, M., *After Empire: Multiethnic Societies and Nation-Building: The Soviet Union, and the Russian, Habsburg and Ottoman Empires* (Boulder, 1997).

Barkey, K., *Empire of Difference: The Ottomans in Comparative Perspective* (Cambridge, 2008).

Barnard, T.P., 'The Timber Trade in Pre-Modern Siak,' *Indonesia*, 65 (1998), 87–96.

Barnard, T.P., 'Texts, Raja Ismail, and Violence: Siak and the Transformation of Malay Identity in the Eighteenth Century,' *Journal of Southeast Asian Studies*, 32 (2001), 331–42.

Barnard, T.P., *Multiple Centres of Authority: Society and Environment in Siak and Eastern Sumatra, 1684–1827* (Leiden, 2003).

Barr, J., *Peace Came in the Form of a Woman: Indians and Spaniards in the Texas Borderlands* (Chapel Hill, 2007).

Barrett, T.M., 'Lines of Uncertainty: The Frontiers of the North Caucasus,' *Slavic Review*, 114 (1995), 578–601.

Barrett, T.M., *On the Edge of Empire: The Terek Cossacks and the North Caucasus Frontier, 1700–1860* (Boulder, 1999).

Barth, F., *Ethnic Groups and Boundaries* (Oslo, 1969).

Bashford, A., 'Quarantine and the Imagining of the Australian Nation,' *Health*, 2 (1998), 387–402.

Bashford, A., 'At the Border: Contagion, Immigration, Nation,' *Australian Historical Studies*, 33 (2002), 344–58.

Baud, M. and van Schendel, W., 'Towards a Comparative History of Borderlands,' *Journal of World History*, 8 (1997), 211–42.

Baycroft, T., *Culture, Identity and Nationalism: French Flanders in the Nineteenth and Twentieth Centuries* (Woodbridge, 2004).

Bayley, W.A., *Border City: History of Albury New South Wales* (Albury, 1976).

Belich, J., *Replenishing the Earth: The Settler Revolution and the Rise of the Anglo-World, 1783–1939* (Oxford, 2009).

Bell, D., 'Nation-Building and Cultural Particularism in Eighteenth-Century France: The Case of Alsace,' *Eighteenth-Century Studies*, 21 (1988), 472–90.

Benton, L., *A Search for Sovereignty: Law and Geography in European Empires, 1400–1900* (Cambridge, 2010).

Benton-Cohen, K., *Borderline Americans: Racial Division and Labor War in the Arizona Borderlands* (Cambridge, MA, 2009).

Berg, F.J., 'The Swahili Community of Mombasa, 1500–1900,' *Journal of African History*, 9 (1968), 35–56.

Berg, F.J. and Walter, B.J., 'Mosques, Population, and Urban Development in Mombasa,' *Hadith*, 1 (1968), 47–100.

Bigham, D.E., *The Indiana Territory, 1800–2000* (Indianapolis, 2001).

Bigham, D.E., *On Jordan's Banks: Emancipation and Its Aftermath in the Ohio River Valley* (Lexington, 2006).

Bjork, J.E. *Neither German nor Pole: Catholicism and National Indifference in a Central European Borderland* (Ann Arbor, 2008).

Blackburn, K., 'Mapping Aboriginal Nations: The "Nation" Concept of Late Nineteenth Century Anthropologists in Australia,' *Aboriginal History*, 26 (2002), 131–58.

Blackhawk, N., *Violence over the Land: Indians and Empires in the Early American West* (Cambridge, MA, 2006).

Block, D., *Mission Culture on the Upper Amazon: Native Tradition, Jesuit Enterprise, and Secular Policy in Moxos, 1660–1880* (Lincoln, NE, 1994).

Boeck, B.J., *Imperial Boundaries: Cossack Communities and Empire-Building in the Age of Peter the Great* (Cambridge, 2009).

Bolton, H.E., *The Spanish Borderlands: A Chronicle of Old Florida and the Southwest* (New Haven, 1921).

Brantley, C., *The Giriama and Colonial Resistance in Kenya, 1800–1920* (Berkeley, 1981).

Breyfogle, N.B., *Heretics and Colonizers: Forging Russia's Empire in the South Caucasus* (Ithaca, NY, 2005).

Breyfogle, N.B., Schrader, A., and Sunderland, W. (eds), *Peopling the Russian Periphery: Borderland Colonization in Eurasian History* (London, 2007).

Brockington, L.G., *Blacks, Indians and Spaniards in the Eastern Andes: Reclaiming the Forgotten in the Colonial Mizque, 1550–1782* (Lincoln, NE, 2006).

Broers, M., *Europe under Napoleon, 1799–1815* (London, 1996).

Broers, M., *Napoleon's Other War* (Oxford, 2010).

Brooks, J.F., *Captives and Cousins: Slavery, Kinship, and Community in the Southwest Borderlands* (Chapel Hill, 2002).

Broun, D. et al. (eds), *Image and Identity: The Making and Re-Making of Scotland through the Ages* (Edinburgh, 1988).

Brown, K., *A Biography of No Place: From Ethnic Borderland to Soviet Heartland* (Cambridge, MA, 2004).

Brubaker, R., *Nationalism Reframed: Nationhood and Nationalism in the New Europe* (Cambridge, 1996).

Bushnell, A.T., *Situado and Sabana: Spain's Support System for the Presidio and Mission Provinces of Florida* (Athens, GA, 1994).

Campbell, J., *Middle Passages: African American Journeys to Africa, 1787–2005* (New York, 2006).

Canney, D.L., *Africa Squadron: The US Navy and the Slave Trade, 1842–1861* (Washington DC, 2006).

Cayton, A.R.L., *The Frontier Republic: Ideology and Politics in the Ohio Country, 1780–1825* (Kent, OH, 1986).

Cayton, A.R.L., *Frontier Indiana* (Bloomington, 1996).

Cayton, A.R.L., *Ohio: The History of a People* (Columbus, 2002).

Cayton, A.R.L. and Onuf, P.S., *The Midwest and the Nation: Rethinking the History of an American Region* (Bloomington, 1990).

Clegg III, C.A., *The Price of Liberty: African Americans and the Making of Liberia* (Chapel Hill, 2004).

Cole, J.W. and Wolf, E.R., *The Hidden Frontier: Ecology and Ethnicity in an Alpine Valley* (2nd ed., Berkeley, 1999 [1974]).

Colley, L., 'Britishness and Otherness: An Argument,' *Journal of British Studies*, 31 (1992), 309–29.

Colley, L., *Britons: Forging the Nation 1707–1837* (2nd ed., New Haven, 2009 [1992]).

Colls, R. (ed.), *Northumbria: History and Identity 547–2000* (Chichester, 2007).

Cozzens, P. (ed.), *Eyewitnesses to the Indian Wars, 1865–1890, Vol. I: The Struggle for Apacheria* (Mechanicsburg, 2001).

Cragoe, M., *Culture, Politics and National Identity in Wales, 1832–1886* (Oxford, 2004).

Cross, M.B., *Treasures in the Trunk: Quilts of the Oregon Trail* (Nashville, 1993).

Curzon of Kedleson, Lord, *Frontiers: The Romanes Lecture 1907* (Oxford, 1907).

Daniels, C. and Kennedy, M.V. (eds), *Negotiated Empires: Centers and Peripheries in the Americas, 1500–1820* (London, 2002).

Davis, J.A., *Naples and Napoleon: Southern Italy and the European Revolutions, 1780–1860* (Oxford, 2006).

Davis, S.L. and Prescott, J.R.V., *Aboriginal Frontiers and Boundaries in Australia* (Carlton, 1992).

Day, D., *Smugglers and Sailors: The Customs History of Australia 1788–1901* (Canberra, 1992).

de Blij, H.J. *Mombasa: An African City* (Evanston, 1968).

de la Teja, J.F. and Frank, R. (eds), *Choice, Persuasion and Coercion. Social Control on Spain´s North American Frontiers* (Albuquerque, 2005).

Deeds, S.M., *Defiance and Deference in Mexico's Colonial North* (Austin, 2003).

DeLay, B., *War of a Thousand Deserts: Indian Raids and the US–Mexican War* (New Haven, 2008).

Diener, A.C. and Hagen, J. (eds), *Borderlines and Borderlands* (Lanham, 2010).

Donnan, H. and Wilson, T.M., *Borders: Frontiers of Identity, Nation and State* (Oxford, 1999).

Donnan, H. and Wilson, T.M. (eds), *Border Approaches: Anthropological Perspectives on Frontiers* (Lanham, 1994).

Edmonds, P., *Urbanizing Frontiers: Indigenous Peoples and Settlers in 19th-Century Pacific Rim Cities* (Vancouver, 2010).

Ekberg, C.J., *French Roots in the Illinois Country: The Mississippi Frontier in Colonial Times* (Urbana, 1998).

Ellis, G., *Napoleon's Continental Blockade: The Case of Alsace* (London, 1981).

Eslinger, E., 'The Shape of Slavery on the Kentucky Frontier, 1775–1800,' *Register of the Kentucky Historical Society*, 92 (1994), 1–23.

Etcheson, N., *The Emerging Midwest: Upland Southerners and the Political Culture of the Old Northwest, 1787–1861* (Bloomington, 1996).

Etherington, N. (ed.), *Mapping Colonial Conquest: Australia and Southern Africa* (Crawley, 2007).

Evans, R.J.W., 'Frontiers and National Identities in Central-European History,' in Evans (ed.), *Austria, Hungary, and the Habsburgs: Essays on Central Europe, c.1683–1867* (Oxford, 2006), pp. 114–33.

Faragher, J.M., *Rereading Frederick Jackson Turner* (New Haven, 1998).

Faragher, J.M. and Stansell, C., 'Women and their Families on the Overland Trail to California and Oregon, 1842 to 1867,' *Feminist Studies*, 2 (1975), 150–66.

Fitzpatrick, B., 'The Big Man's Frontier and Australian Farming,' *Agricultural History*, 21 (1947), 8–12.

Fleisher, M.L., *Kuria Cattle Raiders: Violence and Vigilantism on the Tanzania/Kenya Frontier* (Ann Arbor, 2000).

Flynn, D.K., '"We Are the Border": Identity, Exchange, and the State along the Bénin–Nigeria Border,' *American Ethnologist*, 24 (1997), 311–30.

Ford, L., *Settler Sovereignty: Jurisdiction and Indigenous People in America and Australia, 1788–1836* (Cambridge, MA, 2010).

Ganson, B., *The Guaraní under Spanish Rule in the Río de la Plata* (Stanford, 2003).

Ganter, R., *Mixed Relations: Asian–Aboriginal Contact in North Australia* (Crawley, 2006).

Gara, L., *The Liberty Line: The Legend of the Underground Railroad* (Lexington, 1967).

Grab, A., *Napoleon and the Transformation of Europe* (Basingstoke, 2003).

Gray, S.E. and Cayton, A.R.L., *The American Midwest* (Bloomington, 2001).

Graybill, A.R., *Policing the Great Plains: Rangers, Mounties, and the North American Frontier, 1875–1910* (Lincoln, NE, 2007).

Graybill, A.R., *The Red and the White: A Family Saga of the American West* (New York, 2013).

Griffen, W.B., *The Apaches at War and Peace: The Janos Presidios, 1750–1858* (Albuquerque, 1988).

Griffler, K.P., *Front Line of Freedom: African Americans and the Forging of the Underground Railroad in the Ohio Valley* (Lexington, 2004).

Grivno, M., *Gleanings of Freedom: Free and Slave Labor along the Mason–Dixon Line, 1790–1860* (Urbana, 2011).

Gruenwald, K.M., *River of Enterprise: The Commercial Origins of Regional Identity in the Ohio Valley, 1790–1850* (Bloomington, 2002).

Gruenwald, K.M., 'Space and Place on the Early American Frontier: The Ohio Valley as Region, 1790–1850,' *Ohio Valley History*, 4 (2004), 31–48.

Gudmestad, R.H., *A Troublesome Commerce: The Transformation of the Interstate Slave Trade* (Baton Rouge, 2003).

Guy, D. and Sheridan, T.E. (eds), *Contested Ground: Comparative Frontiers on the Northern and Southern Edges of the Spanish Empire* (Tucson, 1998).

Haefeli, E. et al., 'Responses: Borders and Borderlands,' *American Historical Review*, 104 (1999), 1221–39.

Hallgarth, S.A., 'Women Settlers on the Frontier: Unwed, Unreluctant, Unrepentant,' *Women's Studies Quarterly*, 17 (1989), 23–34.

Hämäläinen, P., *The Comanche Empire* (New Haven, 2008).

Hämäläinen P. and Johnson, B.H. (eds), *Major Problems in the History of North American Borderlands* (Boston, MA, 2012).

Hämäläinen, P. and Truett, S., 'On Borderlands,' *Journal of American History*, 98 (2011), 338–61.

Hammond, J.C., *Slavery, Freedom, and Expansion in the Early American West* (Charlottesville, 2007).

Harper, N.D., 'Frontier and Section: A Turner "Myth"?,' *Historical Studies*, 5 (1952), 135–53.

Harper, T.N., 'The Politics of the Forest in Colonial Malaya,' *Modern Asian Studies*, 31 (1997), 1–29.

Harrold, S., *Border War: Fighting Over Slavery before the Civil War* (Chapel Hill, 2010).

Helvenston, S.I., 'Ornament or Instrument? Proper Roles for Women on the Kansas Frontier,' *Kansas Quarterly*, 18 (1986), 35–49.

Hertslet, E., *The Map of Africa by Treaty* (3rd ed., London, 1967).

Hinderaker, E., *Elusive Empires: Constructing Colonialism in the Ohio Valley, 1673–1800* (New York, 1997).

Hinderaker, E. and Mancall, P.C., *At the Edge of Empire: The Backcountry in British North America* (Baltimore, 2003).

Hirst, J., *The Sentimental Nation: The Making of the Australian Commonwealth* (South Melbourne, 2000).

Horne, G., *Black and Brown: African Americans and the Mexican Revolution* (New York, 2005).

Horton, M., Brown, H.W., and Mudida, N., *Shanga: The Archaeology of a Muslim Trading Community on the Coast of East Africa* (London, 1996).

Hudson, J.B., *Fugitive Slaves and the Underground Railroad in the Kentucky Borderland* (Jefferson, 2002).

Hutterer, K.L. (ed.), *Economic Exchange and Social Interaction in Southeast Asia* (Ann Arbor, 1977).

Jackle, J.A., *Images of the Ohio Valley: A Historical Geography of Travel, 1740–1860* (New York, 1977).

Jasanoff, M., *Edge of Empire: Lives, Culture, and Conquest in the East, 1750–1850* (New York, 2005).

Jeffrey, J.R., *Frontier Women: 'Civilizing' the West? 1840–1880* (New York, 1998).

Jenkins, G.H. (ed.), *The Welsh Language and Its Social Domains 1801–1911* (Cardiff, 2000).

John, B., *Pembrokeshire* (Newton Abbot, 1976).

Johnson, B.H., *Revolution in Texas* (New Haven, 2003).

Johnson, B.H. and Graybill, A.R. (eds), *Bridging National Borders in North America: Transnational and Comparative Histories* (Durham, NC, 2010).

Johnson, B.H. and Gusky, J., *Bordertown* (New Haven, 2008).

Johnson, C.S., *Bitter Canaan: The Story of the Negro Republic* (New Brunswick, 1987).

Johnson, D.E. and Michaelsen, S. (eds), *Border Theory: The Limits of Cultural Politics* (Minneapolis, 1997).

Jones, I.G., *Mid-Victorian Wales* (Cardiff, 1992).

Judson, P., *Guardians of the Nation: Activists on the Language Frontiers of Imperial Austria* (Cambridge, MA, 2007).

Kelley, S., '"Mexico in His Head": Slavery and the Texas–Mexico Border, 1810–1860,' *Journal of Social History*, 37 (2004), 709–23.

Khodarkovsky, M., *Russia's Steppe Frontier: The Making of a Colonial Empire, 1500–1800* (Bloomington, 2002).

Kidd, C., 'Race, Empire, and the Limits of Nineteenth-Century Scottish Nationhood,' *Historical Journal*, 46 (2003), 873–92.

King, J., *Budweisers into Czechs and Germans: A Local History of Bohemian Politics, 1848–1948* (Princeton, 2002).

Kinsey, J.H., *Strange Empire: A Narrative of the Northwest* (New York, 1952).

Knapman, G., 'Mapping an Ancestral Past: Discovering Charles Richards' Maps of Aboriginal South-Eastern Australia,' *Australian Aboriginal Studies*, 1 (2011), 19–34.

Kohut, Z., *Russian Centralism and Ukrainian Autonomy: Imperial Absorption of the Hetmanate 1760s–1830s* (Cambridge, MA, 1988).

Kolodny, A., *The Land before Her: Fantasy and Experience of the American Frontiers, 1630–1860* (Chapel Hill, 1984).

Kopytoff, I. (ed.), *The African Frontier* (Bloomington, 1987).

Kopytoff, I. and Miers, S. (eds), *Slavery in Africa: Historical and Anthropological Perspectives* (Madison, 1977).

Kuzniewski, A.J., *Faith and Fatherland: The Polish Church War in Wisconsin, 1896–1918* (Notre Dame, 1980).

La Rosa Corzo, G., *Runaway Slave Settlements in Cuba* (Chapel Hill, 1988).

LaDow, B., *The Medicine Line: Life and Death on a North American Borderland* (New York, 2001).

Langfur, H., *The Forbidden Lands: Colonial Identity, Frontier Violence, and the Persistence of Brazil's Eastern Indians, 1750–1830* (Stanford, 2006).

Lantzef, G.V. and Pierce, R.A., *Eastward to Empire: Exploration and Conquest on the Russian Open Frontier to 1750* (Montreal, 1973).

Lawson, R., 'The Bush Ethos and Brisbane in the 1890's,' *Historical Studies*, 15 (1972), 276–83.

LeDonne, J., 'The Frontier in Modern Russian History,' *Russian History*, 19 (1992), 143–54.

Lewis, B. and Braude, B., *Christians and Jews in the Ottoman Empire* (New York, 2000).

Liebenow, J.G., *Liberia: The Evolution of Privilege* (Ithaca, NY, 1969).

Limerick, P.N., *The Legacy of Conquest: The Unbroken Past of the American West* (New York, 1987).

Loriaux, M., *European Union and the Deconstruction of the Rhineland Frontier* (Cambridge, 2008).

Lyons, M. and Russell, P. (eds), *Australia's History* (Sydney, 2005).

Maier, C., 'Consigning the Twentieth Century to History: Alternative Narratives for the Modern Era,' *American Historical Review*, 105 (2000), 807–33.

Maier, H.M.J., *In the Center of Authority: The Malay Hikayat Merong Mahawangsa* (Ithaca, NY, 1988).

Marsden, W., *The History of Sumatra* (Oxford, 1986).

Mayne, A. and Atkinson, S. (eds), *Outside Country: Histories of Inland Australia* (Kent Town, 2011).

Mazower, M., *The Balkans* (London, 2000).

McCrady, D., *Living with Strangers: The Nineteenth-Century Sioux and the Canadian–American Borderlands* (Lincoln, NE, 2006).

McEwan, A.C., *International Boundaries of East Africa* (Oxford, 1971).

McIntosh, J., *The Edge of Islam: Power, Personhood, and Ethnoreligious Boundaries on the Kenya Coast* (Durham, NC, 2009).

Meinig, D.W., *The Shaping of America: A Geographical Perspective on 500 Years of History* (4 vols, New Haven, 1986).

Midgal, J.S. (ed.), *Boundaries and Belonging: State and Societies in the Struggle to Shape Identities and Local Practices* (New York, 2004).

Miles, W.F.S., *Hausaland Divided: Colonialism and Independence in Nigeria and Niger* (Ithaca, NY, 1994).

Miller, L.A., *The Border and Beyond: Camooweal 1884–1984* (Camooweal, 1984).

Misra, S., *Becoming a Borderland: The Politics of Space and Identity in Colonial Northeastern India* (New Delhi, 2011).

Montejano, D., *Anglos and Mexicans in the Making of Texas, 1836–1986* (Austin, 1987).

Moon, D., 'Peasant Migration and the Settlement of Russia's Frontiers, 1550–1897,' *Historical Journal*, 40 (1997), 859–93.

Moorhead, M., *The Apache Frontier: Jacobo Ugarte and Spanish–Indian Relations in Northern New Spain, 1769–1791* (Norman, 1968).

Mora-Torres, J., *The Making of the Mexican Border: The State, Capitalism, and Society in Nuevo León, 1848–1910* (Austin, 2001).

Morgan, K.O., *Wales 1880–1980* (Oxford, 1982).

Morieux, R., 'Diplomacy from Below and Belonging: Fishermen and Cross-Channel Relations in the Eighteenth Century,' *Past and Present*, 202 (2009), 83–125.

Morton, F., *Children of Ham: Freed Slaves and Fugitive Slaves on the Kenya Coast, 1873 to 1907* (Boulder, 1990).

Moses, W.J., *Liberian Dreams: Back-to-Africa Narratives from the 1850s* (University Park, PA, 1998).

Mullins, S., *Torres Strait: A History of Colonial Occupation and Culture Contact 1864–1897* (Rockhampton, 1994).

Murdock, C., *Changing Places: Society, Culture, and Territory in the Saxon–Bohemian Borderlands, 1870–1946* (Ann Arbor, 2011).

Murphy, R.C., *Guestworkers in the German Reich: A Polish Community in Wilhelmine Germany* (New York, 1983).

Myres, S.L. (ed.), *Ho for California! Women's Overland Diaries from the Huntington Library* (San Marino, CA, 1980).

Norris, R., *The Emergent Commonwealth: Australian Federation: Expectations and Fulfilment 1889–1910* (Carlton, 1975).

Nugent, P., *Smugglers, Secessionists and Loyal Citizens on the Ghana–Toga Frontier: The Life of the Borderlands since 1914* (Athens, NY, 2002).

Nugent, P. and Asiwaju, A.I. (eds), *African Boundaries* (London, 1996).

Nurse, D. and Hinnebusch, T., *Swahili and Sabaki: A Linguistic History* (Berkeley, 1993).

Onuf, P.S., *Statehood and Union* (Bloomington, 1987).

Påhlsson, C., *The Northumbrian Burr: A Sociolinguistic Study* (Lund, 1972).

Patch, R.W., *Maya Revolt and Revolution in the Eighteenth Century* (New York, 2002).

Pearson, M.N., *Port Cities and Intruders: The Swahili Coast, India, Portugal in the Early Modern Era* (Baltimore, 1998).

Pearson, M.N., 'Littoral Society: The Concept and the Problems,' *Journal of World History*, 17 (2006), 353–73.

Pennay, B., *Federation at the Border: A Thematic History and Survey of Places Related to Federation in the Albury and Corowa District* (Albury, 1997).

Peterson, J., *Province of Freedom: A History of Sierra Leone, 1787–1870* (London, 1969).

Power, D. and Standen, N. (eds), *Frontiers in Question: Eurasian Borderlands, 700–1700* (New York, 1999).

Radding, C., *Landscapes of Power and Identity: Comparative Histories in the Sonoran Desert and the Forests of Amazonia from Colony to Republic* (Durham, NC, 2005).

Radding, C., 'Sonora-Arizona: The *común*, Local Governance, and Defiance in Colonial Sonora,' in de la Teja, J.F. and Frank, R. (eds), *Choice, Persuasion, and Coercion: Social Control on Spain's North American Frontiers* (Albuquerque, 2005), pp. 179–99.

Raj, K., 'Circulation and the Emergence of Modern Mapping: Great Britain and Early Colonial India, 1764–1820,' in Markovits, C., Pouchepadass, J., and Subrahmanyam, S. (eds), *Society and Circulation: Mobile People and Itinerant Cultures in South Asia, 1750–1950* (Delhi, 2003), pp. 23–54.

Readman, P., '"The Cliffs Are Not Cliffs": The Cliffs of Dover and National Identities in Britain, c.1750–c.1900,' *History*, 99 (2014).

Rebert, P., *La Gran Línea: Mapping the United States–Mexico Boundary, 1849–1857* (Austin, 2001).

Rees, T., *Arc of the Medicine Line: Mapping the World's Longest Undefended Border across the Western Plains* (Lincoln, NE, 2007).

Reid, A.M. and Lane, P.J. (eds), *African Historical Archaeologies* (New York, 2004).

Reid, R.L., *Always a River: The Ohio River and the American Experience* (Bloomington, 1991).

Restall, M. (ed.), *Beyond Black and Red: African–Native Relations in Colonial Latin America* (Albuquerque, 2005).

Reynolds, H., *Aboriginal Sovereignty: Reflections on Race, State and Nation* (St Leonards, 1996).

Reynolds, H., *North of Capricorn: The Untold Story of Australia's North* (Crows Nest, 2003).

Rowe, M., *From Reich to State: The Rhineland in the Revolutionary Age, 1780–1830* (Cambridge, 2003).

Rugeley, T., *Rebellion Now and Forever: Mayas, Hispanics, and Caste War Violence in Yucatan, 1800–1880* (Stanford, 2009).

Russell, D., *Looking North: Northern England and the National Imagination* (Manchester, 2004).

Saeger, J.S., *Chaco Mission Frontier: The Guaycuruan Experience* (Tucson, 2000).

Sahlins, P., *Boundaries: The Making of France and Spain in the Pyrenees* (Berkeley, 1989).

Sahlins, P., 'Natural Frontiers Revisited: France's Boundaries since the Seventeenth Century,' *American Historical Review*, 95 (1990), 1423–51.

Sanneh, L., *Abolitionists Abroad: American Blacks and the Making of Modern West Africa* (Cambridge, 1999).

Santiago, M., *The Jar of Severed Hands: Spanish Deportation of Apache Prisoners of War, 1770–1810* (Norman, 2011).

Schlissel, L., *Women's Diaries of the Westward Journey* (New York, 1982).

Scott, J.C., *The Art of Not Being Governed: An Anarchist History of Upland Southeast Asia* (New Haven, 2009).

Shick, T.W., *Behold the Promised Land: A History of Afro-American Settler Society in Nineteenth-Century Liberia* (Baltimore, 1977).

Simpson, M., *Trafficking Subjects: The Politics of Mobility in Nineteenth-Century America* (Minneapolis, 2005).

Skinner, B., *The Western Front of the Eastern Church: Uniate and Orthodox Conflict in 18th-Century Poland, Ukraine, Belarus, and Russia* (DeKalb, 2009).

Sopher, D., *The Sea Nomads: A Study of the Maritime Boat People of Southeast Asia* (Singapore, 1977).

Spear, T.T., *The Kaya Complex: A History of the Mijikenda Peoples of the Kenya Coast to 1900* (Nairobi, 1978).

St John, R., *Line in the Sand: A History of the Western US–Mexico Border* (Princeton, 2011).

Stanley, G.F.G., *Mapping the Frontier: Charles Wilson's Diary of the Survey of the 49th Parallel, 1858–1862, While Secretary of the British Boundary Commission* (Toronto, 1970).

Staudenraus, P.J., *The African Colonization Movement, 1816–1865* (New York, 1980).

Stauter-Halsted, K., *The Nation in the Village: The Genesis of Peasant National Identity in Austrian Poland, 1848, 1914* (Ithaca, NY, 2001).

Sunderland, W., *Taming the Wild Field: Colonization and Empire on the Russian Steppe* (Ithaca, NY, 2004).

Sundiata, I., *Brothers and Strangers: Black Zion, Black Slavery, 1914–1940* (Durham, NC, 2004).

Sutton, P., *Country: Aboriginal Boundaries and Land Ownership in Australia* (Canberra, 1995).

Tagliacozzo, E., *Secret Trades, Porous Borders: Smuggling and States along a Southeast Asian Frontier, 1865–1915* (New Haven, 2005).

Tarling, N., *Anglo–Dutch Rivalry in the Malay World, 1780–1824* (Cambridge, 1962).

Taylor, A., *The Divided Ground: Indians, Settlers, and the Northern Borderland of the American Revolution* (New York, 2007).

Taylor, N.M., *Frontiers of Freedom: Cincinnati's Black Community, 1802–1868* (Athens, OH, 2005).

Thaden, E.C., *Russia's Western Borderlands, 1710–1870* (Princeton, 1984).

Thom, D.J., *The Niger–Nigeria Boundary, 1890–1906* (Athens, NY, 1975).

Thompson, J., *Cortina: Defending the Mexican Name in Texas* (College Station, TX, 2007).

Thongchai, W., *Siam Mapped: The History of a Geo-Body of a Nation* (Honolulu, 1994).

Torpey, J., *The Invention of the Passport* (Cambridge, 2000).

Torrans, T., *Forging the Tortilla Curtain: Cultural Drift and Change along the United States–Mexico Border from the Spanish Era to the Present* (Ft Worth, 2000).

Truett, S. and Young, E. (eds), *Continental Crossroads: Remapping US–Mexico Borderlands History* (Durham, NC, 2004).

Turner, F.J., *The Frontier in American History* (Huntington, NY, 1976 [1920]).

van Schendel, W., *The Bengal Borderland* (London, 2005).

van Sickle, E.S., 'Reluctant Imperialists: The US Navy and Liberia, 1819–1845,' *Journal of the Early Republic*, 31 (2011), 107–34.

Weber, D.J., *The Spanish Frontier in North America* (New Haven, 1992).

Weber, D.J., *Bárbaros: Spaniards and their Savages in the Age of Enlightenment* (New Haven, 2005).

Weber, D.J. and Rausch, J.M. (eds), *Where Cultures Meet: Frontiers in Latin American History* (Wilmington, 1994).

Weber, E., *Peasants into Frenchmen* (London, 1977).

Weeks, T.R., *Nation and State in Late Imperial Russia: Nationalism and Russification on the Western Frontier, 1863–1914* (DeKalb, 1996).

West-Pavlov, R. and Wawrzinek, J. (eds), *Frontier Skirmishes: Literary and Cultural Debates in Australia after 1992* (Heidelberg, 2010).

White, R., *The Middle Ground: Indians, Empires, and Republics in the Great Lakes Region, 1650–1815* (Cambridge, 1991).

Whitten, A.J. et al., *The Ecology of Sumatra* (Yogyakarta, 1984).

Wieczynski, J.L., *The Russian Frontier: The Impact of Borderlands upon the Course of Early Russian History* (Charlottesville, 1976).

Wigmore, G., 'Before the Railroad: From Slavery to Freedom in the Canadian–American Borderland,' *Journal of American History*, 98 (2011), 437–54.

Wiley, B.I., *Slaves No More: Letters from Liberia, 1833–1869* (Lexington, 1980).

Willis, J., *Mombasa, the Swahili, and the Making of the Mijikenda* (New York, 1993).

Wilson, T. and Donnan, H. (eds), *Border Identities* (Cambridge, 1998).

Woloch, I., *The New Regime: Transformations of the French Civic Order, 1789–1820s* (New York, 1994).

Wolters, O.W., *Early Indonesian Commerce: A Study of the Origins of Srivijaya* (Ithaca, NY, 1967).

Woolf, S., *Napoleon's Integration of Europe* (London, 1991).

Wunder, J. and Hämäläinen, P., 'Of Lethal Places and Lethal Essays,' *American Historical Review*, 104 (1999), 1229–35.

Yaremko, J., 'Colonial Wars and Indigenous Geopolitics: Aboriginal Agency, the Cuba–Florida–Mexico Nexus, and the Other Diaspora,' *Canadian Journal of Latin American and Caribbean Studies*, 35 (2010), 165–96.

Young, E., *Catarino Garza's Revolution on the Texas–Mexico Border* (Durham, NC, 2004).

Zahra, T., *Kidnapped Souls: National Indifference and the Battle for Children in the Bohemian Lands, 1900–1948* (Ithaca, NY, 2008).

Index

Printed and bound in the United States of America